Uncertainty Analysis with High Dimensional Dependence Modelling

Uncertainty Analysis with High Dimensional Dependence Modelling

Dorota Kurowicka and Roger Cooke
Delft University of Technology, The Netherlands

John Wiley & Sons, Ltd

Other Wiley Editorial Offices

John Wiley & Sons Inc., 111 River Street, Hoboken, NJ 07030, USA

Jossey-Bass, 989 Market Street, San Francisco, CA 94103-1741, USA

Wiley-VCH Verlag GmbH, Boschstr. 12, D-69469 Weinheim, Germany

John Wiley & Sons Australia Ltd, 42 McDougall Street, Milton, Queensland 4064, Australia

John Wiley & Sons (Asia) Pte Ltd, 2 Clementi Loop #02-01, Jin Xing Distripark, Singapore 129809

John Wiley & Sons Canada Ltd, 22 Worcester Road, Etobicoke, Ontario, Canada M9W 1L1

Wiley also publishes its books in a variety of electronic formats. Some content that appears
in print may not be available in electronic books.

Library of Congress Cataloging-in-Publication Data

Kurowicka, Dorota, 1967-
    Uncertainty analysis : mathematical foundations and applications / Dorota Kurowicka
and Roger Cooke.
        p. cm. – (Wiley series on statistics in practice)
    Includes bibliographical references and index.
    ISBN-13: 978-0-470-86306-0
    ISBN-10: 0-470-86306-4 (alk. paper)
        1. Uncertainty (Information theory) – Mathematics. I. Cooke, Roger, 1942– II. Title. III.
Statistics in practice.

    Q375.K87 2006
    003'.54 – dc22

                                                                                        2005057712

British Library Cataloguing in Publication Data

A catalogue record for this book is available from the British Library

ISBN-13: 978-0-470-86306-0 (HB)
ISBN-10: 0-470-86306-4 (HB)

Typeset in 10/12pt Times by Laserwords Private Limited, Chennai, India
Printed and bound in Great Britain by TJ International, Padstow, Cornwall
This book is printed on acid-free paper responsibly manufactured from sustainable forestry
in which at least two trees are planted for each one used for paper production.

Contents

Preface   **ix**

1 Introduction   **1**
   1.1   Wags and Bogsats   1
   1.2   Uncertainty analysis and decision support: a recent example   4
   1.3   Outline of the book   9

2 Assessing Uncertainty on Model Input   **13**
   2.1   Introduction   13
   2.2   Structured expert judgment in outline   14
   2.3   Assessing distributions of continuous univariate uncertain quantities   15
   2.4   Assessing dependencies   16
   2.5   Unicorn   20
   2.6   Unicorn projects   20

3 Bivariate Dependence   **25**
   3.1   Introduction   25
   3.2   Measures of dependence   26
         3.2.1   Product moment correlation   26
         3.2.2   Rank correlation   30
         3.2.3   Kendall's tau   32
   3.3   Partial, conditional and multiple correlations   32
   3.4   Copulae   34
         3.4.1   Fréchet copula   36
         3.4.2   Diagonal band copula   37
         3.4.3   Generalized diagonal band copula   41
         3.4.4   Elliptical copula   42
         3.4.5   Archimedean copulae   45
         3.4.6   Minimum information copula   47
         3.4.7   Comparison of copulae   49
   3.5   Bivariate normal distribution   50
         3.5.1   Basic properties   50
   3.6   Multivariate extensions   51
         3.6.1   Multivariate dependence measures   51

        3.6.2   Multivariate copulae   53
        3.6.3   Multivariate normal distribution   53
    3.7   Conclusions   54
    3.8   Unicorn projects   55
    3.9   Exercises   61
    3.10 Supplement   67

4   High-dimensional Dependence Modelling   81
    4.1   Introduction   81
    4.2   Joint normal transform   82
    4.3   Dependence trees   86
        4.3.1   Trees   86
        4.3.2   Dependence trees with copulae   86
        4.3.3   Example: Investment   90
    4.4   Dependence vines   92
        4.4.1   Vines   92
        4.4.2   Bivariate- and copula-vine specifications   96
        4.4.3   Example: Investment continued   98
        4.4.4   Partial correlation vines   99
        4.4.5   Normal vines 101
        4.4.6   Relationship between conditional rank and partial correla-
                tions on a regular vine 101
    4.5   Vines and positive definiteness 105
        4.5.1   Checking positive definiteness 105
        4.5.2   Repairing violations of positive definiteness 107
        4.5.3   The completion problem 109
    4.6   Conclusions 111
    4.7   Unicorn projects 111
    4.8   Exercises 115
    4.9   Supplement 116
        4.9.1   Proofs 116
        4.9.2   Results for Section 4.4.6 127
        4.9.3   Example of fourvariate correlation matrices 129
        4.9.4   Results for Section 4.5.2 130

5   Other Graphical Models   131
    5.1   Introduction 131
    5.2   Bayesian belief nets 131
        5.2.1   Discrete bbn's 132
        5.2.2   Continuous bbn's 133
    5.3   Independence graphs 141
    5.4   Model inference 142
        5.4.1   Inference for bbn's 143
        5.4.2   Inference for independence graphs 144
        5.4.3   Inference for vines 145

5.5   Conclusions 150
5.6   Unicorn projects 150
5.7   Supplement 157

6   Sampling Methods   **159**
6.1   Introduction 159
6.2   (Pseudo-) random sampling 160
6.3   Reduced variance sampling 161
       6.3.1   Quasi-random sampling 161
       6.3.2   Stratified sampling 164
       6.3.3   Latin hypercube sampling 166
6.4   Sampling trees, vines and continuous bbn's 168
       6.4.1   Sampling a tree 168
       6.4.2   Sampling a regular vine 169
       6.4.3   Density approach to sampling regular vine 174
       6.4.4   Sampling a continuous bbn 174
6.5   Conclusions 180
6.6   Unicorn projects 180
6.7   Exercise 184

7   Visualization   **185**
7.1   Introduction 185
7.2   A simple problem 186
7.3   Tornado graphs 186
7.4   Radar graphs 187
7.5   Scatter plots, matrix and overlay scatter plots 188
7.6   Cobweb plots 191
7.7   Cobweb plots local sensitivity: dike ring reliability 195
7.8   Radar plots for importance; internal dosimetry 199
7.9   Conclusions 201
7.10   Unicorn projects 201
7.11   Exercises 203

8   Probabilistic Sensitivity Measures   **205**
8.1   Introduction 205
8.2   Screening techniques 205
       8.2.1   Morris' method 205
       8.2.2   Design of experiments 208
8.3   Global sensitivity measures 214
       8.3.1   Correlation ratio 215
       8.3.2   Sobol indices 219
8.4   Local sensitivity measures 222
       8.4.1   First order reliability method 222
       8.4.2   Local probabilistic sensitivity measure 223
       8.4.3   Computing $\frac{\partial E(X|g_o)}{\partial g_o}$ 225

8.5 Conclusions 227
8.6 Unicorn projects 228
8.7 Exercises 230
8.8 Supplement 236
    8.8.1 Proofs 236

9 Probabilistic Inversion   **239**
9.1 Introduction 239
9.2 Existing algorithms for probabilistic inversion 240
    9.2.1 Conditional sampling 240
    9.2.2 PARFUM 242
    9.2.3 Hora-Young and PREJUDICE algorithms 243
9.3 Iterative algorithms 243
    9.3.1 Iterative proportional fitting 244
    9.3.2 Iterative PARFUM 245
9.4 Sample re-weighting 246
    9.4.1 Notation 246
    9.4.2 Optimization approaches 247
    9.4.3 IPF and PARFUM for sample re-weighting probabilistic
        inversion 248
9.5 Applications 249
    9.5.1 Dispersion coefficients 249
    9.5.2 Chicken processing line 252
9.6 Convolution constraints with prescribed margins 253
9.7 Conclusions 255
9.8 Unicorn projects 256
9.9 Supplement 258
    9.9.1 Proofs 258
    9.9.2 IPF and PARFUM 263

10 Uncertainty and the UN Compensation Commission   **269**
10.1 Introduction 269
10.2 Claims based on uncertainty 270
10.3 Who pays for uncertainty 272

Bibliography   **273**

Index   **281**

Preface

This book emerges from a course given at the Department of Mathematics of the Delft University of Technology. It forms a part of the program on Risk and Environmental Modelling open to graduate students with the equivalent of a Bachelor's degree in mathematics. The students are familiar with undergraduate analysis, statistics and probability, but for non-mathematicians this familiarity may be latent. Therefore, most notions are 'explained in-line'. Readers with a nodding acquaintance with these subjects can follow the thread. To keep this thread visible, proofs are put in supplements of the chapters in which they occur. Exercises are also included in most chapters.

The real source of this book is our experience in applying uncertainty analysis. We have tried to keep the applications orientation in the foreground. Indeed, the whole motivation for developing generic tools for high dimensional dependence modelling is that decision makers and problem owners are becoming increasingly sophisticated in reasoning with uncertainty. They are making demands, which an analyst with the traditional tools of probabilistic modelling cannot meet. Put simply, our point of view is this: a joint distribution is specified by specifying a sampling procedure. We therefore assemble tools and techniques for sampling and analysing high dimensional distributions with dependence. These same tools and techniques form the design requirements for a generic uncertainty analysis program. One such program is UNcertainty analysis wIth CORrelatioNs (UNICORN). A fairly ponderous light version may be downloaded from http://ssor.twi.tudelft.nl/ risk/. UNICORN projects are included in each chapter to give hands on experience in applying uncertainty analysis.

The people who have contributed substantially to this book are too numerous to list, but certainly include Valery Kritchallo, Tim Bedford, Daniel Lewandowski, Belinda Chiera, Du Chao, Bernd Kraan and Jolanta Misiewicz.

1

Introduction: Uncertainty Analysis and Dependence Modelling

1.1 Wags and Bogsats

'...whether true or not [it] is at least probable; and he who tells nothing exceeding the bounds of probability has a right to demand that they should believe him who cannot contradict him'. Samuel Johnson, author of the first English dictionary, wrote this in 1735. He is referring to the Jesuit priest Jeronimo Lobo's account of the unicorns he saw during his visit to Abyssinia in the 17th century (Shepard (1930) p. 200).

Johnson could have been the apologist for much of what passed as decision support in the period after World War II, when think tanks, forecasters and expert judgment burst upon the scientific stage. Most salient in this genre is the book *The Year 2000* (Kahn and Wiener (1967)) in which the authors published 25 'even money bets' predicting features of the year 2000, including interplanetary engineering and conversion of humans to fluid breathers. Essentially, these are statements without pedigree or warrant, whose credibility rests on shifting the burden of proof. Their cavalier attitude toward uncertainty in quantitative decision support is representative of the period. Readers interested in how many of these even money bets the authors have won, and in other examples from this period, are referred to (Cooke (1991), Chapter 1).

Quantitative models pervade all aspects of decision making, from failure probabilities of unlaunched rockets, risks of nuclear reactors and effects of pollutants on health and the environment to consequences of economic policies. Such quantitative models generally require values for parameters that cannot be measured or

Uncertainty Analysis with High Dimensional Dependence Modelling   D. Kurowicka and R. Cooke
© 2006 John Wiley & Sons, Ltd

assessed with certainty. Engineers and scientists sometimes cover their modesty with churlish acronyms designating the source of ungrounded assessments. 'Wags' (wild-ass guesses) and 'bogsats' (bunch of guys sitting around a table) are two examples found in published documentation.

Decision makers, especially those in the public arena, increasingly recognize that input to quantitative models is uncertain and demand that this uncertainty be quantified and propagated through the models.

Initially, it was the modellers themselves who provided assessments of uncertainty and did the propagating. Not surprisingly, this activity was considered secondary to the main activity of computing 'nominal values' or 'best estimates' to be used for forecasting and planning and received cursory attention.

Figure 1.1 shows the result of such an in-house uncertainty analysis performed by the National Radiological Protection Board (NRPB) and The Kernforschungszentrum Karlsruhe (KFK) in the late 1980s (Crick et al. (1988); Fischer et al. (1990)). The models in question predict the dispersion of radioactive material in the atmosphere following an accident in a nuclear reactor. The figure shows predicted lateral dispersion under stable conditions, and also shows wider and narrower plumes, which the modellers are 90% certain will enclose an actual plume under the stated conditions.

It soon became evident that if things were uncertain, then experts might disagree, and using one expert-modeller's estimates of uncertainty might not be sufficient. Structured expert judgment has since become an accepted method for quantifying models with uncertain input. 'Structured' means that the experts are identifiable, the assessments are traceable and the computations are transparent. To appreciate the difference between structured and unstructured expert judgment, Figure 1.2 shows the results of a structured expert judgment quantification of the same uncertainty pictured in Figure 1.1 (Cooke (1997b)). Evidently, the picture of uncertainty emerging from these two figures is quite different.

One of the reasons for the difference between these figures is the following: The lateral spread of a plume as a function of down wind distance x is modelled, per stability class, as

$$\sigma(x) = Ax^B.$$

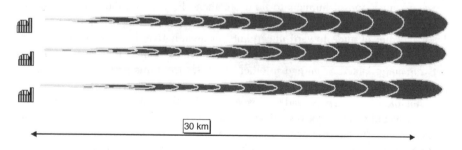

Figure 1.1 5%, 50% and 95% plume widths (stability D) computed by NRPB and KFK.

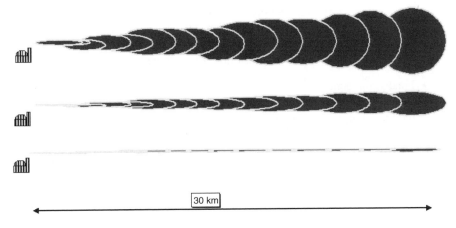

Figure 1.2 5%, 50% and 95% plume widths (stability D) computed by the EU-USNRC Uncertainty Analysis of accident consequence codes.

Both the constants A and B are uncertain as attested by spreads in published values of these coefficients. However, these uncertainties *cannot* be independent. Obviously if A takes a large value, then B will tend to take smaller values. Recognizing the implausibility of assigning A and B as *independent* uncertainty distributions, and the difficulty of assessing a joint distribution on A and B, the modellers elected to consider B as a constant; that is, as known with certainty.[1]

The differences between these two figures reflect a change in perception regarding the goal of quantitative modelling. With the first picture, the main effort has gone into constructing a quantitative deterministic model to which uncertainty quantification and propagation are added on. In the second picture, the model is essentially about capturing uncertainty. Quantitative models are useful insofar as they help us resolve and reduce uncertainty. Three major differences in the practice of quantitative decision support follow from this shift of perception.

- First of all, the representation of uncertainty via expert judgment, or some other method is seen as a scientific activity subject to methodological rules every bit as rigorous as those governing the use of measurement or experimental data.

- Second, it is recognized that an essential part of uncertainty analysis is the analysis of dependence. Indeed, if all uncertainties are independent, then their propagation is mathematically trivial (though perhaps computationally

---

[1]This is certainly not the only reason for the differences between Figures 1.1 and 1.2. There was also ambivalence with regard to what the uncertainty should capture. Should it capture the plume uncertainty in a single accidental release, or the uncertainty in the average plume spread in a large number of accidents? Risk analysts clearly required the former, but meteorologists are more inclined to think in terms of the latter.

challenging). Sampling and propagating independent uncertainties can easily be trusted to the modellers themselves. However, when uncertainties are dependent, things become much more subtle, and we enter a domain for which the modellers' training has not prepared them.

- Finally, the domains of communication with the problem owner, model evaluation, and so on, undergo significant transformations once we recognize that the main purpose of models is to capture uncertainty.

1.2   Uncertainty analysis and decision support: a recent example

A recent example serves to illustrate many of the issues that arise in quantifying uncertainty for decision support. The example concerns transport of *Campylobacter* infection in chicken processing lines. The intention here is not to understand *Campylobacter* infection, but to introduce topics covered in the following chapters. For details on *Campylobacter*, see Cooke et al. (Appearing); Van der Fels-Klerx et al. (2005); Nauta et al. (2004).

Campylobacter contamination of chicken meat may be responsible for up to 40% of *Campylobacter*-associated gastroenteritis and for a similar proportion of deaths. A recent effort to rank various control options for *Campylobacter* contamination has led to the development of a mathematical model of a processing line for chicken meat (these chickens are termed 'broilers').

A typical broiler processing line involves a number of phases as shown in Figure 1.3. Each phase is characterized by transfers of *Campylobacter* colony forming units from the chicken surface to the environment, from the environment back to the surface and from the faeces to the surface (until evisceration), and the destruction of the colonies. The general model, applicable with variations in each processing phase, is shown in Figure 1.4.

Given the number of *Campylobacter* on and in the chickens at the inception of processing, and given the number initially in the environment, one can run the model with values for the transfer coefficients and compute the number of *Campylobacter* colonies on the skin of a broiler and in the environment at the end of each phase. Ideally, we would like to have field measurements or experiments

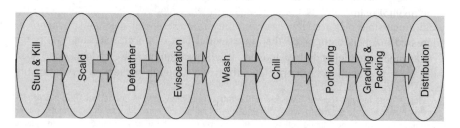

Figure 1.3   Broiler chicken processing line.

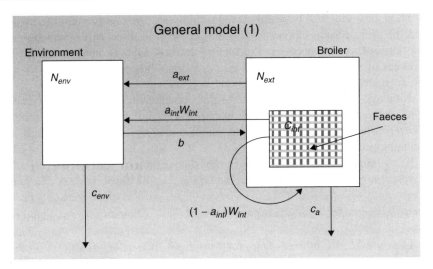

Figure 1.4 Transfer coefficients in a typical phase of a broiler chicken processing line.

to determine values for the coefficients in Figure 1.4. Unfortunately, these are not feasible. Failing that, we must quantify the *uncertainty* in the transfer coefficients, and propagate this uncertainty through the model to obtain uncertainty distributions on the model output.

This model has been quantified in an expert judgment study involving 12 experts (Van der Fels-Klerx et al. (2005)). Methods for applying expert judgments are reviewed in Chapter 2. We may note here that expert uncertainty assessments are regarded as statistical hypotheses, which may be tested against data and combined with a view to optimizing performance of the resulting 'decision maker'.

The experts have detailed knowledge of processing lines, but owing to the scarcity of measurements, they have no direct knowledge of the transfer mechanisms defined by the model. Indeed, use of environmental transport models is rather new in this area, and unfamiliar. Uncertainty about the transfer mechanisms can be large, and, as in the dispersion example discussed in the preceding text, it is unlikely that these uncertainties could be independent. Combining possible values for transfer and removal mechanism independently would not generally yield a plausible picture. Hence, uncertainty in one transfer mechanism cannot be addressed independently of the rest of the model.

Our quantification problem has the following features:

• There are no experiments or measurements for determining values.

• There is relevant expert knowledge, but it is not directly applicable.

• The uncertainties may be large and may not be presumed to be independent, and hence dependence must be quantified.

These obstacles will be readily recognized by anyone engaged in mathematical modelling for decision support beyond the perimeter of direct experimentation and measurement. As the need for quantitative decision support rapidly outstrips the resources of experimentation, these obstacles must be confronted and overcome. The alternative is regression to wags and bogsats.

Although experts cannot provide useful quantification for the transfer coefficients, they are able to quantify their uncertainty regarding the number of *Campylobacter* colonies on a broiler in the situation described below taken from the elicitation protocol:

At the beginning of a new slaughtering day, a thinned flock is slaughtered in a 'typical large broiler chicken slaughterhouse'. ... We suppose every chicken to be externally infected with 10^5 Campylobacters per carcass and internally with 10^8 Campylobacters per gram of caecal content at the beginning of each slaughtering stage. ...

Question A1: All chickens of the particular flock are passing successively through each slaughtering stage. How many Campylobacters (per carcass) will be found after each of the mentioned stages of the slaughtering process each time on the first chicken of the flock?

Experts respond to questions of this form, for different infection levels, by stating the 5%, 50% and 95% quantiles, or percentiles, of their uncertainty distributions. If distributions on the transfer coefficients in Figure 1.4 are given, then distributions per processing phase for the number of *Campylobacter* per carcass (the quantity assessed by the experts) can be computed by Monte Carlo simulation: We sample a vector of values for the transfer coefficients, compute a vector of *Campylobacter* per carcass and repeat this until suitable distributions are constructed. We would like the distributions over the assessed quantities computed in this way to agree with the quantiles given by the combined expert assessments. Of course we could guess an initial distribution over the transfer coefficients, perform this Monte Carlo computation and see if the resulting distributions over the assessed quantities happen to agree with the experts' assessments. In general they will not, and this trial-and-error method is quite unlikely to produce agreement. Instead, we start with a diffuse distribution over the transfer coefficients, and adapt this distribution to fit the requirements in a procedure called 'probabilistic inversion'.

More precisely, let X and Y be n- and m-dimensional random vectors, respectively, and let G be a function from $\mathbb{R}^n$ to $\mathbb{R}^m$. We call $x \in \mathbb{R}^n$ an inverse of $y \in \mathbb{R}^m$ under G if $G(x) = y$. Similarly, we call X a probabilistic inverse of Y under G if $G(X) \sim Y$, where $\sim$ means 'has the same distribution as'. If $\{Y|Y \in C\}$ is the set of random vectors satisfying constraints C, then we say that X is an element of the probabilistic inverse of $\{Y|Y \in C\}$ under G if $G(X) \in C$. Equivalently, and more conveniently, if the distribution of Y is partially specified, then we say that X is a probabilistic inverse of Y under G if $G(X)$ satisfies the partial specification of Y. In the current context, the transfer coefficients in Figure 1.4 play the role of X, and the assessed quantities play the role of Y.

In our *Campylobacter* example, the probabilistic inversion problem may now be expressed as follows: Find a joint distribution over the transfer coefficients such that the quantiles of the assessed quantities agree with the experts' quantiles. If more than one such joint distribution exists, pick the least informative of these. If no such joint distribution exists, pick a 'best-fitting' distribution, and assess its goodness of fit.

Probabilistic inversion techniques are the subject of Chapter 9.

In fact, the best fit produced with the model in Figure 1.4 was not very good. It was not possible to find a distribution over the transfer coefficients, which, when pushed through the model, yielded distributions matching those of the experts. On reviewing the experts' reasoning, it was found that the 'best' expert (see Chapter 2) in fact recognized two types of transfer from the chicken skin to the environment. A rapid transfer applied to *Campylobacter* on the feathers, and a slow transfer applied to *Campylobacter* in the pores of the skin. When the model was extended to accommodate this feature, a satisfactory fit was found. The second model, developed after the first probabilistic inversion, is shown in Figure 1.5.

Distributions resulting from probabilistic inversion typically have dependencies. In fact, this is one of the ways in which dependence arises in uncertainty analysis. We require tools for studying such dependencies. One simple method is to simply compute rank correlations. Notions of correlation and their properties are discussed in Chapter 3. For now it will suffice simply to display in Table 1.1 the rank correlation matrix for the transfer coefficients in Figure 1.5, for the scalding phase.

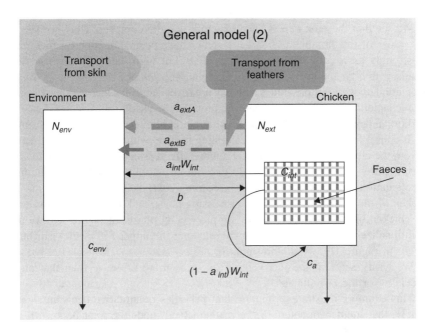

Figure 1.5 Processing phase model after probabilistic inversion.

Table 1.1   Rank correlation matrix of transfer coefficients, scalding phase.

Variable	a_{extA}	a_{extB}	ca	b	ce	$aint$
a_{extA}	1.00	0.17	−0.60	−0.04	0.03	0.00
a_{extB}	0.17	1.00	−0.19	−0.10	−0.06	0.00
ca	−0.60	−0.19	1.00	0.01	0.02	0.00
b	−0.04	−0.10	0.01	1.00	0.02	0.00
ce	0.03	−0.06	0.02	0.02	1.00	0.00
$aint$	0.00	0.00	0.00	0.00	0.00	0.00

Table 1.1 shows a pronounced negative correlation between the rapid transfer from the skin (a_{extA}) and evacuation from the chicken (ca), but other correlations are rather small. Correlations of course do not tell the whole story. Chapter 7 discusses visual tools for studying dependence in high-dimensional distributions. One such tool is the cobweb plot. In a cobweb plot, variables are represented as vertical lines. Each sample realizes one value of each variable. Connecting these values by line segments, one sample is represented as a jagged line intersecting all the vertical lines. Plate 1 shows 2000 such jagged lines and gives a picture of the joint distribution. In this case, we have plotted the quantiles, or percentiles, or ranks of the variables rather than the values themselves. The negative rank correlation between a_{extA} and ca is readily visible if the picture is viewed in colour: The lines hitting low values of a_{extA} are red, and the lines hitting values of ca are also red.

We see that the rank dependence structure is quite complex. Thus, we see that low values of the variable ce (c_{env}, the names have been shortened for this graph) are strongly associated with high values of b, but high values of ce may occur equally with high and low values of b. Correlation (rank or otherwise) is an average association over all sample values and may not reveal complex interactions. In subsequent chapters, we shall see how cobweb plots can be used to study dependence and conditional dependence. One simple illustration highlights their use in this example.

Suppose, we have a choice of accelerating the removal from the environment ce or from the chicken ca; which would be more effective in reducing *Campylobacter* transmission? To answer this, we add two output variables: $a1$ (corresponding to the elicitation question given in the preceding text) is the amount on the *first* chicken of the flock as it leaves the processing phase and $a2$ is the amount on the *last* chicken of the flock as it leaves the processing phase. In Figure 1.6, we have conditionalized the joint distribution by selecting the upper 5% of the distribution for ca; in Figure 1.7, we do the same for ce.

We easily see that the intervention on ce is more effective than that on ca, especially for the last chicken.

This example illustrates a feature that pervades quantitative decision support, namely, that input parameters of the mathematical models cannot be known with certainty. In such situations, mathematical models should be used to capture and propagate uncertainty. They should not be used to help a bunch of guys sitting

Samples selected: 497

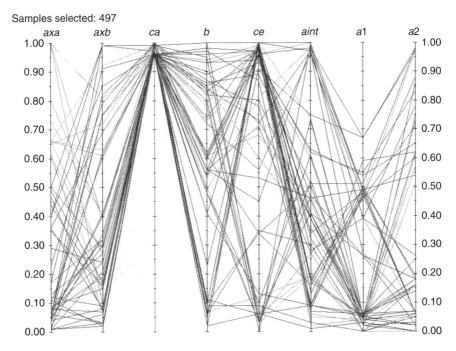

Figure 1.6 Cobweb plot conditional on high *ca*.

around a table make statements that should be believed if they cannot be contradicted. In particular, it shows the following:

- Expert knowledge can be brought to bear in situations where direct experiment or measurement is not possible, namely, by quantifying expert uncertainty on variables that the models should predict.

- By utilizing techniques like probabilistic inversion in such situations, models become vehicles for capturing and propagating uncertainty.

- Configured in this way, expert input can play an effective role in evaluating and improving models.

- Models quantified with uncertainty, rather than wags and bogsats, can provide meaningful decision support.

1.3 Outline of the book

This book focuses on techniques for uncertainty analysis, which are generally applicable. Uncertainty distributions may *not* be assumed to conform to any parametric form. Techniques for specifying, sampling and analysing high-dimensional distributions should therefore be non-parametric. Our goal is to present the mathematical

Samples selected: 500

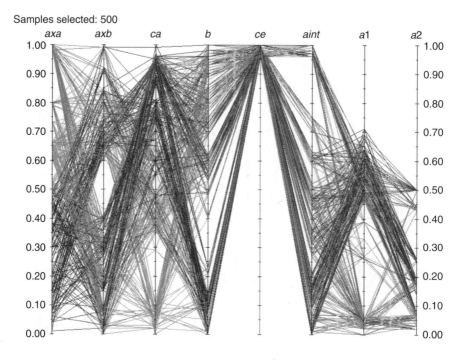

Figure 1.7 Cobweb plot conditional on high *ce*.

concepts that are essential in understanding uncertainty analysis and to provide the practitioners with tools they will need in applications.

Some techniques, in particular those associated with bivariate dependence modelling, are becoming familiar to a wide range of users. Even this audience will benefit from a presentation focused on applications in higher dimensions. Subjects like the minimal information, diagonal band and elliptical copula will probably be new. Good books are available for bivariate dependence: Dall'Aglio et al. (1991); Doruet Mari and Kotz (2001); Joe (1997); Nelsen (1999). High-dimensional dependence models, sampling methods, post-processing analysis and probabilistic inversion will be new to non-specialists, both mathematicians and modellers.

The focus of this book is not how to assess dependencies in high-dimensional distributions, but what to do with them once we have them. That being said, the uncertainty, which gets analysed in uncertainty analysis is often the uncertainty of experts, and expert judgment deserves brief mention. Expert judgment is treated summarily in Chapter 2. Chapter 2 also introduces the uncertainty analysis, package UNICORN. Each chapter contains UNICORN projects designed to sensitize the reader to issues in dependence modelling and to step through features of the program. The projects in Chapter 2 provide a basic introduction to UNICORN and are strongly recommended. Chapter 3 treats bivariate dependence, focusing on techniques that are useful in higher dimensions. The UNICORN projects in Chapter 3 introduce the concepts of *aleatory* and *epistemic* uncertainty.

With regard to dependence in higher dimensions, much is not known. For example, we do not know whether an arbitrary correlation matrix is also a rank correlation matrix.[2] We do know that characterizing dependence in higher dimensions via product moment correlation matrices is *not* the way to go. Product moment correlations impose unwelcome constraints on the one-dimensional distributions. Further, correlation matrices must be positive definite, and must be completely specified. In practice, data errors, rounding errors or simply vacant cells lead to intractable problems with regard to positive definiteness. We must design other friendlier ways to let the world tell us, and to let us tell computers, as to which high-dimensional distribution to calculate. We take the position that graphical models are the weapon of choice. These may be Markov trees, vines, independence graphs or Bayesian belief nets. For constructing sampling routines capable of realizing richly complex dependence structures, we advocate regular vines. They also allow us to move beyond discrete Bayesian belief nets without defaulting to the joint normal distribution. Much of this material is new and only very recently available in the literature: Bedford and Cooke (2001a); Cowell et al. (1999); Pearl (1988); Whittaker (1990). Chapter 4 is devoted to this.

Chapter 5 studies graphical models, which have interesting features, but are not necessarily generally applicable in uncertainty analysis. Bayesian belief nets and independence graphs are discussed. The problem of inferring a graphical model from multivariate data is addressed. The theory of regular vines is used to develop non-parametric continuous Bayesian belief nets. Chapter 6 discusses sampling methods. Particular attention is devoted to sampling regular vines and Bayesian belief nets.

Problems in measuring, inferring and modelling high-dimensional dependencies are mirrored at the end of the analysis by problems in communicating this information to problem owners and decision makers. Here graphical tools come to the fore in Chapter 7.

Chapter 8 addresses the problem of extracting useful information from an uncertainty analysis. This is frequently called *sensitivity analysis* (Saltelli et al. (2000)). We explore techniques for discovering which input variables contribute significantly to the output.

Chapter 9 takes up probabilistic inversion. Inverse problems are as old as probability itself, but their application in uncertainty analysis is new. Again, this material is only very recently available in the literature.

The concluding chapter speculates on the future role of uncertainty analysis in decision support.

Each chapter contains mathematical exercises and projects. The projects can be performed with the uncertainty analysis package UNICORN (UNcertainty analysis wIth CORrelatioNs), a light version of which can be downloaded free at

---

[2]We have recently received a manuscript from H. Joe that purports to answer this question in the negative for dimensions greater than four.

http://ssor.twi.tudelft.nl/~risk/. These projects are meant to sensitize the reader to reason with uncertainty, and modelling dependence. Many of these can also be done with popular uncertainty analysis packages that are available as spread sheet add-ons, such as Crystal Ball and @Risk. However, these packages do not support features such as multiple copula, vine modelling, cobweb plots, iterated and conditional sampling and probabilistic inversion. All projects can be performed with UNICORN Light, and step-by-step guidance is provided. Of course, the users can program these themselves.

In conclusion, we summarize the mathematical issues that arise in 'capturing' uncertainty over model input, propagating this uncertainty through a mathematical model, and using the results to support decision making. References to the relevant chapters are given below:

1. The standard product moment (or Pearson) correlation cannot be assessed independently of the marginal distributions, whereas the rank (or Spearman) correlation can (Chapter 3).

2. We cannot characterize the set of rank correlation matrices. We do know that the joint normal distribution realizes a 'thin' set of rank correlation matrices (Chapter 4).

3. There is no general algorithm for extending a partially specified matrix to a positive definite matrix (Chapter 4).

4. Even if we have a valid rank correlation matrix, it is not clear how we should define and sample a joint distribution with this rank correlation matrix. These problems motivate the introduction of regular vines (Chapter 4; sampling is discussed in Chapter 6).

5. Given sufficient multivariate data, how should we infer a graphical model, or conditional independence structure, which best fits the data (Chapter 5)?

6. After obtaining a simulated distribution for the model input and output, how can we analyse the results graphically (Chapter 7) and how can we characterize the importance of various model inputs with respect to model output (Chapter 8)?

7. How can we perform probabilistic inversion (Chapter 9)?

This book assumes knowledge of basic probability, statistics and linear algebra. We have put proofs and details in mathematical supplements for each chapter. In this way, the readers can follow the main line of reasoning in each chapter before immersing themselves in mathematical details.

2

Assessing Uncertainty on Model Input

2.1 Introduction

The focus of this book is not how to obtain uncertainty distributions, but rather what to do with them once we have them. Quantifying uncertainty is the subject of a vast literature; this chapter can only hope to point the reader in the right direction. More detailed attention is given to the assessment of dependence. Even here, however, the treatment must be cursory. Rank correlation and minimal information are formally introduced in Chapter 3; this chapter appeals to an intuitive understanding of these notions. The apoplexed may elect to surf this chapter and return in earnest after digesting Chapter 3.

As indicated in the previous chapter, the quantification and propagation of uncertainty become essential in precisely those situations where quantitative modelling cannot draw upon extensive historical, statistical or measurement data. Of course modellers should always use such data sources whenever possible. As data becomes sparse, modellers are increasingly forced to evaluate and combine disparate data sources not perfectly tailored to the purpose at hand. As the judgmental element increases, the task of interpreting and synthesizing scanty evidence exceeds the modeller's competence and must be turned over to domain experts. We have entered the arena of expert judgment.

It seems paradoxical at first, but upon reflection it is a truism: If quantification must rely on expert judgment, the experts will not agree. We do not use expert judgment to assess the speed of light in a vacuum; if asked, all experts would give the same answer. The speed of light has been measured to everyone's satisfaction and its value, 2.998×10^8 m/s, is available in standard references. At the other extreme, expert judgment is not applied to assess the possibility of the existence of a god. Theology is not science; there are no relevant experiments or measurements. Between these two extremes, there is a wide domain where sparse data must be synthesized and adapted for the purposes at hand by experts. By the nature of the case, different experts will do this in different ways, and it is exactly in these cases that the quantification and propagation of uncertainty is, or should be, the primary goal of quantitative modelling.

Structured expert judgment is increasingly accepted as scientific input in quantitative modelling. A salient example is found in recent court proceedings of the United Nations Claims Commission investigating damages from the 1990 Gulf War. Structured expert judgment was used to assess the health effect on the Kuwaiti population of Iraq's firing the Kuwaiti oil wells. The fires burned for nearly nine months in 1990–1991, and the plumes from these fires contained fine particulate matter that is believed to have adverse health effects. Although most reputable experts acknowledge the noxious effects of this form of pollution, the exact dose-response relation is a matter of substantial uncertainty. The Kuwaiti government based its claims on the combined uncertainty distributions of six top international experts on these dose-response relations. The concentrations to which the population was exposed were assessed on the basis of extensive measurements. The lawyers for Iraq raised many pertinent questions regarding the use of structured expert judgment to assess damages in tort law. The judges should decide whether the claimant's burden of proof has been met. Because of its very high profile, and because of the significant scientific effort underlying the claim, this case promises to have profound repercussions. The concluding chapter discusses this case in detail and speculates on its consequences.

2.2   Structured expert judgment in outline

Methods for assessing and combining expert uncertainty are available in the literature.[1] This material will not be rehearsed here. Instead, we itemize the main features of the structured expert judgment method as employed by the present authors (Cooke (1991)).

- A group of experts are selected.

- Experts are elicited individually regarding their uncertainty over the results of possible measurements or observations within their domain of expertise.

- Experts also assess variables within their field, the true values of which are known post hoc.

- Experts are treated as statistical hypotheses and are scored with regard to statistical likelihood (sometimes called 'calibration') and informativeness.

---

[1]A very partial set of references is: Bedford and Cooke (2002); Budnitz et al. (1997); Clemen and Winkler (1995, 1999); Clemen et al. (1995); Cooke (1991); Cooke and Goossens (2000a); Garthwaite et al. (appearing); Granger Morgan and Henrion (1990); Hogarth (1987); Hora and Iman (1989); O'Hagan and Oakley (2004).

- The scores are combined to form weights. These weights are constructed to be 'strictly proper scoring rules' in an appropriate asymptotic sense: experts receive their maximal expected long-run weight by, and only by, stating their true degrees of belief. With these weights, statistical accuracy strongly dominates informativeness – one cannot compensate poor statistical performance by very high information.

- The likelihood and informativeness scores are used to derive performance-based weighted combinations of the experts' uncertainty distributions.

The key feature of this method is the performance-based combination of expert uncertainty distributions. People with a palliative approach to expert judgment find this unsettling, but extensive experience overwhelmingly confirms that experts actually like this since indeed the performance measures are entirely objective.

This experience also shows that in the wide majority of cases, the performance-based combination of expert judgment gives more informative and statistically more accurate results than either the best expert or the 'equal weight' combination of expert distributions (Cooke (2004); Cooke and Goossens (2000b); Goossens et al. (1998)). Upon reflection, it is evident that equal weighting has a very serious drawback. As the number of experts increases, the equal weight combination typically becomes increasingly diffuse until it represents no one's belief and is useless for decision support. This is frequently seen as the number of experts exceeds, say, eight. The viability of equal weighting is maintained only by sharply restricting the number of experts who will be treated equally, leaving others outside the process. It appeals to a sort of *one-man-one-vote* (for registered voters) consensus ideal. Science, on the other hand, is driven by *rational* consensus. Ultimately, consensus is the equilibrium of power; in science, it is not the power of the ballot but the power of arguments that counts.

Does expert performance on the calibration variables (whose values are or become known) predict performance on the other variables? This question is usually impossible to answer for obvious reasons. If we could learn the true values of the variables of interest, we would not be doing expert judgment in the first place. Such evidence as exists points to 'yes' (van Overbeek (1999); Xu (2002)). Of course one can divide the calibration variables into two subsets and look at correlations of scores. This gives a favourable answer, but is not really answering the question (Bier (1983)).

2.3 Assessing distributions of continuous univariate uncertain quantities

We are concerned with cases in which the uncertain quantity can assume values in a continuous range of real numbers. An expert is confronted with an uncertain quantity, say X, and is asked to specify information about his subjective distribution over possible values of X. The assessment may take a number of different forms.

The expert may specify his cumulative distribution function, or density or mass function (whichever is appropriate). Alternatively, the analyst may require only partial information about the distribution. This partial information might be the mean and standard deviation, or it might be several quantiles of his distribution. For r in $[0, 1]$, the rth quantile is the smallest number x_r such that the expert's probability for the event $\{X \leq x_r\}$ is equal to r. The 50% quantile is the median of his distribution. Typically, $(5\%, 50\%, 95\%)$ or $(5\%, 25\%, 50\%, 75\%, 95\%)$ quantiles are queried. Distributions are fitted to the elicited quantiles such that:

1. The densities agree with the expert's quantile assessments;

2. The densities are minimally informative with respect to a background measure (usually either uniform or loguniform), given the quantile constraints (Kullback (1967)).

The actual questions put to the experts must be prepared very carefully to avoid unnecessary ambiguity and to get the experts to conditionalize on the proper background information. By way of example, the following is an elicitation question for the experts who participated in the Kuwait damage claim project.

Question	Setting	Exposure (Effect Interval)	Change	Pollutant	Composition	Baseline
2	EU	Long-term	$1~\mu g/m^3$	$PM_{2.5}$	Ambient	20 ug/m3

What is your estimate of the true, but unknown, percent change in the total annual, non-accidental mortality rate in the adult European population resulting from a permanent $1~\mu g/m^3$ reduction in long-term annual average $PM_{2.5}$ (from a population-weighted baseline concentration of 20 $\mu g/m^3$) throughout the EU? To express the uncertainty associated with the concentration-response relationship, please provide the 5th, 25th, 50th, 75th, and 95th percentiles of your estimate.

5% :_____   25%:_____   50% :_____   75%:_____   95%:_____

2.4   Assessing dependencies

As mentioned in the previous chapter, dependencies between input variables may be induced by probabilistic inversion. Briefly summarizing, suppose the range of some function is observable, but the domain of the function is not observable. Suppose that we want an uncertainty distribution over the domain, and that we have an uncertainty distribution over the range. We then invert the function probabilistically by finding a distribution over the domain, which, when pushed through this function, best matches the range-distribution. As in the example in Chapter 1, the best matching distribution will generally be the 'push through' of a distribution on the domain having dependencies.

Sometimes dependencies will be imposed by functional constraints on the input variables. Suppose, for example, we have yearly data on the numbers of registered voters for the Liberal (L), Conservative (C) and Other (O) parties, with the number of voters (N) constant. In a randomly chosen year we may represent the uncertainty that a random individual is Liberal, Conservative or Other as the distributions of L, C and O. However, these distributions cannot be independent since we must have $L + O + C = N$.

When input uncertainty is directly assessed by experts, the experts' uncertainties regarding these inputs may be the source of dependence. We focus here on the question, how should we extract information about dependencies from the experts?

One obvious strategy is to ask the experts directly to assess a (rank) correlation coefficient.[2] However, even trained statisticians have difficulty with this type of assessment. Two approaches have been found to work satisfactorily in practice. The choice between these depends on whether the dependence is *lumpy* or *smooth*. Consider uncertain quantities X and Y. If Y has the effect of switching various processes on or off, which influence X, then the dependence of X on Y is called *lumpy*. In this case, the best strategy is to elicit conditional distributions for X given the switching values of Y, and to elicit the probabilities for Y. This might arise, for example, if corrosion rates for underground pipes are known to depend on soil type (sand, clay, peat), where the soil type itself is uncertain. In other cases the dependence may be *smooth*. For example, uncertainties in biological half-lives of cesium in dairy and beef cattle are likely to be smoothly dependent. The biological half-life of cesium in dairy cattle is uncertain within a certain range. For each value in that range, there is a conditional distribution for the biological half-life of cesium in beef cattle given the dairy cattle value. We may expect this conditional distribution to change only slightly when we slightly vary the value in dairy cattle.

When the analyst suspects a potential smooth dependence between (continuous) variables X and Y, experts first assess their marginal distributions. They are then asked:

Suppose Y were observed in a given case and its values were found to lie above the median value for Y; what is your probability that, in this same case, X would also lie above its median value?

This probability can vary between 0 and 1. If X and Y are completely rank-correlated, its value is 1; and if they are completely anti-correlated, its value is 0; and if they are independent its value is $\frac{1}{2}$.

---

[2]We glide over mathematical issues that receive ample attention in the following chapters. Recapitulating the discussion at the end of Chapter 1: (1) The standard product moment (or Pearson) correlation cannot be assessed independently of the marginal distributions; the rank (or Spearman) correlation can. (2) It is not known which correlation matrices can be rank correlation matrices. (3) There is no general algorithm for extending a partially specified matrix to a positive definite matrix. (4) Even if we have a valid rank correlation matrix, it is not clear how we should define and sample a joint distribution with this rank correlation matrix. These problems motivate the introduction of regular vines in Chapter 4. Vines use rank and conditional rank correlations, and the results of this discussion apply.

This technique was used extensively in a joint EU-US study on the consequences of an accident in a nuclear power plant (Cooke and Goossens (2000b); Kraan and Cooke (2000a,b)).

Experts quickly became comfortable with this assessment technique and provided answers which were meaningful to them and to the project staff. If F_X, F_Y are the (continuous, invertible) cumulative distribution functions of X and Y respectively, the experts thus assess

$$\pi_{\frac{1}{2}}(X, Y) = P\left(F_X(X) > \frac{1}{2} \mid F_Y(Y) > \frac{1}{2}\right).$$

Consider all joint distributions for (X, Y) having margins F_X, F_Y, having rank correlation $r(X, Y)$, and having minimum information relative to the distribution with independent margins. For each $r \in [-1, 1]$, there corresponds a unique value for $\pi_{\frac{1}{2}}(X, Y) \in [0, 1]$. Hence, we may characterize the minimal informative distribution with rank correlation $r(X, Y)$ in terms of $\pi_{\frac{1}{2}}(X, Y)$, or more generally as $r(\pi_q)$ where for each $q \in (0, 1)$

$$\pi_q(X, Y) = P(F_X(X) > q \mid F_Y(Y) > q).$$

We illustrate the procedure. The functions $r(\pi_q)$ have been computed with the uncertainty package UNICORN. The results for $q = -0.9, \ldots, 0.9$ are shown in Figure 2.1.

When a single expert assesses $\pi_{\frac{1}{2}}(X, Y)$, then we simply use the minimum information joint distribution with rank correlation $r(\pi_{\frac{1}{2}})$ found from Figure 2.1. Suppose, for example, that $\pi_{\frac{1}{2}} = 0.8$. Start on the horizontal axis with $q = 0.5$ and travel up until reaching 0.8 on the vertical axis. The line for $r = 0.7$ passes very close to the point $(0.5, 0.8)$, hence $r(\pi_{\frac{1}{2}}) \approx 0.7$.

When we consider a weighted combination of experts distributions, a complication arises. Since the medians for X and Y will not be the same for all experts, the conditional probabilities $\pi_{\frac{1}{2}}(X, Y)$ cannot be combined via the linear pooling. However, the marginal distributions can be pooled, resulting in a 'Decision Makers' (DM) cumulative distribution functions $F_{X,DM}$ and $F_{Y,DM}$ for X and Y. Let $x_{DM,50}$ and $y_{DM,50}$ denote the medians for DM's distribution for X and Y. We can compute the conditional probabilities

$$P_e(X > x_{DM,50} \mid Y > y_{DM,50}) = \frac{P_e(X > x_{DM,50} \cap Y > y_{DM,50})}{P_e(Y > y_{DM,50})}$$

for each expert e. Since the probabilities on the right-hand side are defined over the same events for all experts, they can be combined via the linear pool. This yields a value for $\pi_{\frac{1}{2}}$ for DM, for which we can find the corresponding r.

For more details, we refer to sources cited in the preceding text. This discussion is intended only to convey an idea how input distributions for an uncertainty analysis may be obtained. The following chapters discuss in detail how this information is used.

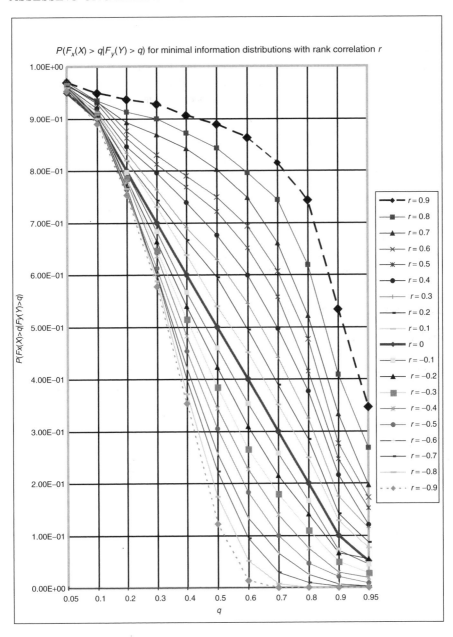

Figure 2.1 π_q for minimal information distributions with rank correlation r.

2.5   Unicorn

UNICORN is a stand-alone uncertainty analysis package. Dependence modelling is vine- and tree-based coupling of univariate random variables, though joint distributions in Ascii format are also supported. An extended formula parser is available, as well as a min-cost-flow network solver. Post-processing satellite programs provide for

- report generation

- graphics

- sensitivity analysis and

- probabilistic inversion.

A light version may be downloaded from http://ssor.twi.tudelft.nl/~risk/, which is suitable for doing all the projects in this book. These projects do not explore all facets of UNICORN, a tutorial downloaded with the program explores other features. A Pro version may be ordered from the same URL.

2.6   Unicorn projects

Project 2.1 Copulae

This project introduces different copulae and shows how to create a simple dependence model and study its properties graphically. A copula is a joint distribution on the unit square with uniform margins. The four copulae introduced here are supported in UNICORN and described in Chapter 3.

- *Launch UNICORN and click on the icon to create a new file. The random variable input panel appears.*

- *Click on the ADD button; a variable named V 1 is created with a uniform distribution on the interval [0,1] as default. We will accept this default. Other distributions can be assigned from the distribution list box. Click a second time on ADD, and accept the default. Note that information on the uniform distribution is visible on-screen: the density function, the parameters, main quantiles and moments.*

- *From the MODEL menu select Dependence, or simply click on the Dependence icon. This dependence input panel appears. From the dependence menu, select ADD NEW, and choose dependence tree. V 1 and V 2 are listed as available nodes. Double click on V 1, then on V 2. A tree is created with V 1 as root. The rank correlation between V 1 and V 2 is 0 by default. The rank correlation can be changed either with the slider which appears when the value 0 in the tree is clicked, or by entering a number between −1 and 1 in the correlation box, while the value in the tree is highlighted. Make the correlation 0.7.*

- *The diagonal band copula is chosen by default from the copula list box.*

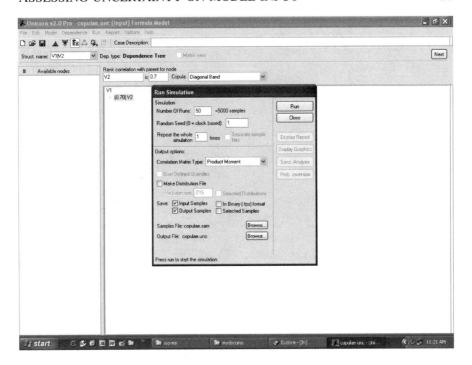

Figure 2.2   Simulate panel.

- *Now go to the RUN menu and select Simulate, or click on the Simulate icon. The panel shown in Figure 2.2 appears. One 'Run' is 100 samples. Choose 50 (5000 samples). Choose to save the input samples (there are no output samples here, as there are no User Defined Functions) and hit RUN. The simulation is almost instantaneous. The post-processing options are now available.*

- *Hit GRAPHICS to call the graphics package. When the UNIGRAPH package opens, hit the icon for scatter plots. The number of samples is controlled by a slider, and the dot size and colour are adjusted in the options menu. Your scatter plot with 5000 samples should look like the first picture in Figure 2.3.*

- *Go back to the dependence panel and repeat these steps using the successively the elliptical, Frank's and the Minimum Information copulae. The results should appear as in Figure 2.3. Note the close resemblance between Frank's copulae and the minimum information copula.*

Project 2.2 Bivariate normal and normals with copulae
This project compares the bivariate normal distribution and the result of two normal variables with the above four copulae. Unicorn includes the normal copula, but in this exercise you build your own bivariate normal distribution.

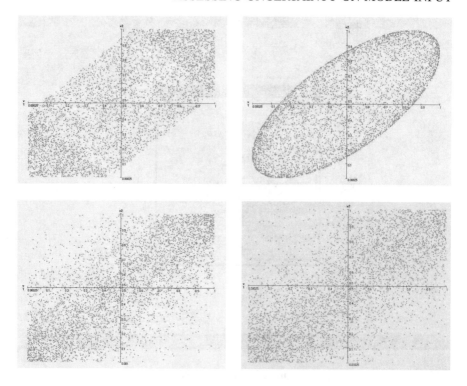

Figure 2.3 Diagonal band, elliptical, Frank's and Minimum Information copulae
with correlation 0.7.

- *Create a file with 4 standard normal variables (mean 0, unit standard deviation), named* $n1, n2, n3$ *and* $n4$.

- *Go to the formula panel, create a User Defined Function (UDF) named 'Y' and type '0.51*n1+0.49*n2' in the formula definition box.* $(n1, Y)$ *is a bivariate normal vector with correlation* $\rho(n1, Y) = 0.72$; *the rank correlation is 0.7 (see Chapter 3).*

- *Join* $n3$ *and* $n4$ *with rank correlation 0.7 using the minimum information copula; hit RUN and go to graphics and plot the scatter plots of* $(n1, Y)$ *and* $(n3, n4)$. *The results should look like the first two graphs in Figure 2.4.*

- *Go back to the dependence panel and choose Frank's copula, and make the scatter plot of* $(n3, n4)$ *again; the result should look like the third plot in Figure 2.4. Now do the same with the elliptical and diagonal band copulae. Note that the minimum information and Frank's copula produce scatter plots very similar to the bivariate normal plot, whereas the last two are quite different.*

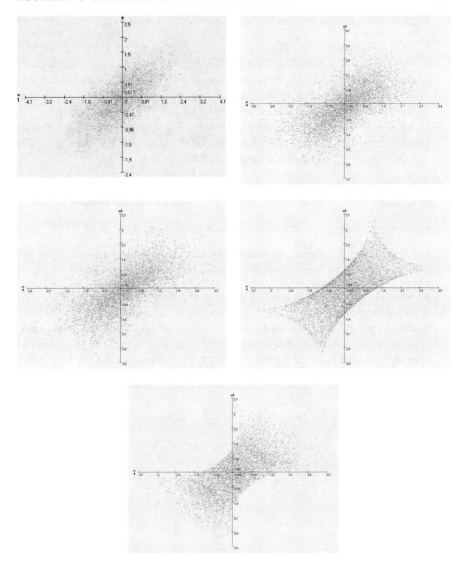

Figure 2.4   From upper left to lower right bivariate normal and normals with minimum information, Frank's, the elliptical and the diagonal band copula; all with rank correlation 0.7.

Project 2.3 Dependence Graph

The graph in Figure 2.1 is based on the probability that one uniform variate exceeds its qth percentile given that a second uniform variate has exceeded its qth percentile. This project shows how to compute these probabilities; we choose q = 0.5.

- *Create a case with two uniform [0,1] variables, $V1$ and $V2$.*

- *We are interested in the distribution of V1 conditional on $V2 \geq 0.5$. To perform conditional sampling with UNICORN, we must have a User Defined Function (UDF) named 'condition' taking values 0 or 1; if the value is 1 the sample is accepted, otherwise it is rejected. Create a UDF named 'condition' and enter its definition as 'i1{0.5,V2,1}'. This returns the value 1 if $0.5 \leq V2 \leq 1$, and returns zero otherwise. The report is trained to display the percentiles of UDFs, so we need a UDF equal to $V1$. Create a second UDF named U, and define it simply as $V1$.*

- *In the Dependence panel, make a dependence tree connecting $V1$ to $V2$. Assign a rank correlation 0.7 and choose the minimal information copula.*

- *Go to the Sample panel, choose 50 runs (5000 samples) and hit RUN.*

- *Click on Display Report. A panel appears with which you can configure the report. Check the box Output percentiles, and click on Generate.*

- *When the report appears, scroll down to the percentiles of the UDF U. The unconditional median of $V1$ is 0.5. From the percentiles of U, we see that there is about 20% probability that U is less or equal to 0.5. There is thus about 80% probability that V1 is ≥ 0.5, given that $V2 \geq 0.5$. Compare this to the curve in Figure 2.1 for $r = 0.7$. This curve crosses the line $q = 0.5$ at about the value 0.8.*

3

Bivariate Dependence

3.1 Introduction

The theory and history of bivariate dependence is rich and easily warrants a separate volume (Doruet Mari and Kotz (2001); Joe (1997)). In this chapter we restrict attention to notions and techniques which translate to higher dimensions and lend themselves to dependence modelling in uncertainty analysis.

The central notion in the study of bivariate dependence is correlation. The celebrated *product moment correlation coefficient* was apparently invented by Francis Galton in 1885, and thrown into its modern form by Karl Pearson in 1895 (Doruet Mari and Kotz (2001), p. 26), for which reason it is also called the *Pearson correlation*. Correlation was invented to further another of Galton's inventions, eugenics. Eugenics is the science of improving the human race through selective breeding ... an abiding concern of the English aristocracy. Galton applied correlation and statistical techniques to answer all sorts of questions, from the efficacy of prayer to the nature of genius (Gould (1981), p. 75). The psychologist Charles Spearman picked up the lance in 1904 and developed tools, most notably the Spearman or *rank correlation*, to isolate and study the illusive *g-factor*. The g-factor is hereditary intelligence, that putative quantity that is fostered by good breeding but not alas by naive plans of social reformers. Neither better education, nor higher wages, nor improved sanitation can enable the rabble to puncture the g-ceiling of bad breeding. The threat of working class g's to the English gentry was nothing compared to the g's from southern and eastern Europe washing onto the shores of the United States. Spearman's ideas had a profound impact on the American Immigration Restriction Act, which Spearman applauded in these terms: "The general conclusion emphasized by nearly every investigator is that as regards 'intelligence', the Germanic stock has on the average a marked advantage over the South European" (cited in Gould (1981), p. 271). Indeed, the results of intelligence tests administered to immigrants to the US in 1912 showed that 83% of the Jews,

Uncertainty Analysis with High Dimensional Dependence Modelling   D. Kurowicka and R. Cooke
© 2006 John Wiley & Sons, Ltd

80% of the Hungarians, 79% of the Italians, and fully 87% of the Russians were feeble-minded (Kamin (1974), p. 16).

3.2 Measures of dependence

Bivariate dependence is concerned with two related questions:

- How do we measure dependence between two random variables, and

- In which bivariate distributions do we measure them?

The first question leads us to concepts like correlation, rank correlation, Kendall's rank correlation, conditional correlation and partial correlation. The second question leads to the introduction of copulae.

The concept of *independence* is fundamental in probability theory. We can say that two events A and B are independent when the occurrence of one of them has no influence on the probability of the other. Otherwise they are *dependent*. More precisely,

Definition 3.1 (Independence) *Random variables* $X_1, \ldots, X_n$ *are* independent *if for any intervals* $I_1, \ldots, I_n$,

$$\mathbf{P}\{X_1 \in I_1, and \ldots X_n \in I_n\} = \prod_{i=1}^{n} \mathbf{P}\{X_i \in I_i\}.$$

Saying that variables are *not* independent does not say much about their joint distribution. What is the nature of this dependence? How dependent are they? How can we measure the dependence? These questions must be addressed in building a dependence model.

In this section, we present the most common measures of dependence based on linear or monotone relationship that are used later in this book. There are many other concepts of dependence proposed in the literature. For richer exposition and references to original papers, see for example, Doruet Mari and Kotz (2001); Joe (1997); Nelsen (1999).

3.2.1 Product moment correlation

The product moment correlation, also called linear or Pearson correlation is defined as follows:

Definition 3.2 (Product moment correlation) *The* **product moment correlation** *of random variables* X, Y *with finite expectations* $E(X), E(Y)$ *and finite variances* $\sigma_X^2, \sigma_Y^2, is$

$$\rho(X, Y) = \frac{E(XY) - E(X)E(Y)}{\sigma_X \sigma_Y}.$$

The product moment correlation can be also defined in terms of regression coefficients as follows: Let us consider X and Y with means zero. Let b_{XY} minimize

$$E\left((X - b_{XY}Y)^2\right);$$

and b_{YX} minimize

$$E\left((Y - b_{YX}X)^2\right);$$

Proposition 3.1

$$\rho(X, Y) = sgn\,(b_{XY})\,\sqrt{b_{XY}b_{YX}}.$$

In other words, the product moment correlation of X and Y is the appropriately signed geometric mean of the best linear predictor of X given Y and the best linear predictor of Y given X.

If we are given N pairs of samples (x_i, y_i) from the random vector (X, Y), we calculate the sample or population product moment correlation as follows:

$$\rho(X, Y) = \frac{\sum_{i=1}^{N}(x_i - \overline{X})(y_i - \overline{Y})}{\sqrt{\sum_{i=1}^{N}(x_i - \overline{X})^2}\sqrt{\sum_{i=1}^{N}(y_i - \overline{Y})^2}},$$

where $\overline{X} = \frac{1}{N}\sum_{i=1}^{N} x_i$ and $\overline{Y} = \frac{1}{N}\sum_{i=1}^{N} y_i$.

The product moment correlation is standardized:

Proposition 3.2 *For any random variables X, Y, with finite means and variances:*

$$-1 \leq \rho(X, Y) \leq 1;$$

In the propositions below, we see that in case of independent random variables the product moment correlation is equal to zero. It is not the case, however, that correlation zero implies independence. This can be immediately seen by taking X to be any random variable with $E(X) = E(X^3) = 0$, $E(X^2) > 0$. Take $Y = X^2$. The correlation between X and Y is equal to zero but they are obviously not independent. In the case of perfect linear dependence, that is, $Y = aX + b$, where $a \neq 0$ we get $\rho(X, Y) = 1$ if $a > 0$ and $\rho(X, Y) = -1$ if $a < 0$.

Proposition 3.3 *If X and Y are independent, then $\rho(X, Y) = 0$.*

Proposition 3.4 *For a, b real numbers*

(i) $\rho(X, Y) = \rho(aX + b, Y)$, $a > 0$;

(ii) $\rho(X, Y) = -\rho(aX + b, Y)$, $a < 0$;

(iii) *if $\rho(X, Y) = 1$ then for some $a > 0, b \in \mathbb{R}$, $X = aY + b$;*

In the proposition below, it is shown how correlation is affected by taking a mixture of joint distributions with identical univariate first and second moments.

Proposition 3.5 *If F and G are two bivariate distribution functions with identical univariate first and second moments, and $H = aF + (1 - a)G$, $0 \le a \le 1$, then*

$$\rho(H) = a\rho(F) + (1 - a)\rho(G),$$

where $\rho(H)$ denotes the product moment correlation of two variables with distribution H.

The product moment correlation is very popular because it is often straightforward to calculate. For uncertainty analysis it has few disadvantages, however.

a. The product moment correlation is not defined if the expectations and variances of X and Y are not finite (e.g. Cauchy distribution).

b. The product moment correlation is not invariant under non-linear strictly increasing transformations.

c. The possible values of the product moment correlation depend on marginal distributions (Theorem 3.1, below).

In the propositions below, these facts are established. Throughout, if X is a random variable, F_X denotes the cumulative distribution function of X.

Proposition 3.6 *For given X, Y let*

$$Rho(X, Y) = \{a| \text{ there exist } W, Z \text{ with } F_X = F_W \text{ and } F_Y = F_Z,$$

$$\text{and } \rho(W, Z) = a\}.$$

Then Rho is an interval, in general a strict subinterval of $[-1,1]$.

We can see in the example below that $Rho(X, Y)$ is a strict subinterval of $[-1, 1]$ that depends on the marginal distributions. Thus, given marginal distributions F_X and F_Y for X and Y, not all product moment correlations between -1 and 1 can be attained.

Example 3.1 *(Embrechts et al. (2002)) Let X and Y be random variables with support $[0, \infty]$, so that $F_X(x) = F_Y(y) = 0$ for all $x, y < 0$. Let the right endpoints of F_X and F_Y be infinite: the suprema of the sets $\{x \mid F_X(x) < 1\}$, $\{y \mid F_Y(y) < 1\}$ are infinite. Assume that $\rho(X, Y) = -1$, which would imply $Y = aX + b$, with $a < 0$ and $b \in \mathbb{R}$. It follows that for all $y < 0$*

$$F_Y(y) = \mathbf{P}(Y \le y) = \mathbf{P}(X \ge (y - b)/a)$$

$$\ge \mathbf{P}(X > (y - b)/a) = 1 - F_X((y - b)/a) > 0,$$

which contradicts the assumption $F_Y(y) = 0$ for $y < 0$. $\square$

Lemma 3.1 shows how the product moment correlation can be expressed in terms of the joint cumulative distribution function of random vectors (X, Y). This lemma is essential in the proof of Theorem 3.1.

Lemma 3.1 *Let (X, Y) be a vector of continuous random variables with support $\Omega = [a, b] \times [c, d]$ (a, b, c, d can be infinite). Let $F(x, y)$ be the joint cumulative distribution function of (X, Y) with cumulative marginal distribution functions $F_X(x)$ and $F_Y(y)$. Assume $0 < \sigma_X^2, \sigma_Y^2 < \infty$ then*

$$\rho(X, Y) = \frac{1}{\sigma_X \sigma_Y} \iint_\Omega F(x, y) - F_X(x) F_Y(y) \, dx \, dy.$$

We can now determine the set of product moment correlations, which are consistent with given margins. We say that random variables X and Y are *comonotonic* if there is a strictly increasing function G such that $X = G(Y)$ except on a set of measure zero; X and Y are *countermonotonic* if X and $-Y$ are comonotonic.

Theorem 3.1 (Hoeffding) *(Hoeffding (1940)) Let (X, Y) be a random vector with marginal distributions F_X and F_Y, and with $0 < \sigma_X^2, \sigma_Y^2 < \infty$; then*

a. *The set of all possible correlations is a closed interval $[\rho_{min}, \rho_{max}]$ and $\rho_{min} < 0 < \rho_{max}$.*

b. *The extremal correlation $\rho = \rho_{min}$ is attained if and only if X and Y are countermonotonic; similarly, the extremal correlation $\rho = \rho_{max}$ is attained if X and Y are comonotonic.*

From Hoeffding's theorem, we get that $\rho_{min} < 0 < \rho_{max}$. There are examples for which this interval is very small (Embrechts et al. (2002)).

Example 3.2 *Let $X = e^Z$, $Y = e^{\sigma W}$, where Z, W are standard normal variables. Then X and Y are lognormally distributed, and:*

$$\lim_{\sigma \to \infty} \rho_{min}(X, Y) = \lim_{\sigma \to \infty} \rho_{max}(X, Y) = 0.$$

Proof.
From Hoeffding's theorem ρ_{max}, ρ_{min}, are attained by taking $W = Z$, $W = -Z$ respectively. We calculate using the properties of the lognormal distribution. Since $E(Y) = e^{0.5\sigma^2}$ and $Var(Y) = e^{\sigma^2}(e^{\sigma^2} - 1)$ then

$$\rho_{max}(X, Y) = \frac{e^{0.5(1+\sigma)^2} - e^{0.5} e^{0.5\sigma^2}}{\sqrt{e(e-1)e^{\sigma^2}(e^{\sigma^2-1})}}$$

$$= \frac{e^\sigma - 1}{\sqrt{(e-1)(e^{\sigma^2} - 1)}}.$$

Similarly,

$$\rho_{min}(X, Y) = \frac{e^{-\sigma} - 1}{\sqrt{(e - 1)(e^{\sigma^2} - 1)}}.$$

Calculation of limits of ρ_{min} and ρ_{max} concludes the proof. $\square$

3.2.2 Rank correlation

The Spearman, or rank correlation was introduced by Spearman (1904). K. Pearson regarded the introduction of rank correlation as 'a retrograde step... I cannot therefore look upon the correlation of ranks as conveying any real idea of the correlation of variates, unless we have a means of passing from the correlation of ranks to the value of the correlation of the variates...' (Pearson (1904), p. 2). To this end, Pearson proved in Pearson (1904) the result in Proposition 3.25.

Definition 3.3 (Rank correlation) *The* **rank correlation** *of random variables X, Y with cumulative distribution functions F_X and F_Y is*

$$\rho_r(X, Y) = \rho(F_X(X), F_Y(Y)).$$

We will denote the rank correlation as ρ_r or sometimes simply r.

The population version of the rank correlation can be defined as proportional to the probability of concordance minus the probability of discordance for two vectors (X_1, Y_1) and (X_2, Y_2), where (X_1, Y_1) has distribution F_{XY} with marginal distribution functions F_X and F_Y and X_2, Y_2 are independent with distributions F_X and F_Y. Moreover (X_1, Y_1), (X_2, Y_2) are independent (Joe (1997)):

$$\rho_r = 3 \left(\mathbf{P}[(X_1 - X_2)(Y_1 - Y_2) > 0] - \mathbf{P}[(X_1 - X_2)(Y_1 - Y_2) < 0] \right). \quad (3.1)$$

If we are given N pairs of samples (x_i, y_i) for the random vector (X, Y), then to calculate rank correlation we must first replace the value of each x_i by the value of its rank among the other x_i's in the sample, that is $1, 2, \ldots, N$. If the x_i are all distinct, then each integer will occur precisely once. If some of x_i's have identical values, then we assign to these 'ties' the mean of ranks that they would have had if their values were slightly different (for more information about how to deal with ties see Press et al. (1992)). We apply the same procedure for the y_i's. Let R_i be the rank of x_i among the other x's, S_i be the rank of y_i among the other y's. Then the rank correlation is defined to be a product moment correlation of ranks and can be calculated as follows:

$$\rho_r(X, Y) = \frac{\sum_{i=1}^{N} (R_i - \overline{R})(S_i - \overline{S})}{\sqrt{\sum_{i=1}^{N} (R_i - \overline{R})^2} \sqrt{\sum_{i=1}^{N} (S_i - \overline{S})^2}},$$

where $\overline{R} = \frac{1}{N} \sum_{i=1}^{N} R_i$ and $\overline{S} = \frac{1}{N} \sum_{i=1}^{N} S_i$.

From Proposition 3.7 and Definition 3.3, there is a clear relationship between the product moment and the rank correlation. The rank correlation is a correlation of random variables transformed to uniform random variables. Hence we get immediately that rank correlation is symmetric and takes values from the interval $[-1,1]$.

Proposition 3.7 *If X is a random variable with a continuous invertible cumulative distribution function F_X, then $F_X(X)$ has the uniform on $[0,1]$ distribution, denoted by $U(0, 1)$.*

Rank correlation is independent of marginal distributions and invariant under non-linear strictly increasing transformations.

Proposition 3.8

a. *If $G : \mathbb{R} \to \mathbb{R}$ is a strictly increasing function, then*

$$\rho_r(X, Y) = \rho_r(G(X), Y);$$

b. *If $G : \mathbb{R} \to \mathbb{R}$ is a strictly decreasing function, then*

$$\rho_r(X, Y) = -\rho_r(G(X), Y);$$

c. *If $\rho_r(X, Y) = 1$, then there exists a strictly increasing function $G : \mathbb{R} \to \mathbb{R}$ such that*

$$X = G(Y).$$

In contrast to the product moment correlation, the rank correlation always exists and does not depend on marginal distributions.

Proposition 3.9 *For given X, Y having continuous invertible distribution functions let*

$$Rnk(X, Y) = \{a| \text{ there exist } W, Z \text{ with } F_X = F_W \text{ and } F_Y = F_Z,$$

$$\text{and } \rho_r(W, Z) = a\}.$$

Then $Rnk = [-1,1]$.

By definition, product moment and rank correlations are equal for uniform variables, but in general they are different. The following proposition shows how different they can be (Kurowicka (2001)).

Proposition 3.10 *Let $X_k = U^k$, $k \geq 1$, then*

$$\rho(X_1, X_k) = \frac{\sqrt{3(2k + 1)}}{k + 2} \to 0 \text{ as } k \to \infty;$$

$$\rho_r(X_1, X_k) = 1.$$

For the joint normal distribution, the relationship between rank and product moment correlations is known (Section 3.5, Proposition 3.25).

The rank correlation can be expressed in terms of copula (Section 3.4, Proposition 3.26).

3.2.3 Kendall's tau

Let (X_1, Y_1) and (X_2, Y_2) be two independent pairs of random variables with joint distribution function F and marginal distributions F_X and F_Y. Kendall's rank correlation, also called Kendall's tau (Kendall (1938)) is given by

$$\tau = \mathbf{P}[(X_1 - X_2)(Y_1 - Y_2) > 0] - \mathbf{P}[(X_1 - X_2)(Y_1 - Y_2) < 0]. \qquad (3.2)$$

To calculate Kendall's tau from data, we do not have to rank the data. If we have N data points (x_i, y_i), we consider all $\frac{N(N-1)}{2}$ unordered pairs of data points. We call the pair *concordant* if the ordering of the two x's is the same as the ordering of the two y's. We call a pair *discordant* if the ordering of the two x's is opposite from the ordering of the two y's. If there is a tie in either of the two x's or the two y's, then the pair is neither concordant nor discordant. If the tie is x's, we will call the pair an 'extra- y pair'. If the tie is in the y's, we will call the pair 'extra- x pair'. If the tie is in both the x's and the y's, we do not call the pair anything at all. The Kendall's τ is the combination of counts:

$$\tau = \frac{concord - discord}{\sqrt{concord + discord + extra\text{-}y}\sqrt{concord + discord + extra\text{-}x}}.$$

We can easily see that τ is symmetric and normalized to the interval $[-1, 1]$. It is also straightforward to verify the following proposition:

Proposition 3.11 *Let X and Y be random variables with continuous distributions. If X and Y are independent, then $\tau(X, Y) = 0$.*

Kendall's tau can be expressed in terms of the copula (Section 3.4, Proposition 3.26). This entails that Kendall's tau is invariant under continuous, increasing transformations. The relationship between product moment correlation and Kendall's tau for variables joined by the normal distribution is known (Proposition 3.26, Section 3.5).

3.3 Partial, conditional and multiple correlations

Partial correlation A partial correlation can be defined in a way similar to product moment correlation in terms of partial regression coefficients. Consider variables X_i with zero mean and standard deviations $\sigma_i = 1$, $i = 1, \ldots, n$. Let the numbers $b_{12;3,\ldots,n}, \ldots, b_{1n;2,\ldots,n-1}$ minimize

$$E\left((X_1 - b_{12;3,\ldots,n}X_2 - \ldots - b_{1n;2,\ldots,n-1}X_n)^2\right).$$

Definition 3.4 (Partial correlation)

$$\rho_{12;3,\ldots,n} = sgn(b_{12;3,\ldots,n})\sqrt{b_{12;3,\ldots,n}b_{21;3,\ldots,n}}.$$

Equivalently, we could define the partial correlation as

$$\rho_{12;3,...,n} = -\frac{C_{12}}{\sqrt{C_{11}C_{22}}},$$

where $C_{i,j}$ denotes the (i, j)th cofactor of the correlation matrix; that is, the determinant of the submatrix is gotten by removing row i and column j.

The partial correlation $\rho_{12;3,...,n}$ can be interpreted as the correlation between the orthogonal projections of X_1 and X_2 on the plane orthogonal to the space spanned by $X_3, \ldots, X_n$. Partial correlations can be computed from correlations with the following recursive formula (Yule and Kendall (1965)) (an illustrative calculation for three variables is given in the Supplement):

$$\rho_{12;3,...,n} = \frac{\rho_{12;3,...,n-1} - \rho_{1n;3,...,n-1} \cdot \rho_{2n;3,...,n-1}}{\sqrt{1 - \rho_{1n;3,...,n-1}^2}\sqrt{1 - \rho_{2n;3,...,n-1}^2}}. \tag{3.3}$$

Conditional correlation

Definition 3.5 (Conditional correlation) *The **conditional correlation** of Y and Z given X*

$$\rho_{YZ|X} = \rho(Y|X, Z|X)$$
$$= \frac{E(YZ \mid X) - E(Y \mid X)E(Z \mid X)}{\sigma(Y \mid X)\sigma(Z \mid X)}$$

is the product moment correlation computed with the conditional distribution of Y and Z given X.

For joint normal distribution, partial and conditional correlations are equal (Proposition 3.29). In general, however, partial and conditional correlations are not equal and the difference can be large:[1]

Proposition 3.12 *If*

a. *X is distributed uniformly on the interval $[0, 1]$,*

b. *Y, Z are conditionally independent given X,*

c. *$Y|X$ and $Z|X$ are distributed uniformly on $[0, X^k]$, $k > 0$,*

then

$$|\rho_{YZ|X} - \rho_{YZ;X}| = \frac{3k^2(k-1)^2}{4(k^4 + 4k^2 + 3k + 1)};$$

which converges to $\frac{3}{4}$ as $k \to \infty$.

Since Y and Z are conditionally independent given X, their conditional correlation is zero. Their partial correlation, however, is not zero. This is in sharp contrast to the situation described in Proposition 3.3.

---

[1]This example was suggested by P. Groeneboom, and published in Kurowicka (2001).

Multiple correlation

Definition 3.6 (Multiple correlation) *The **multiple correlation** $R_{1\{2,...,n\}}$ of variable X_1 with respect to $X_2, \ldots, X_n$ is given by:*

$$1 - R_{1\{2,...,n\}}^2 = \frac{D}{C_{11}}, \tag{3.4}$$

where D is the determinant of the correlation matrix and C_{11} is the $(1, 1)$ cofactor. It is the correlation between X_1 and the best linear predictor of X_1 based on $X_2, \ldots, X_n$.

In Kendall and Stuart (1961) it is shown that $R_{1\{2,...,n\}}$ is non-negative, is invariant under permutation of $\{2, \ldots, n\}$ and satisfies:

$$1 - R_{1\{2,...,n\}}^2 = (1 - R_{1\{3,...,n\}}^2)(1 - \rho_{1,2;3...n}^2)$$
$$= (1 - \rho_{1,n}^2)(1 - \rho_{1,n-1;n}^2)(1 - \rho_{1,n-2;n-1,n}^2) \ldots (1 - \rho_{1,2;3...n}^2).$$

Further, it is easy to show that (see Exercise 3.7)

$$D = \left(1 - R_{1\{2,...,n\}}^2\right)\left(1 - R_{2\{3,...,n\}}^2\right) \ldots \left(1 - R_{n-1\{n\}}^2\right). \tag{3.5}$$

Of course $R_{n-1\{n\}} = \rho_{n-1,n}$.

3.4 Copulae

The notion of 'copula' was introduced to separate the effect of dependence from the effect of marginal distributions in a joint distribution. A *copula* is simply a distribution on the unit square with uniform marginal distributions. (Here, we will also consider copulae as distributions on $[-\frac{1}{2}, \frac{1}{2}]^2$ with uniform margins on $[-\frac{1}{2}, \frac{1}{2}]$. This transformation simplifies calculations since then variables have means zero.) Copulas are then functions that join or 'couple' bivariate distribution functions to their marginal distribution functions.

Definition 3.7 *(Sklar (1959)) Random variables X and Y are **joined by copula** C if their joint distribution can be written*

$$F_{XY}(x, y) = C(F_X(x), F_Y(y)).$$

Every continuous bivariate distribution can be represented in terms of a copula. Moreover, we can always find a unique copula that corresponds to a given continuous joint distribution. For example, if Φ_ρ is the bivariate normal cdf with correlation ρ (see Section 3.5) and Φ^{-1} the inverse of the standard univariate normal distribution function then

$$\mathbf{C}_\rho(u, v) = \Phi_\rho\left(\Phi^{-1}(u), \Phi^{-1}(v)\right)$$

$u, v \in [0, 1]$ is called normal copula (see Figure 3.1).

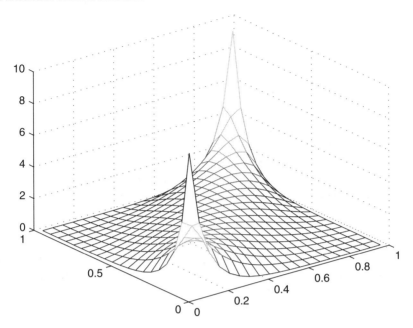

Figure 3.1   A density function of the normal copula with correlation 0.8.

If $f_{X,Y}$ and f_X, f_Y denote the nonzero density and marginal densities of (X, Y) with distribution function F_{XY}, then

$$c(F_X(x), F_Y(y)) = \frac{f_{XY}(x, y)}{f_X(x) f_Y(y)}.$$

is the copula density of C. Hence the density can be written in terms of the copula density and the marginal densities as:

$$f_{XY}(x, y) = c(F_X(x), F_Y(y)) f_X(x) f_Y(y).$$

It is easy to construct a distribution $C_I(U, V)$ with uniform margins such that U and V are independent:

$$C_I(u, v) = uv \quad (u, v) \in [0, 1]^2.$$

The rank correlation and Kendall's tau can be expressed in terms of the copula C (Nelsen (1999)).

Proposition 3.13 *If X and Y are continuous random variables with joint distribution function F and margins F_X and F_Y, let C denote the copula, that is, $F(x, y) = C(F_X(x), F_Y(y))$, then Kendall's tau and the rank correlation for X and Y can be* *expressed as:*

$$\tau = 4 \int_{[0,1]^2} C(s,t) dC(s,t) - 1,$$

$$\rho_r = 12 \int_{[0,1]^2} std C(s,t) - 3 = 12 \int_{[0,1]^2} C(s,t) \, ds \, dt - 3.$$

An overview of copulae can be found in e.g. Dall'Aglio et al. (1991); Doruet Mari and Kotz (2001); Joe (1997); Nelsen (1999). We present here only few families of copulae, which will be used in subsequent chapters. We are particularly interested in families of copulae that have the *zero independence* property: If the copula has zero correlation, then it is the independent copula.

3.4.1 Fréchet copula

The simplest copulae are the Fréchet copula C_L and C_U (Fréchet (1951)). They are also called *Fréchet bounds* because it can be shown that for any copula C

$$C_L \le C \le C_U.$$

C_L and C_U are bivariate distributions such that mass is spread uniformly on the main diagonal or anti-diagonal, respectively. We get for $(u, v) \in [0, 1]^2$

$$C_L(u, v) = \max(u + v - 1, 0),$$

$$C_U(u, v) = \min(u, v).$$

In Figure 3.2, the Fréchet copula C_L and C_U and the independent copula are shown.

Copulae C_U and C_L describe complete positive and negative dependence, respectively.

Proposition 3.14 *If U, V are joined by copula C_U (C_L), then $\rho_r(U, V) = 1$ (−1).*

A mixture of Fréchet copulae is a copula for which the mass is concentrated on the diagonal and anti-diagonal depending on parameter $A \in [0, 1]$:

$$C_A(u, v) = (1 - A)C_L(u, v) + AC_U(u, v)$$

for $(u, v) \in [0, 1]^2$.

By Proposition 3.5 all correlation values from the $[-1,1]$ interval can be realized in this way. We can construct a mixture of the Fréchet copula with correlation zero (take $A = \frac{1}{2}$). Of course variables joined by this mixture will not be independent. The Fréchet copulae are easy to simulate and give mathematical insight. However, as they concentrate mass on a set of Lebesque measure zero, they are not interesting for applications.

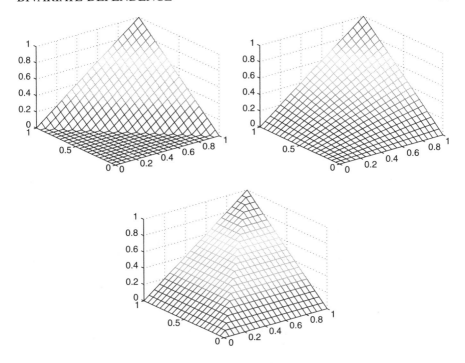

Figure 3.2 Starting from the left C_L, C_I and C_U.

3.4.2 Diagonal band copula

One natural generalization of Fréchet copula is the *diagonal band* copula introduced in Cooke and Waij (1986). In contrast to Fréchet copula, for positive correlation the mass is concentrated on the diagonal band with vertical bandwidth $\beta = 1 - \alpha$. Mass is distributed uniformly on the inscribed rectangle and is uniform but 'twice as thick' in the triangular corners (see Figure 3.3). We can easily verify that the height of the density function on the rectangle is equal to $\frac{1}{2\beta}$ and on the triangles $\frac{1}{\beta}$. For negative correlation, the band is drawn between the other corners. For positive correlations the density b_α of the diagonal band distribution is

$$b_\alpha(u, v) = \frac{1}{2(1 - \alpha)} \left(\mathbb{I}_{[\alpha-1, 1-\alpha]}(u - v) + \mathbb{I}_{[0, 1-\alpha]}(u + v) + \mathbb{I}_{[1+\alpha, 2]}(u + v) \right),$$

where $0 \leq \alpha \leq 1$, $0 \leq u, v \leq 1$ and $\mathbb{I}_A$ denotes indicator function of A. For negative correlations, the mass of the diagonal band density is concentrated in a band along the diagonal $v = 1 - u$. We then have bandwidth $\beta = 1 + \alpha$, $-1 \leq \alpha \leq 0$.

From this construction, we can easily see that the diagonal band distribution has uniform margins, is symmetric in u and v and the correlation between variables joined by this copula depends on bandwidth β. Moreover, for $\alpha = 0$ the variables are independent. For $\alpha = 1$, they are completely positively correlated and for

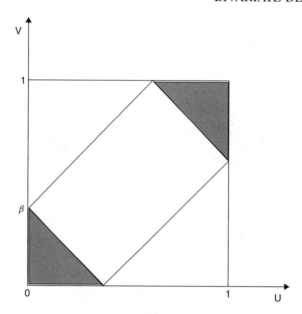

Figure 3.3 A diagonal band copula.

$\alpha = -1$ they are completely negatively correlated. The relationship between the parameter of the diagonal band copula and a correlation is given below (Cooke and Waij (1986)).

Proposition 3.15 *Let (U, V) be distributed according to $b_\alpha(u, v)$ with vertical bandwidth β and correlation ρ. Then the following relations hold*

$$\beta = 1 - |\alpha| = \frac{2}{3} - \frac{4}{3} \sin\left(\frac{1}{3} \arcsin\left(\frac{27}{16}|\rho| - \frac{11}{16}\right)\right), \qquad (3.6)$$

$$\rho = \text{sign}(\alpha)\left((1 - |\alpha|)^3 - 2(1 - |\alpha|)^2 + 1\right). \qquad (3.7)$$

β given by (3.6) is one of the three real solutions of (3.7). The two other solutions

$$\frac{2}{3} \pm \frac{2\sqrt{3}}{3} \cos\left(\frac{1}{3} \arcsin\left(\frac{27}{16}|\rho| - \frac{11}{16}\right)\right) + \frac{2}{3} \sin\left(\frac{1}{3} \arcsin\left(\frac{27}{16}|\rho| - \frac{11}{16}\right)\right)$$

do not yield values for $\beta \in [0, 1]$.

The density of the diagonal band copula with correlation 0.8 is shown in Figure 3.4.

Simulation of correlated variables with the diagonal band copula is very simple. First we sample $u \sim U(0, 1)$. If $u \in (\beta, 1 - \beta)$, then the conditional distribution $V|U = u$ is uniform on the interval $(u - \beta, u + \beta)$. Sampling from this distribution, we obtain v. If u is smaller than β, then the conditional distribution given u is

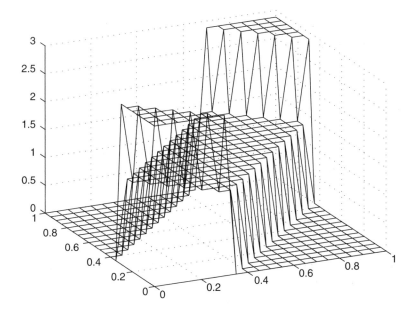

Figure 3.4 A density of the diagonal band copula with correlation 0.8.

piece-wise uniform: uniform on $(0, \beta - u)$ and on $(-u + \beta, u + \beta)$ with density on the first interval twice that on the second interval. The same holds mutatis mutandis when u is greater than $1 - \beta$.

For the diagonal band copula, we can find the conditional distribution and its inverse.

Proposition 3.16 *Let (U, V) be distributed according to $b_\alpha(u, v)$, $-1 \leq \alpha \leq 1$. Then the conditional and inverse conditional distribution of $V|U$ are the following:*

$$
F_{V|U}(v|u; \alpha) = \begin{cases} \frac{1}{1-\alpha} v & u < 1 - \alpha \text{ and } v < -u + 1 - \alpha \\ \frac{1}{1-\alpha}(v - \alpha) & u > \alpha \text{ and } v > -u + 1 + \alpha \\ 1 & u < \alpha \text{ and } v > u + 1 - \alpha \\ 0 & v < u - 1 + \alpha \\ \frac{1}{2(1-\alpha)}(v - u + 1 - \alpha) & \text{otherwise} \end{cases}
$$

and

$$
F_{V|U}^{-1}(t|u; \alpha) = \begin{cases} (1 - \alpha)t & u < 1 - \alpha \text{ and } t < 1 - \frac{u}{1-\alpha} \\ (1 - \alpha)t + \alpha & u > \alpha \text{ and } t > \frac{1-u}{1-\alpha} \\ 2(1 - \alpha)t + u - 1 + \alpha & \text{otherwise} \end{cases}
$$

for $0 \leq \alpha \leq 1$. Moreover, for $-1 \leq \alpha \leq 0$,

$$
F_{V|U}(v|u; \alpha) = F_{V|U}(v|1 - u; -\alpha),
$$

$$
F_{V|U}^{-1}(t|u; \alpha) = F_{V|U}^{-1}(t|1 - u; -\alpha).
$$

Having inverse cumulative distribution functions, the simulation of correlated uniform variables (U, V) with the diagonal band copula can be done with the following algorithm:

$$u = u_1,$$
$$v = F_{V|U}^{-1}(u_2|u, \alpha), \tag{3.8}$$

where u_1, u_2 are realizations of two independent uniformly distributed random variables U_1, U_2.

To simulate correlated variables X and Y with marginal distribution functions F_X and F_Y and rank correlation r_{XY} using the diagonal band copula, we simply sample U, V with the diagonal band with α corresponding to r_{XY} and apply the transformations:

$$x = F_X^{-1}(u), \quad y = F_Y^{-1}(v).$$

The main advantage with the diagonal band distribution is that it can be easily computed. This is convenient in modelling correlated events.

Example 3.3 *Let events A_1, A_2 each have probability 0.1. Find the diagonal band distribution such that*

$$\mathbf{P}(A_1, A_2) = 0.075.$$

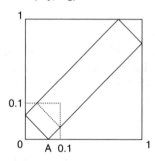

Note that if A_1 and A_2 were independent we should have $P(A_1 \cap A_2) = 0.01$. We want to find α such that probability of the square $[0, 0.1]^2$ is equal to 0.075. Using the properties of the diagonal band copula, we get

$$\mathbf{P}(A_1, A_2) = 0.075 = \frac{(1-\alpha)^2}{2} \frac{1}{1-\alpha} + \frac{(1-\alpha)^2}{2} \frac{1}{2(1-\alpha)}$$

$$+ (0.1 - (1-\alpha))\sqrt{2}(1-\alpha)\sqrt{2}\frac{1}{2(1-\alpha)}$$

$$= 0.1 - \frac{1-\alpha}{4};$$

so

$$\alpha = 1 - 0.1 = 0.9.$$

Using formula (3.7), we find that to realize $\mathbf{P}(A_1, A_2) = 0.075$ we must take diagonal band copula with correlation 0.981. $\square$

3.4.3 Generalized diagonal band copula

The construction of diagonal band copula can be seen as putting the uniform density on $[-\beta, \beta]$ and translating this along the diagonal, always folding the mass that lies outside the unit square inwards. It's easier to picture than to describe, see Figure 3.5. This procedure can be generalized by putting a non-uniform density $G(z)$ along lines perpendicular to the U axis such that $G(0)$ lies on the diagonal. In this way, a band of densities $G(z)$ will be constructed (see Figure 3.5).

This construction was presented in Bojarski (2001) and Ferguson (1995). We follow here the notation used in Ferguson (1995).

Let Z be absolutely continuous with density $G(z)$ for $z \in [0, 1]$. The density function of the generalized diagonal band copula is

$$bg(u, v) = \frac{1}{2}[G(|u - v|) + G(1 - |1 - u - v|)] \text{ for } 0 < u, v < 1. \quad (3.9)$$

Proposition 3.17 *The generalized diagonal band distribution has uniform margins and has rank correlation given by*

$$\rho_r = 1 + 4E(Z^3) - 6Var(Z),$$

where Z has density G.

When Z puts all its mass on 1, we get correlation -1, when Z has the uniform distribution on [0,1], then U and V are independent, and when Z puts all its mass on 0, correlation 1 is realized. The generalized diagonal copula generated with beta distribution with parameters 3,3 is shown in Figure 3.6.

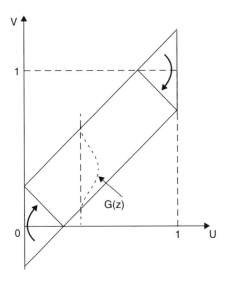

Figure 3.5 Construction of the generalized diagonal band copula.

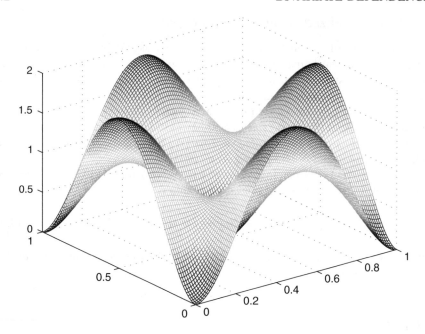

Figure 3.6   Generalized diagonal band copulae with $G(z) = \text{beta}(3, 3)$.

From (3.9) the conditional distribution of V given U follows:

$$(V|U = u) = \begin{cases} |Z - u| & \text{with probability } 1/2 \\ 1 - |1 - Z - u| & \text{with probability } 1/2. \end{cases}$$

This suggests a simple sampling method for this distribution. Sample U from a uniform distribution on $(0,1)$, choose Z independently from G and toss independently a fair coin. On heads, let $V = |Z - U|$ and on tails let $V = 1 - |1 - Z - U|$.

In Figure 3.7, we present three more examples of density functions of the generalized diagonal band distribution. We can see how interesting shapes can be obtained.

It has recently been shown that mixtures of diagonal band copulae are a strict subclass of the generalized diagonal band copulae (Lewandowski (2004)).

3.4.4   Elliptical copula

The elliptical copula[2] is absolutely continuous and can realize any correlation value in $(-1, 1)$ (Kurowicka et al. (2000)). In constructing this copulae properties of elliptically contoured and rotationally invariant random vectors were used

---

[2]The name of this copula was chosen to acknowledge the distribution that was used in constructing this copula. The uniform distribution on the ellipsoid in $\mathbb{R}^3$ was projected to two dimensions. There is, however, another usage according to which elliptical copulae correspond to elliptically contoured distributions.

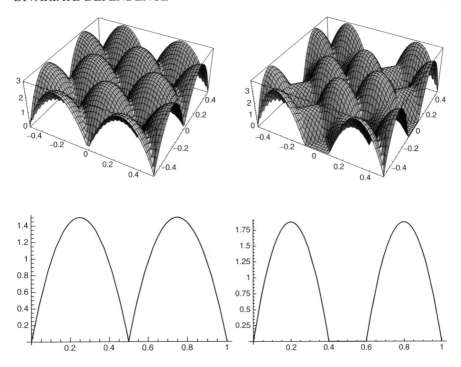

Figure 3.7 Density functions of the generalized diagonal copulae. Below each density is the corresponding $G(z)$ distribution.

(Hardin (1982); Misiewicz (1996)). A density function of the elliptical copula with correlation $\rho \in (-1, 1)$ is the following

$$f_\rho(u, v) = \begin{cases} \dfrac{1}{\pi\sqrt{\frac{1}{4}(1-\rho^2)-u^2-v^2+2\rho uv}} & (u, v) \in B \\ 0 & (u, v) \notin B, \end{cases}$$

where

$$B = \left\{ (u, v) \mid u^2 + \left(\frac{v - \rho u}{\sqrt{1-\rho^2}} \right)^2 < \frac{1}{4} \right\}.$$

Figure 3.8 depicts a graph of the density function of the elliptical copula with correlation $\rho = 0.8$.

In the proposition below, some properties of the elliptical copula are studied. We can find a closed form for the inverse conditional distribution, which is very important for simulation (algorithm 3.8).

Proposition 3.18 *Let U, V be uniform on $[-\frac{1}{2}, \frac{1}{2}]$. If U, V are joined by the elliptical copula with correlation ρ then*

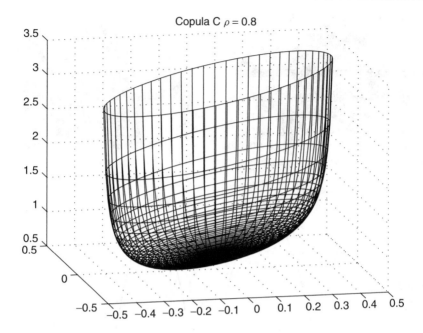

Figure 3.8   A density function of an elliptical copula with correlation 0.8.

a. $E(V|U = u) = \rho U,$

b. $Var(V|U = u) = \frac{1}{2}(1 - \rho^2)\left(\frac{1}{4} - U^2\right),$

c. for $\rho u - \sqrt{1 - \rho^2}\sqrt{\frac{1}{4} - u^2} \leq v \leq \rho u + \sqrt{1 - \rho^2}\sqrt{\frac{1}{4} - u^2}$

$$F_{V|U}(v|u) = \frac{1}{2} + \frac{1}{\pi}\arcsin\left(\frac{v - \rho u}{\sqrt{1 - \rho^2}\sqrt{\frac{1}{4} - u^2}}\right),$$

d. for $-\frac{1}{2} \leq t \leq \frac{1}{2}$

$$F_{V|U}^{-1}(t|u) = \sqrt{1 - \rho^2}\sqrt{\frac{1}{4} - u^2}\sin(\pi t) + \rho u.$$

The elliptical copula inherits some attractive properties of the normal distribution. For example, it has linear regression (property (a) in the preceding text), conditional correlations are constant (see below) and are equal to partial correlations.

Proposition 3.19 *Let X, Y, Z be uniform on $[-\frac{1}{2}, \frac{1}{2}]$ and let X, Y and X, Z be joined by elliptical copula with correlations ρ_{XY} and ρ_{XZ} respectively and assume that the conditional copula for YZ given X does not depend on X; then the conditional correlation $\rho_{YZ|X}$ is constant in X and*

$$\rho_{YZ;X} = \rho_{YZ|X}.$$

The elliptical copula does not enjoy the zero independence property. For correlation zero, mass is concentrated in a disk.

3.4.5 Archimedean copulae

Another very popular family are the Archimedean copulae. Properties of this family were studied by many authors for example, Doruet Mari and Kotz (2001); Genest and MacKay (1986); Joe (1997); Nelsen (1999). A function $\varphi : (0, 1] \to [0, \infty)$, which is convex, strictly decreasing with a positive second derivative such that $\varphi(1) = 0$ is called a *generator*. If we define the inverse (or quasi inverse) by

$$\varphi^{-1}(x) = \begin{cases} \varphi^{-1}(x) & \text{for } 0 \le x \le \varphi(0) \\ 0 & \text{for } \varphi(0) < x < \infty, \end{cases}$$

then an Archimedean copula can be defined as follows.

Definition 3.8 (Archimedean copula) *Copula $C(u, v)$ is **archimedean** with generator φ if*

$$C(u, v) = \varphi^{-1}[\varphi(u) + \varphi(v)]. \tag{3.10}$$

Since the second derivative φ'' of φ exists it is possible to find a density function c of C:

$$c(u, v) = -\frac{\varphi''(C)\varphi'(u)\varphi'(v)}{(\varphi'(C))^3}$$

The following elementary properties of Archimedean copula can be easily checked (Genest and MacKay (1986)).

Proposition 3.20 *Let $C(u,v)$ be given by (3.10), then*

a. $C(u,v)=C(v,u)$.

b. C *is a copula.*

c. U,V *joined by C are independent if $\varphi(x) = -c \log(x)$, where $c > 0$.*

There are many families of Archimedean copulae but only Frank's family (Frank (1979)) has a property of reflection symmetry, that is, $c(u, v) = c(1 - u, 1 - v)$. We find this property very important from an application point of view, so only Frank's family of Archimedean copula will be discussed further. Frank's copula has one parameter θ:

$$C(u, v; \theta) = -\frac{1}{\theta} \log \left(1 + \frac{(e^{-\theta u} - 1)(e^{-\theta v} - 1)}{(e^{-\theta} - 1)} \right)$$

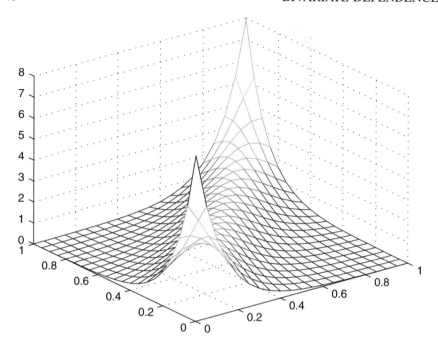

Figure 3.9 Density function of the Frank's Archimedean copulae with parameter $\theta = 7.9026$, and correlation 0.8.

with generating function

$$\varphi(x) = -\ln \frac{e^{-\theta x} - 1}{e^{-\theta} - 1}.$$

When $\theta \to \infty$ $(\theta \to -\infty)$, then Frank's copula corresponds to C_U (C_L). $\theta \to 0$ gives independent copula. Figure 3.9 shows Franks copula density with parameter $\theta = 7.9026$ and correlation 0.8.

The following proposition shows a simple way to calculate Kendall's tau for Archimedean copulae (Genest and MacKay (1986)).

Proposition 3.21 *Let U and V be random variables with an Archimedean copula C with the generating function φ. Then the Kendall's tau for U and V is given by*

$$\tau(U, V) = 1 + 4 \int_0^1 \frac{\varphi(t)}{\varphi'(t)} \, dt.$$

Proposition 3.22 *(Joe (1997)) Let U and V be random variables with the Frank's Archimedean copula $C(u, v; \theta)$, then the density, conditional distribution of $V|U$*

and inverse of the conditional distribution are as follows:

$$c(u, v; \theta) = \theta(1 - e^{-\theta})e^{-\theta(u+v)}/[1 - e^{-\theta} - (1 - e^{-\theta u})(1 - e^{-\theta v})]^2,$$

$$C_{V|U}(v|u; \theta) = e^{-\theta u}[(1 - e^{-\theta})(1 - e^{-\theta v})^{-1} - (1 - e^{-\theta u})]^{-1},$$

$$C_{V|U}^{-1}(t|u; \theta) = -\theta^{-1}\log\{1 - (1 - e^{-\theta})/[(t^{-1} - 1)e^{-\theta u} + 1]\}.$$

3.4.6 Minimum information copula

Intuitively, information is a scale-free measure of the 'concentratedness' in a distribution. It is mathematically more correct to speak of 'relative information' of one distribution with respect to another. Other terms are 'cross entropy', 'relative entropy' and 'directed divergence' (Kullback (1959)).

Definition 3.9 (Relative information) *If f and g are densities with f absolutely continuous with respect to g, then the **relative information** $I(f|g)$ of f with respect to g is*

$$I(f|g) = \int f(x)\log\left(\frac{f(x)}{g(x)}\right) dx.$$

If $(p_1, \ldots, p_n)$ and $(q_1, \ldots, q_n)$ are probability vectors with $q_i > 0$, $i = 1, \ldots n$ then the relative information of p with respect to q is

$$I(p|q) = \sum_{i=1}^{n} p_i ln(p_i/q_i).$$

If $p_i = 0$, then by definition $p_i ln(p_i/q_i) = 0$.

Properties of $I(f|g)$ are that $I(f|g) \geq 0$ and $I(f|g) = 0 \Leftrightarrow f = g$ (see Exercise 3.9). $I(f|g)$ can be interpreted as measuring the degree of 'uniformness' of f with respect to g.

The minimum information copula was introduced and studied in Meeuwissen (1993) and Meeuwissen and Bedford (1997). The construction is based on the fact that for any $\rho_r \in (-1, 1)$ there is a unique bivariate joint distribution satisfying the following constraints:

a. the marginal distributions are uniform on $I = (-1/2, 1/2]$;

b. the rank correlation is $\rho_r \in (-1, 1)$;

c. the distribution has minimal information relative to uniform distribution among all distributions with rank correlation ρ_r.

This is proved by solving the following optimization problem.

minimize
$$\int_I \int_I f(x, y) \log(f(x, y)) \, dx \, dy$$

subject to
$$\int_I f(x, y) \, dx = 1 \quad \forall y \in I,$$

$$\int_I f(x, y) \, dy = 1 \quad \forall x \in I,$$

$$\int_I xy f(x, y) \, dx \, dy = t,$$

$$f(x, y) \geq 0 \quad (x, y) \in I^2,$$

f is a continuous function.

It is shown in Meeuwissen and Bedford (1997) that this optimization problem has a unique solution of the form

$$\kappa(x, \theta)\kappa(y, \theta)e^{\theta xy},$$

where $\kappa : I \times \mathbb{R} \to \mathbb{R}$ is continuous and θ is a parameter corresponding to correlation. Moreover, the solution to the continuous optimization problem can be approximated by a sequence of discrete distributions of the following form (Nowosad (1966))

$$P(x_i, x_j) = \frac{1}{n^2}\kappa_i(\theta)\kappa_j(\theta)e^{\theta x_i x_j},$$

where $x_i = (2i - 1 - n)/n$ for $i = 1, \ldots, n$, and where $P(x_i, x_j)$ is a solution of the discretized problem:

minimize
$$\sum_{i=1}^n \sum_{i=1}^n p_{ij} \log(p_{i,j})$$

subject to
$$\sum_{i=1}^n p_{ij} = \frac{1}{n} \quad j = 1, \ldots, n,$$

$$\sum_{j=1}^n p_{ij} = \frac{1}{n} \quad i = 1, \ldots, n,$$

$$\sum_{i=1}^n \sum_{i=1}^n x_i y_j p_{ij} = \theta,$$

$$p_{ij} \geq 0,$$

where $(x_i, x_j) = ((2i - 1 - n)/n, (2j - 1 - n)/n)$ and $p_{ij} = P(x_i, x_j)$.

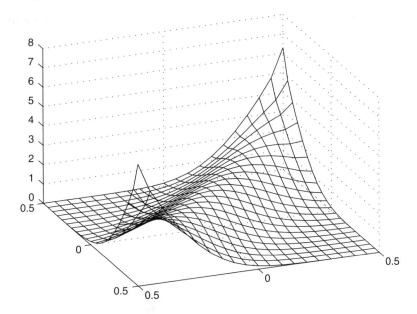

Figure 3.10   A density function of minimum information copula with correlation 0.8 (taking values on $[-0.5, 0.5]^2$).

The minimum information copula is attractive because it realizes a specified rank correlation by 'adding as little information as possible' to the product of the margins. Its main disadvantage is that it does not have a closed functional form. All calculations with this copula must involve numerical approximations.

An immediate consequence of this result is the following: given the arbitrary continuous marginal distributions F_X, F_Y, and $\rho_r \in (-1, 1)$, there is a unique joint distribution having these margins with rank correlation ρ_r, and having minimal information with respect to the independent distribution with margins F_X and F_Y.

3.4.7   Comparison of copulae

In this section, we compare the information values for the above copulae as functions of the rank correlation (Lewandowski (2004)).

The relative information values presented in Table 3.1 have been calculated numerically by first generating a given copula density on a grid of 500 by 500 cells, and then calculating the relative information based on the discrete approximation of its density. We used this method because there is no closed form expression for the density of the minimum information copula. The order of the copulae in the table reflects their performance in terms of the relative information in comparison with the minimum information copula with given rank correlation. Columns B-E express the relative information as a percentage of the relative information of the minimum information copula for the indicated correlation value. Table 3.1

Table 3.1   The relative information of the minimum information copula for a given rank correlation (A) and percent increment for other copulae.

Rank correlation	Copula				
r	A	B	C	D	E
0.1	0.00498	0.80%	5.02%	538.76%	11013.25%
0.2	0.02016	0.94%	5.16%	247.37%	2721.58%
0.3	0.04630	1.10%	5.62%	150.60%	1186.33%
0.4	0.08489	1.35%	7.03%	99.41%	648.72%
0.5	0.13853	1.69%	7.54%	69.40%	399.72%
0.6	0.21212	2.02%	6.19%	51.03%	263.74%
0.7	0.31526	2.35%	4.78%	37.39%	180.75%
0.8	0.47140	2.45%	3.64%	27.45%	124.70%
0.9	0.75820	2.18%	2.76%	18.58%	81.85%
0.95	1.06686	1.45%	1.91%	14.24%	60.51%
0.99	1.82640	1.16%	0.85%	8.84%	37.26%

A - Minimum information copula
B - Frank's copula
C - Generalized diagonal band copula generated with the triangle distribution
D - Diagonal band copula
E - Elliptical copula

shows that Frank's copula would be an attractive substitute for the minimal information copula: It does not add much information to the product of margins, it enjoys the zero independence property and it admits closed form expressions for the conditional and inverse conditional distributions. The relationship between the parameter of Frank's copula and the correlation must be found using numerical integration.

3.5   Bivariate normal distribution

The normal distribution is broadly used in practice. We recall the basic definitions and properties of this distribution.

3.5.1   Basic properties

The *Normal density* with parameters μ and $\sigma > 0$ is:

$$f_{\mu,\sigma}(x) = \frac{1}{\sqrt{2\pi}\sigma} \exp\left[-\frac{(x-\mu)^2}{2\sigma^2}\right], \quad -\infty < x < \infty.$$

and distribution function

$$F_{\mu,\sigma}(x) = \int_{-\infty}^{x} f_{\mu,\sigma}(s)\, ds.$$

The normal distribution with $\mu = 0$ and $\sigma = 1$ is called the *standard normal distribution*.

Any linear combination of normal variables is normal. If X has a normal distribution with parameters μ and σ, then $\frac{X-\mu}{\sigma}$ has a standard normal distribution.

The standard bivariate normal distribution is a joint distribution of two standard normal variables X and Y and has a density function

$$f_\rho(x, y) = \frac{1}{2\pi\sqrt{1-\rho^2}} \exp\left[-\frac{x^2 - 2\rho x y + y^2}{2(1-\rho^2)}\right], \quad -\infty < x, y < \infty.$$

The parameter ρ is equal to the product moment correlation of X and Y.

It is straightforward to show the following two propositions.

Proposition 3.23 *If (X, Y) has the standard normal distribution then $\rho(X, Y) = 0$ is a necessary and sufficient condition for X and Y to be independent.*

Proposition 3.24 *If (X, Y) has the standard bivariate normal distribution with correlation ρ, then*

a. *the marginal distributions of X and Y are standard normal;*

b. *the conditional distribution of Y given $X = x$ is normal with mean ρx and standard deviation $\sqrt{1 - \rho^2}$.*

We can find a relationship between the product moment, rank correlation and Kendall's tau.

Proposition 3.25 *(Pearson (1904)) Let (X, Y) be random vectors with joint normal distribution, then*

$$\rho(X, Y) = 2\sin\left(\frac{\pi}{6}\rho_r(X, Y)\right).$$

Proposition 3.26 *Let (X, Y) be random vectors with joint normal distribution then*

$$\tau(X, Y) = 2\arcsin\left(\rho(X, Y)\right)/\pi;$$

3.6 Multivariate extensions

3.6.1 Multivariate dependence measures

The bivariate dependence measures can be extended to the multivariate case by grouping all pairwise measures of dependence in the form of a matrix, say $\mathbb{M}$. If the pairwise measures are the product moment correlations, then the matrix is called the product moment correlation matrix. For n variables, there are $\binom{n}{2}$ off-diagonal terms in this matrix; for 100 variables that translates to 4950 correlations. Moreover, the matrix must be positive definite; that is, for all $x \in \mathbb{R}^n \setminus \{\mathbb{O}\}$, $x^T \mathbb{M} x > 0$. Equivalently, a matrix is positive definite if all its eigenvalues are positive. This constraint makes elements of the matrix $\mathbb{M}$ algebraically dependent. Changing one element generally requires changing others as well. In the example below, it is shown how elements of $\mathbb{M}$ are constrained when all off-diagonal terms of the matrix are equal.

Example 3.4 *Let $\mathbb{M}$ be $n \times n$ symmetric matrix with ones on the main diagonal and the same off-diagonal terms $m_{ij} = x$ when $i \neq j$. Then $\mathbb{M}$ is positive definite if, and only if, $x \in (-\frac{1}{n-1}, 1)$.*

Proof.
It is easy to see that $\mathbb{M}$ is of the form

$$\mathbb{M} = I + x(E - I) = (1 - x)I + xE$$

where I is the $n \times n$ identity matrix and E is an $n \times n$ matrix of ones. We know that a matrix is positive definite if, and only if, all its eigenvalues are positive. It is easy to check that the eigenvalues of $\mathbb{M}$ are $(1 - x) + xn$ (multiplicity 1) and $(1 - x)$ (multiplicity $n - 1$). Solving for which the values of x the eigenvalues are positive concludes the proof. $\square$

The constraint of positive definiteness is quite strong. To illustrate, we present results of a simple numerical experiment. Sampling 10,000 symmetric matrices with ones on the main diagonal and off-diagonal terms taken uniformly from $(-1,1)$ we found that for dimension 3, 6079 were positive definite, for dimension 4, 1811 were positive definite, for dimension 5, 206, for dimension 6, only 11, and for dimension 7, none of the 10,000 matrices were positive definite. In practical problems some correlations will be deemed important, others less so. Nonetheless, they must *all* be determined, and in such a way as to satisfy positive definiteness. If certain cells are left unspecified, then we create a so-called *completion problem*: extend the partially specified matrix so as to become positive definite. Some information on the completion problem is presented in Chapter 4.

From the proposition below, we see that all positive definite matrices can be obtained as a product of a lower triangular matrix L and its transpose L^T. This could be a way to specify a correlation matrix. If experts are to assess the matrix L, then its terms must be given an intuitive interpretation.

Proposition 3.27 *If $C \in \mathbb{M}$ is positive definite, then we may write $C = LL^T$, where L is lower triangular (Cholesky decomposition).*

In the proposition below, we can see that the covariance matrix is positive definite.

Proposition 3.28 *Let* $\mathbb{X}^T = [X_1, \ldots, X_n]$ *be a random vector with mean* $\mathbb{O}$. *Then*

a. $Cov(\mathbb{X}) = E(\mathbb{X}\mathbb{X}^T)$.

b. *If B is* $n \times n$ *matrix, then* $Cov(B\mathbb{X}) = B Cov(\mathbb{X}) B^T$.

c. *if* $A \in \mathbb{R}^n$, *then* $A^T Cov(\mathbb{X}) A \geq 0$ *with equality holding only if* $P(X_i = 0) = 1$ *for* $i = 1, \ldots, n$ *(in other words,* $Cov(\mathbb{X})$ *is positive definite).*

Another way to specify a correlation matrix is to use the partial correlation vine introduced in section 4.4.4.

3.6.2 Multivariate copulae

The concept of copula, that is, bivariate distribution with uniform one dimensional margins can be extended to the multivariate case (called *n-copula*)[3]. Many families of multivariate copulas are found in Dall'Aglio et al. (1991); Doruet Mari and Kotz (2001); Joe (1997); Nelsen (1999). All multivariate distributions with continuous margins such as the normal have their corresponding multivariate copula. There are also *n*-copulae that are generalizations of the Archimedean copulae. Formula (3.11) shows the *n*-dimensional extension of the Frank's copula.

$$C(u_1, \ldots, u_n; \theta) = -\theta^{-1} \log \left(1 - \frac{\prod_{i=1}^{n}(1 - e^{-\theta u_i})}{(1 - e^{-\theta})^{n-1}} \right) \qquad (3.11)$$

n-Archimedean, and n-copulae are limited in the correlation structures, which they can realize (Joe (1997)).

3.6.3 Multivariate normal distribution

The multivariate normal distribution is a joint distribution of n normally distributed variables $X_1, X_2, \ldots, X_n$. For all $\underline{x} = [x_1, x_2, \ldots, x_n] \in \mathbb{R}^n$ its joint density function is

$$f(\underline{x}) = \frac{1}{\sqrt{(2\pi)^n |V|}} \exp\left[-\frac{1}{2}(\underline{x} - \underline{\mu}) V^{-1} (\underline{x} - \underline{\mu})^T \right], \qquad (3.12)$$

where $\underline{\mu}$ is a vector, and V is a positive definite symmetric matrix with determinant $|V|$. The parameters $\underline{\mu}$ and V are the expectation vector

---

[3]Many authors define a copula as multidimensional distribution on unit hypercube with uniform marginal distribution.

$(E(X_i) = \mu_i,\ i = 1, 2, \ldots, n)$ and covariance matrix $(V = [v_{ij}],\ v_{ij} = Cov(X_i, X_j),\ i, j = 1, 2, \ldots, n)$, respectively.

For the normal distribution, conditional and partial correlations are equal. We show this here only for the trivariate case but the statement holds for higher dimensions as well (Kendall and Stuart (1961); Yule and Kendall (1965)).

Proposition 3.29 *If (X, Y, Z) has the trivariate joint normal distribution, then*

$$\rho(Y|X, Z|X) = \rho_{YZ;X}.$$

3.7 Conclusions

In this chapter, we have presented some basic concepts of bivariate dependence. We are interested in using such measures to model high-dimensional joint distributions. That means eliciting or otherwise inferring dependence structures and using these to perform Monte Carlo simulations.

The product moment correlation was shown to be easy to calculate and very popular but has many drawbacks from a modelling viewpoint:

1. It does not exist if the first or second moments do not exist;

2. Possible values depend on the marginal distributions; and

3. It is not invariant under non-linear strictly increasing transformations.

For variables joined by the bivariate normal distribution, however, the product moment correlation is a very attractive measure. It can take all values between -1 and 1 and zero correlation implies independence.

The rank correlation is a more flexible measure of dependence. It always exists, does not depend on marginal distributions and takes all values from the interval $[-1, 1]$. This algebraic freedom that the rank correlation gives is very important, for example, when the dependence must be assessed by experts. However, directly asking experts for a rank correlation (or product moment correlation) is not a good idea. It is difficult for experts to assess directly, and its population version (3.1) is a bit complicated. There are methods that indirectly infer the appropriate correlation value (see Chapter 2). Kendall's tau has a simple and intuitive interpretation (see (3.2)). It is widely used in connection with Archimedean copulae, for which it can be easily calculated (Proposition 3.21). Kendall's tau does lead to a tractable constraint for information minimization, but it is less convenient than rank correlation in this regard. To represent multivariate dependence, all of these measures of dependence between pairs of variables must be collected in a correlation matrix. Using a correlation matrix in high-dimensional dependence modelling has one strong disadvantage, namely, it requires specification of a full correlation matrix. It cannot work with a partially specified correlation structure. If extensive data is not available for this purpose, how should we proceed? We can ask experts to assess dependencies in the manner of Chapter 2, but we cannot do this for thousands of correlations. Even if we could, nothing short of a miracle would assure that the numbers obtained would be positive definite. Even if this miracle obliged for each expert individually, would it also oblige for the combination of experts? A new approach is called for, and this is taken up in the next chapter.

3.8   Unicorn projects

Project 3.1 Nominal values

This little project shows that 'nominal values' of random variables when plugged into simple functions may not give a good picture of the function's behaviour when the variables are dependent. Suppose we are interested in the function $G(XY) = XY$, where X and Y are uniformly distributed on [0,2], and $\rho_r(X, Y) = -1$. Suppose we wish to get an idea of G by plugging in the expectations of X and Y. Clearly, $G(EX, EY) = 1$; what is $E(G(X, Y))$?

- *Create a case with two variables, X, Y uniformly distributed on $[0, 2]$.*

- *Go to the function panel and create a user defined function UDF G: $X * Y$.*

- *Go to the dependence panel and build a dependence tree with X and Y with rank correlation -1.*

- *Simulate and generate a report. Click on the option 'Output percentiles'.*

- *Open the report. What is $E(G(X, Y))$? Which percentile of the distribution of G corresponds to $G(EX, EY)$?*

Compare the results with the ones obtained in the case when $\rho_r(X, Y) = 0$ and $\rho_r(X, Y) = 1$.

Project 3.2 Epistemic and aleatory uncertainty: polio incidence

Some applications require distinguishing 'epistemic' from 'aleatory' uncertainty. Consider a class of objects, which we wish to evaluate; it may be days, or people in a population, or spatial grid cells. Some variables may be uncertain, but whatever their values, they effect all members of the class in the same way. Such variables are said to possess epistemic uncertainty. Aleatory variables are also uncertain but their values are chosen independently for each element of the class. Some variables may have both aleatory and epistemic components. Of course from a mathematical viewpoint, 'epistemic' and 'aleatory' simply denote different types of dependence. We illustrate the 'repeat whole simulation' feature and the 'prev' function in UNICORN, and show how these can be used to distinguish these two types of uncertainty.

UNICORN can save the previous sample values of variables and UDFs in memory and use these to compute current values. When a UDF contains the expression 'prev(x)' in its definition, this expression is evaluated as the value of the UDF on the previous sample. 'x' may be a number or an input variable, and is used as a seed value on the first sample. You may also use 'prev([other UDF])' in the definition of a UDF, but then the other UDF must contain a prev expression in its definition, with seed value.

We illustrate with a simplified version of the dynamic Polio transmission model of Edmunds et al. (1999). Each sample represents a time step – in this case days. The transmission rate from the infected to the susceptible population is unknown, and is thought to vary from day to day depending on temperature, humidity and season. The transmission rate is modelled as the sum of an epistemic component [b] which is uncertain, but the same on every day, and an aleatory component [e] which varies from day to day. The number of susceptibles [s] in the current time step is determined by the number of people in the population who are not yet infected, have not previously recovered and have not died. The number of infecteds [i] in the current time step is determined by the number of infecteds and susceptibles in the previous time step and the transmission rate [tr]. A UDF 'day' is introduced to number the days. The precise definitions are given below. The random variables and their parameters for this model are pasted from the UNICORN report: The initial values for i and s are found by solving the equations for equilibrium, and do not concern us here.

Random variable bo: epistemic initial transmission coefficient [1/day]
Distribution type: LogUniform Parameters

$$a = 5E - 7, \ b = 1E - 6.$$

Random variable e: aleatory initial transmission coefficient [1/day]
Distribution type: Beta: Parameters

$$\alpha = 4.000, \ \beta = 4.000, \ a = -5E - 7, \ b = 5E - 7.$$

Random variable ro: initial recovery rate [1/day]
Distribution type: Constant:
$$CNST = 0.100.$$

Random variable mo: initial mortality and birth rate [1/day]
Distribution type: Constant
$$CNST = 5E - 5.$$

Random variable no.: initial population size
Distribution type: Constant

$$CNST = 1000000.000.$$

The UDFs are as follows:
1. r: prev(ro),
2. b: prev(bo),
3. tr: b*(1+0.05*sin(day/365))+e,
4. s: prev((mo+r)/b)*(1−tr*prev(i)−mo)+mo*no,
5. i: prev((b*no−(mo+r))*mo/(b*(mo+r)))*(tr*prev(s)−(r+mo)+1),
6. day: prev(0)+1.

Note that on any day the epistemic component of the transmission, b, is the value used on the previous day, starting with bo. The seed value bo is sampled from a log uniform distribution. If we perform the simulation once, the same value of b is used on each sample, drawn from the distribution of bo. If we repeat the simulation a second time, a different value for bo will be used on every sample. UNICORN is able to perform repeated simulations and store the results in one sample file. The steps are as follows:

- Create a new case with the random variables and UDFs as shown in the preceding text.

- Go to Simulate and for 'Repeat whole simulation' choose 20 repetitions, and do NOT check the box for separate sam files. Choose 10 runs (1000) samples, corresponding to a simulation of 1000 days. We see that statistical jitter caused by the daily variation in the transmission rate gradually dies out as the population reaches equilibrium.

- Go to graphics and make a scatter plot for 'i' and 'day'. If 'day' is on the vertical axis, click its box in the variable selection list until it becomes the horizontal axis. The 'plot samples' slider regulates the number of samples used in the graphic plot. If you move the slider with the mouse or arrow keys to the right, you see the number of samples increases. You also see the number of infecteds against days as days go from 1 to 1000. As you continue sliding, the number of days goes back to 1, and we are following the second simulation. Pushing the slider all the way to the left shows all 20 simulations. The results should look like Figure 3.11.

Project 3.3 NOx Aggregation

This project illustrates some handy tricks in working with large numbers of variables. Although UNICORN Light is restricted to 101 input variables, in fact using nested applications you can effectively have as many variables as you like. This is a stylized version of results published in Cooke (2004). This project also

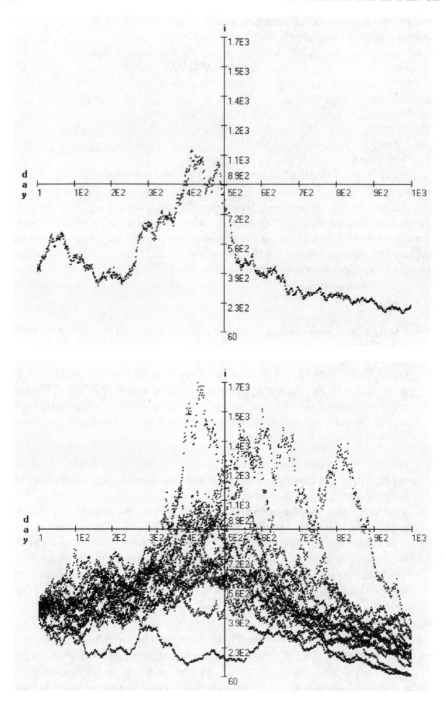

Figure 3.11 One sample path of Polio incidence with aleatory uncertainty, and 20 sample paths showing both aleatory and epistemic uncertainty.

concerns aleatory versus epistemic uncertainty – reflections on this example have persuaded many to take dependence seriously.

Let us review some probability. Suppose the amount of NOx emitted by one automobile in one year is normally distributed with mean 30[kg] and variance 25. The 5% and 95% quantiles of the NOx distribution will be $30 - \sqrt{25} \times 1.65 = 21.75$ and $30 + \sqrt{25} \times 1.65 = 38.25$. Consider now 5000 identical and independent auto's; how much will they emit? Well, the sum of 5000 identical and independent normal variables will be normal[4] with mean $30[Kg] \times 5000 = 150,000[Kg]$ and variance $25 \times 5000 = 125,000$. The 5% and 95% quantiles will be $150,000 - \sqrt{125,000} \times 1.65 = 149,416$, and $150,000 + \sqrt{125,000} \times 1.65 = 150,583$. The uncertainty relative to the mean becomes very small indeed. This is the effect of the central limit theorem: the fluctuations tend to interfere destructively as we add independent variables, and the spread grows much more slowly than the mean.

- *Create a file named NOx, and using Batch, create variables v1 ... v100. The variables remain selected until you assign a distribution; choose the normal distribution with mean 30 and standard deviation 5.*

- *Create a variable called 'dummy' and assign it the Uniform [0,1] distribution. This is for creating a symmetric dependence structure. Move this variable to the first position.*

- *From the Options panel choose 1000 quantiles (this is simply to avoid a large distribution file, see below).*

- *Go to the formula panel and create a UDF*

$$sum1\_50 = v1 + v2 + \cdots + v50.$$

  This requires some typing but with copy paste, it can be done more quickly. After typing $v1 + \ldots v10+$, copy these and paste them after 'v10+', then toggle insert off and just change the indices, and so on. Create a second UDF

$$sum51\_100 = v51 + \cdots + v100.$$

- *Go to the dependence panel and create a dependence tree as follows. Right click on the variables and select all, with right click, add this selection as child nodes. A tree is created with all 100 variables attached to 'dummy'. To rank-correlate the 100 variables to each other with correlation 0.09 (a very weak correlation), we can rank-correlate them all to dummy at 0.3. Select all correlations with shift+scroll down, and change the rank correlation to 0.3. Use the Minimum information copula.*

- *Go to the Simulate panel, choose 1000 samples (10 runs), choose Rank correlation, and check the Make a Distribution File, check selected variables, and select sum1_50. Now run.*

---

[4]Even if the individual auto's are not very normal, the sum will be approximately normal.

- *Generate a report, check Rank Correlation matrix. You do not want all variables in this matrix. In the column and row check lists, first select none, then select sum1_50 and sum51_100. Look at the correlation matrix. The rank correlation between these two UDFs is 0.81! This is worth remembering: If the variables v1...v100 were independent, then of course sum1_50 and sum51_100 would be independent as well. Adding a very weak correlation between the vi's creates a strong correlation between sums of these variables, and the more variables you sum, the stronger the correlation becomes (Exercise 3.14).*

- *Now go back to the variable panel; using shift + ctrl + end select all variables v1...v100. Click on Distribution file, and assign them the distribution file 'sum1_50.dis'. Each of the variables now has the distribution of the sum of 50 variables in the first run.*

- *Go to the formula panel, and change the names of the UDFs to sum1_2500 and sum2501_5000. Add a new UDF, named 'sumall': sum1_2500 + sum2501_5000.*

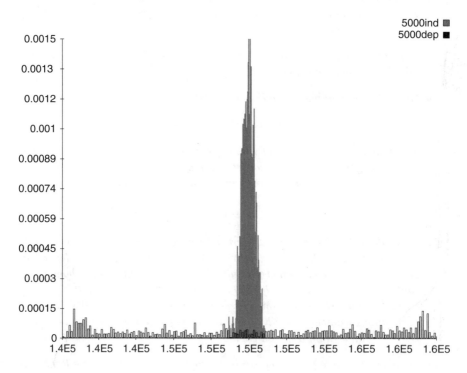

Figure 3.12 Sum of NOx emissions for 5000 autos, with independence and weak dependence between individual auto's.

- *Now simulate as before, 1000 samples, with rank correlation 0.81. Generate the report as before. The 5 and 95% values for the sum of 5000 automobiles is now 139000[Kg] and 161000[Kg]. The rank correlation between sum1_2500 and sum2501_5000 is 0.9.*

This example emphasizes that a very small correlation in the uncertainties of NOx pollution for individual autos causes a very large difference, compared to the assumption of independence as we aggregate over a population of autos. Neglecting dependence at the level of the individual auto can cause large errors at the population level. Compare this with the result of Exercise 3.14 below. Figure 3.12 compares the independent and dependent distribution of emissions from 5000 autos.

Project 3.4 NOx, Choice of copula

This project builds on the previous project to explore the effect of the choice of copula. Use the UDFs Sum1_50 and Sum51_100 with normal variables as in the previous project, but join them with the Minimal information, Frank's, Diagonal band and Elliptical copula's. Look at the scatter plots.

Now create two variables, say A and B and assign them the distribution Sum1_50.dis. Correlate them with rank correlation 0.81 using the same four copulas as above, and look at the scatter plots.

The eight scatter plots are shown in Figure (3.13). Notice that the Minimal information and Frank's copula's yield scatter plots very similar to those with the sums of 50 autos, whereas the Diagonal band and Elliptical copulas do not. This means that when the autos are joined with the Minimal information or Frank's copula's, the joint distribution is well approximated with A and B using the same copula's.

3.9 Exercises

Ex 3.1 *Prove Propositions 3.1, 3.2, 3.3, 3.4, 3.6.*

Ex 3.2 *Find the most negative correlation that can be attained between two exponential variables with parameter λ.*

Ex 3.3 *For events A, B the product moment correlation is defined as*

$$\rho(A, B) = \rho(\mathbb{I}_A, \mathbb{I}_B),$$

where $\mathbb{I}_A$ is the indicator function for A. Show that the following inequalities are sharp:

$$-\left[\frac{P(A')P(B')}{P(A)P(B)}\right]^{1/2} \le \rho(A, B)$$

$$\left[\frac{P(A)P(B')}{P(A')P(B)}\right]^{1/2} \ge \rho(A, B); \ P(A) \le P(B).$$

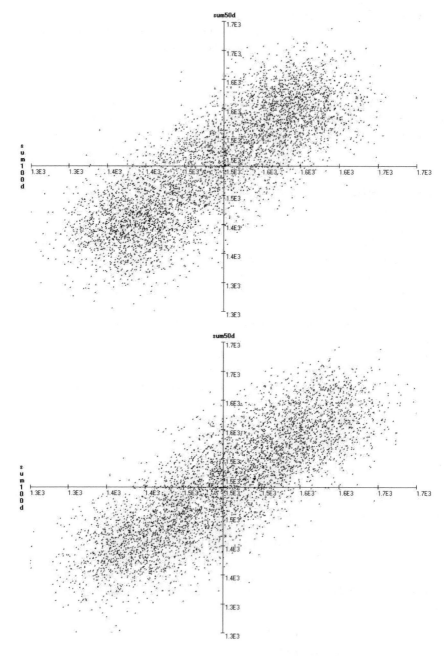

Figure 3.13 Scatter plots of (Sum 1_50) × (Sum 51_100) (top row) with Min. Inf, Frank's, Diagonal band and Elliptical copulas. The bottom row shows scatter plots of two variables with the distribution of the sum of 50 autos, with the copulas as in the top row.

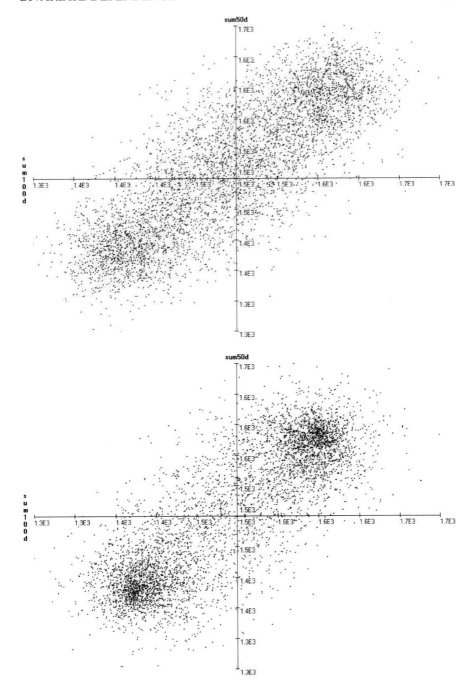

Figure 3.13 (*continued*)

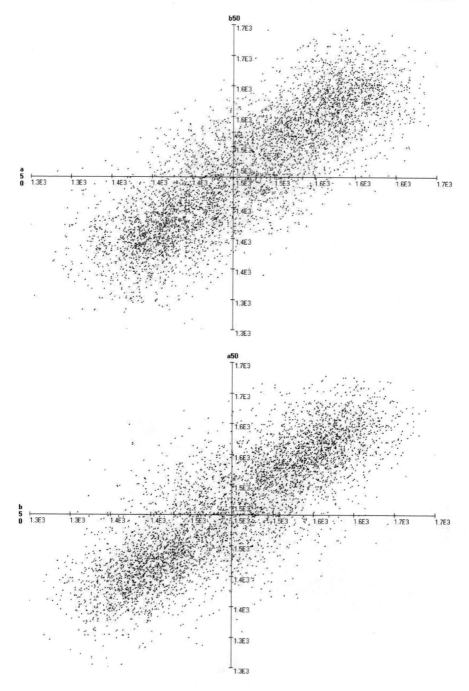

Figure 3.13 (*continued*)

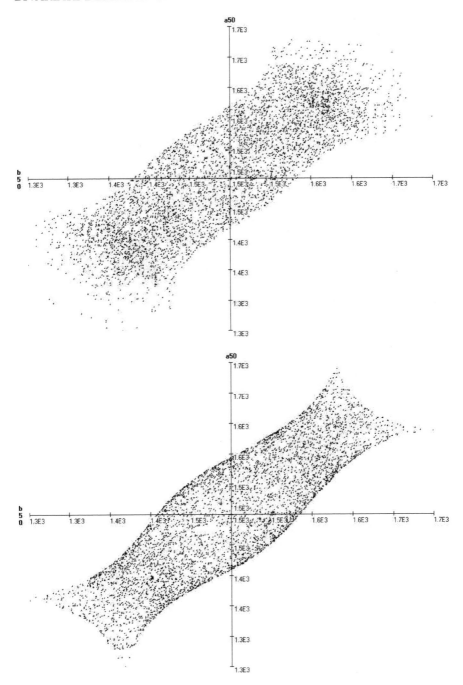

Figure 3.13 (*continued*)

Ex 3.4 *Prove Propositions 3.7, 3.8, 3.9, 3.10, 3.11.*

Ex 3.5 *For events A, B, define $\rho_r(A, B) = \rho(F_{\mathbb{I}_A}, F_{\mathbb{I}_B})$ then:*

$$\rho_r(A, B) = \rho(A, B).$$

Ex 3.6 *Let F be the cdf of a symmetric variable with finite second moment. Show that the full range of product moment correlations in the class of bivariate distributions with both univariate margins equal to F can be realized.*

Ex 3.7 *Show that if D is the determinant of a correlation matrix C, then*

$$D = \left(1 - R^2_{1\{2,\ldots,n\}}\right)\left(1 - R^2_{2\{3,\ldots,n\}}\right)\cdots\left(1 - R^2_{n-1\{n\}}\right);$$

Ex 3.8 *Prove Propositions 3.14, 3.15, 3.16, 3.18, 3.22.*

Ex 3.9 *Let $s = (s_1, \ldots, s_n)$ and $p = (p_1, \ldots, p_n)$ such that $p_i, s_i > 0$ and $\sum_{i=1}^{n} s_i = \sum_{i=1}^{n} p_i = 1$. Show that $I(s|p) = \sum_{i=1}^{n} s_i \log \frac{s_i}{p_i}$ is always non-negative and vanishes only if $s = p$. [Hint, recall that $x - 1 \geq \ln(x)$, with equality if and only if $x = 1$]*

Ex 3.10 *Prove equation 3.5.*

Ex 3.11 *Let (U, V) be distributed according to the distribution f shown in the figure below. The mass is uniformly distributed on the four lines inside the unit square.*

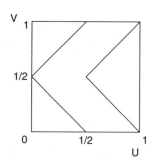

Show that

1. *f is a copula;*

2. *$E(U|V) \equiv 0$ and $E(V|U) \not\equiv 0$;*

3. *$\rho(U, V) = 0$;*

Ex 3.12 *The relative information for absolutely continuous bivariate distributions f and g is defined*

$$I(f|g) = \iint f(x, y) \ln \left(\frac{f(x, y)}{g(x, y)} \right) dx \, dy.$$

Show that the relative information with respect to bivariate uniform distribution u of variables joined by

a. *the elliptical copula g with correlation ρ is equal to*

$$I(g|u) = 1 + \ln 2 - \ln(\pi \sqrt{1 - \rho^2}),$$

The above can be shown using standard integration methods. One may have to calculate the following integral $-\frac{2}{\pi} \int_0^{\frac{\pi}{2}} \log(\cos u) \, du$ *which can be calculated with, for example, Maple as* $\log(2)$.

b. *the diagonal band copula with parameter α is equal to*

$$I(db|u) = -\ln 2^{|\alpha|}(1 - |\alpha|).$$

Show this for example, for the case $\alpha > \frac{1}{2}$.

Ex 3.13 *Prove Propositions 3.20, 3.22, 3.23, 3.24, 3.27, 3.28.*

Ex 3.14 *Let random variables* $X_1, X_2, \ldots, X_n, \ldots X_{2n}$ *have the same distributions F with mean* μ *and variance* σ^2. *Assume that* $(X_i, X_j), i \neq j, i, j = 1, 2, \ldots, 2n$ *have the same distributions H and correlation* $\rho(X_i, X_j) = \rho > 0$. *Show that*

$$\lim_{n \to \infty} \rho(\mathbb{X}_n, \mathbb{X}_{2n}) = 1$$

where $\mathbb{X}_n = \sum_{i=1}^{n} X_i$ *and* $\mathbb{X}_{2n} = \sum_{i=n+1}^{2n} X_i$.

3.10 Supplement

Proposition 3.5 If F and G are two dimensional distribution functions with identical univariate first and second moments, and for $0 \leq a \leq 1$, $H = aF + (1 - a)G$, then

$$\rho(H) = a\rho(F) + (1 - a)\rho(G),$$

where $\rho(H)$ means $\rho(X, Y)$ and joint distribution of X and Y is H.

Proof.

Since $E_F(X) = E_G(X)$ and $Var_F(X) = Var_G(X)$, we can easily show that $E_H(X) = E_F(X) = E_G(X)$ and $Var_H(X) = Var_F(X) = Var_G(X)$. Similarly for Y. Thus,

$$\alpha\rho(F) + (1-\alpha)\rho(G) = \frac{\alpha[E_F(XY) - E_F(X)E_F(Y)]}{\sqrt{Var_F(X)Var_F(Y)}} + \frac{(1-\alpha)[E_G(XY) - E_G(X)E_G(Y)]}{\sqrt{Var_G(X)Var_G(Y)}}.$$

$$= \frac{\alpha[E_F(XY) - E_H(X)E_H(Y)] + (1-\alpha)[E_G(XY) - E_H(X)E_H(Y)]}{\sqrt{Var_H(X)Var_H(Y)}},$$

$$= \frac{\alpha E_F(XY) + (1-\alpha)E_G(XY) - E_H(X)E_H(Y)}{\sqrt{Var_H(X)Var_H(Y)}} = \frac{E_H(XY) - E_H(X)E_H(Y)}{\sqrt{Var_H(X)Var_H(Y)}} = \rho(H). \quad \square$$

Lemma 3.1 Let (X, Y) be a vector of continuous random variables with support $\Omega = [a, b] \times [c, d]$ (a, b, c, d can be infinite). Let $F(x, y)$ be a joint cumulative distribution function of (X, Y) with cumulative marginal distribution functions $F_X(x)$ and $F_Y(y)$. Assume $0 < \sigma_X^2, \sigma_Y^2 < \infty$, then

$$\rho(X, Y) = \frac{1}{\sigma_X \sigma_Y} \iint_\Omega F(x, y) - F_X(x)F_Y(y) \, dx \, dy.$$

Proof.

Notice that it is enough to show that

$$E(XY) - E(X)E(Y) = \iint_\Omega F(x, y) - F_X(x)F_Y(y) \, dx \, dy.$$

We get

$$E(XY) - E(X)E(Y) = \iint_\Omega xy \, dF(x, y) - \int_a^b x \, dF_X(x) \int_c^d y \, dF_Y(y)$$

$$= \iint_\Omega xy \frac{\partial}{\partial x}\left(\frac{\partial}{\partial y}(F(x, y) - F_X(x)F_Y(y))\right) dx \, dy.$$

Recalling that $F(x, b) = F_X(x)$, *and so on*, integration by parts with respect to x and y concludes the proof. $\square$

Theorem 3.1 [Hoeffding] (Hoeffding (1940)) Let (X, Y) be a random vector with marginal distributions F_X and F_Y, and with $0 < \sigma_X^2, \sigma_Y^2 < \infty$; then

a. The set of all possible correlations is a closed interval $[\rho_{min}, \rho_{max}]$ and $\rho_{min} < 0 < \rho_{max}$.

b. The extremal correlation $\rho = \rho_{min}$ is attained if, and only if, X and Y are countermonotonic; similarly the extremal correlation $\rho = \rho_{max}$ is attained if X and Y are comonotonic.

Proof.

Using Lemma 3.1 we see that $\rho(X, Y)$ is maximum (minimum) if $\int_\Omega F(x, y)\, dx\, dy$ is maximum (minimum). Since each joint distribution can be represented in terms of the copula, $F(x, y) = C(F_X(x), F_Y(y))$ and since all copulae are bounded from below and above by C_L and C_U, respectively, then

$$\int_\Omega C_L(F_X(x), F_Y(y))\, dx\, dy \le \int_\Omega F(x, y)\, dx\, dy \le \int_\Omega C_U(F_X(x), F_Y(y))\, dx\, dy.$$

Hence the maximum (minimum) product moment correlations are attained when X and Y are comonotonic (countermonotonic). $\square$

Proof of Formula 3.3 for three variables

Let X_1, X_2, X_3 have mean zero and standard deviations $\sigma_i, i = 1, 2, 3$.
Since $b_{12;3}, b_{13;2}, b_{21;3}, b_{23;1}$ minimize

$$E([X_1 - b_{12;3}X_2 - b_{13;2}X_3]^2);$$
$$E([X_2 - b_{21;3}X_1 - b_{23;1}X_3]^2);$$

then

$$\frac{\partial}{\partial b_{12;3}} E([X_1 - b_{12;3}X_2 - b_{13;2}X_3]^2) = -2E([X_1 - b_{12;3}X_2 - b_{13;2}X_3]X_2) = 0,$$
$$\frac{\partial}{\partial b_{13;2}} E([X_1 - b_{12;3}X_2 - b_{13;2}X_3]^2) = -2E([X_1 - b_{12;3}X_2 - b_{13;2}X_3]X_3) = 0,$$
$$\frac{\partial}{\partial b_{21;3}} E([X_2 - b_{21;3}X_1 - b_{23;1}X_3]^2) = -2E([X_2 - b_{21;3}X_1 - b_{23;1}X_3]X_1) = 0,$$
$$\frac{\partial}{\partial b_{23;1}} E([X_2 - b_{21;3}X_1 - b_{23;1}X_3]^2) = -2E([X_2 - b_{21;3}X_1 - b_{23;1}X_3]X_3) = 0.$$

Hence

$$\begin{cases} Cov(X_1, X_2) - b_{12;3}Var(X_2) - b_{13;2}Cov(X_2, X_3) = 0, \\ Cov(X_1, X_3) - b_{12;3}Cov(X_2, X_3) - b_{13;2}Var(X_3) = 0, \\ Cov(X_1, X_2) - b_{21;3}Var(X_1) - b_{23;1}Cov(X_1, X_3) = 0, \\ Cov(X_2, X_3) - b_{21;3}Cov(X_1, X_3) - b_{23;1}Var(X_3) = 0. \end{cases}$$

Determine

$$b_{13;2} = \frac{Cov(X_1, X_3) - b_{12;3}Cov(X_2, X_3)}{Var(X_3)}$$

and

$$b_{23;1} = \frac{Cov(X_2, X_3) - b_{21;3}Cov(X_1, X_3)}{Var(X_3)}$$

from the second and fourth equations preceding text and substituting in the first and third yields

$$b_{12;3} = \frac{\frac{Cov(X_1, X_3)Cov(X_2, X_3)}{Var(X_3)} - Cov(X_1, X_2)}{\frac{Cov^2(X_2, X_3)}{Var(X_3)} - Var(X_2)},$$

$$b_{21;3} = \frac{\frac{Cov(X_1, X_3)Cov(X_2, X_3)}{Var(X_3)} - Cov(X_1, X_2)}{\frac{Cov^2(X_1, X_3)}{Var(X_3)} - Var(X_1)}.$$

From the above dividing numerator and denominator by $\sqrt{Var(X_1)Var(X_2)}$, we get

$$\rho_{23;1} = \sqrt{b_{12;3}b_{21;3}}$$

$$= \sqrt{\frac{\frac{Cov(X_1,X_3)Cov(X_2,X_3)}{Var(X_3)}-Cov(X_1,X_2)}{\frac{Cov^2(X_2,X_3)}{Var(X_3)}-Var(X_2)} \cdot \frac{\frac{Cov(X_1,X_3)Cov(X_2,X_3)}{Var(X_3)}-Cov(X_1,X_2)}{\frac{Cov^2(X_1,X_3)}{Var(X_3)}-Var(X_1)}}$$

$$= \sqrt{\frac{\frac{Cov(X_1,X_2)}{\sqrt{Var(X_1)Var(X_2)}} - \frac{Cov(X_1,X_3)Cov(X_2,X_3)}{Var(X_3)\sqrt{Var(X_1)Var(X_2)}}}{\left(\frac{Cov^2(X_2,X_3)}{Var(X_3)Var(X_2)}-1\right)\left(\frac{Cov^2(X_1,X_3)}{Var(X_3)Var(X_1)}-1\right)}}$$

$$= \frac{\rho_{12}-\rho_{13}\rho_{23}}{\sqrt{(1-\rho_{23}^2)(1-\rho_{13}^2)}}.$$

Proposition 3.12 If

(a) X is distributed uniformly on an interval $[0, 1]$,

(b) Y, Z are independent given X,

(c) $Y|X$ and $Z|X$ are distributed uniformly on $[0, X^k], k > 0$,

then

$$|\rho_{YZ|X} - \rho_{YZ;X}| = \frac{3k^2(k-1)^2}{4(k^4+4k^2+3k+1)},$$

which converges to $\frac{3}{4}$ as $k \to \infty$.

Proof.
We get

$$E(Y) = E(Z) = E(E(Y|X)) = E(\tfrac{X^k}{2}) = \tfrac{1}{2(k+1)},$$

$$E(Y^2) = E(Z^2) = E(E^2(Y|X)) = E(\tfrac{X^{2k}}{3}) = \tfrac{1}{3(2k+1)},$$

$$Var(Y) = Var(Z) = \tfrac{1}{3(2k+1)} - (\tfrac{1}{2(k+1)})^2,$$

$$E(XY) = E(XZ) = E(E(XY|X)) = E(X(E(Y|X)) = E(\tfrac{X^{k+1}}{2}) = \tfrac{1}{2(k+2)},$$

$$Cov(X, Y) = Cov(X, Z) = E(XY) - E(X)E(Y) = \tfrac{1}{2(k+2)} - \tfrac{1}{2}\tfrac{1}{2(k+1)},$$

$$E(YZ) = E(E(YZ|X)) = E(E(Y|X)E(Z|X)) = E(\tfrac{X^{2k}}{4}) = \tfrac{1}{4(2k+1)},$$

$$Cov(Y, Z) = E(YZ) - E(Y)(E(Z) = \tfrac{1}{4(2k+1)} - \tfrac{1}{4(k+1)^2}.$$

From the above calculations, we obtain

$$\rho_{YZ} = \frac{Cov(Y, Z)}{\sigma_Y\sigma_Z} = \frac{3k^2}{4k^2+2k+1}$$

and

$$\rho_{XY}\rho_{XZ} = \frac{Cov^2(X, Y)}{VarX\, VarY} = \frac{9k^2(2k + 1)}{(k + 2)^2(4k^2 + 2k + 1)}.$$

$$|\rho_{YZ|X} - \rho_{YZ;X}|$$

$$= \frac{\rho_{YZ} - \rho_{XY}\rho_{XZ}}{\sqrt{1 - \rho_{XY}^2}\sqrt{1 - \rho_{XZ}^2}} = \frac{3k^2(k-1)^2}{4(k^4 + 2k^2 + 3k + 1)} \rightarrow \frac{3}{4} \text{ as } k \rightarrow \infty. \quad \square$$

Proposition 3.13 (Nelsen (1999)) Let X and Y be continuous random variables with joint distribution function H and margins F and G, respectively. Let C denote the copula: $H(x, y) = C(F(x), G(y))$; then Kendall's tau and the rank correlation for X and Y can be expressed as:

$$\tau(X, Y) = 4 \int_{[0,1]^2} C(s, t)dC(s, t) - 1,$$

$$\rho_r(X, Y) = 12 \int_{[0,1]^2} st\, dC(s, t) - 3 = 12 \int_{[0,1]^2} C(u, v)\, du\, dv - 3.$$

Proof.
Let (X_1, Y_1) and (X_2, Y_2) be independent copies of (X, Y). Since the variables are continuous $P[(X_1 - X_2)(Y_1 - Y_2) < 0] = 1 - P[(X_1 - X_2)(Y_1 - Y_2) > 0]$ and then

$$\tau(X, Y) = 2P[(X_1 - X_2)(Y_1 - Y_2) > 0] - 1.$$

We get that $P[(X_1 - X_2)(Y_1 - Y_2) > 0] = P[X_1 > X_2, Y_1 > Y_2] + P[X_1 < X_2, Y_1 < Y_2]$. We can calculate these probabilities by integrating over the distribution of one of the vectors (X_1, Y_1) or (X_2, Y_2).

$$P[X_1 > X_2, Y_1 > Y_2] = P[X_2 < X_1, Y_2 < Y_1],$$

$$= \int_{\mathbb{R}^2} P[X_2 < x, Y_2 < y]dC(F(x), G(y)),$$

$$= \int_{\mathbb{R}^2} C(F(x), G(y))dC(F(x), G(y)).$$

Employing the transformation $u = F(x)$ and $v = G(y)$ yields

$$P[X_1 > X_2, Y_1 > Y_2] = \int_{[0,1]^2} C(u, v)dC(u, v).$$

Similarly,

$$P[X_1 < X_2, Y_1 < Y_2] = \int_{[0,1]^2} 1 - u - v + C(u, v)dC(u, v). \quad (3.13)$$

Since C is the joint distribution function of a pair (U, V) of uniform variables on $(0,1)$, then $E(U) = E(V) = \frac{1}{2}$ and (3.13) takes the following form

$$P[X_1 < X_2, Y_1 < Y_2] = 1 - \frac{1}{2} - \frac{1}{2} + \int_{[0,1]^2} C(u, v)dC(u, v)$$

$$= \int_{[0,1]^2} C(u, v)dC(u, v).$$

We then get

$$P[(X_1 - X_2)(Y_1 - Y_2) > 0] = 2 \int_{[0,1]^2} C(u, v)dC(u, v)$$

which concludes the proof for Kendall's tau. To prove the formula for rank correlation notice that

$$\rho_r(X, Y) = \rho(F(x), G(y)) = \rho(U, V) = \frac{E(UV) - 1/4}{1/12} = 12E(UV) - 3$$

$$= 12 \int_{[0,1]^2} uv dC(u, v) - 3.$$

This concludes the proof. $\square$

Proposition 3.17 The generalized diagonal band distribution has uniform margins and has Spearman's correlation given by

$$\rho_r = 1 + 4E(Z^3) - 6Var(Z),$$

where Z is a random variable whose density is the generator G.

Proof.
Notice that $E(X) = E(Y) = \frac{1}{2}$ and $Var(X) = Var(Y) = \frac{1}{12}$, where X and Y are uniform on the $[0,1]$ interval. To evaluate $E(X, Y)$, we first find $E(XY|Z = z)$, where Z has density G. Conditional on $Z = z$, (X, Y) is chosen from the uniform distribution on four line segments that form the boundary of the rectangle with corners $(z, 0)$, $(0, z)$, $(1 - z, 1)$ and $(1, 1 - z)$, hence

$$E(XY|Z = z) = \frac{1}{2} \left[\int_0^z x(z - x)\, dx + \int_z^1 x(x - z)\, dx \right.$$

$$+ \int_0^{1-z} x(x + z)\, dx + \left. \int_{1-z}^1 x(2 - x - z)\, dx \right]$$

$$= \frac{1}{3}z^3 - \frac{1}{2}z^2 + \frac{1}{3}.$$

Taking expectation with respect to Z, multiplying by 12 and subtracting 3 gives the result. $\square$

Proposition 3.19 Let X, Y, Z be uniform on $[-\frac{1}{2}, \frac{1}{2}]$ and let X, Y and X, Z be joined by elliptical copula with correlations ρ_{XY} and ρ_{XZ} respectively and assume that the conditional copula for YZ given X does not depend on X; then the conditional correlation $\rho_{YZ|X}$ is constant in X and

$$\rho_{YZ;X} = \rho_{YZ|X}.$$

Proof.
We calculate the conditional correlation for an arbitrary copula $f(u, v)$ using Proposition 3.18:

$$\rho_{YZ|X} = \frac{E(YZ|X) - E(Y|X)E(Z|X)}{\sigma_{Y|X}\sigma_{Z|X}}.$$

Let $I = [-\frac{1}{2}, \frac{1}{2}]$, then

$$E(YZ|X) = \int_{I^2} F_{Y|X}^{-1}(u) F_{Z|X}^{-1}(v) f(u, v) \, du \, dv$$

$$= \int_{I^2} \rho_{XY}\rho_{XZ} X^2 f(u, v) \, du \, dv$$

$$+ \int_{I^2} \sqrt{1 - \rho_{XY}^2} \rho_{XZ} \sqrt{\frac{1}{4} - X^2} X \sin(\pi u) f(u, v) \, du \, dv$$

$$+ \int_{I^2} \sqrt{1 - \rho_{XZ}^2} \rho_{XY} \sqrt{\frac{1}{4} - X^2} X \sin(\pi v) f(u, v) \, du \, dv$$

$$+ \int_{I^2} \sqrt{1 - \rho_{XY}^2} \sqrt{1 - \rho_{XZ}^2} \left(\frac{1}{4} - X^2\right) \sin(\pi u) \sin(\pi v) f(u, v) \, du \, dv.$$

Since f is a density function of the copulae and

$$\int_I \sin(\pi u) \, du = 0,$$

then we get

$$E(YZ|X) = \rho_{XY}\rho_{XZ} X^2 + \sqrt{1 - \rho_{XY}^2} \sqrt{1 - \rho_{XZ}^2} \left(\frac{1}{4} - X^2\right)$$

$$\int_{I^2} \sin(\pi u) \sin(\pi v) f(u, v) \, du \, dv.$$

Put

$$\mathbf{I} = \int_{I^2} \sin(\pi u) \sin(\pi v) f(u, v) \, du \, dv,$$

then from the above calculations and Proposition 3.18 we obtain

$$\rho_{YZ|X} = \frac{\rho_{XY}\rho_{XZ}X^2 + \sqrt{1 - \rho_{XY}^2}\sqrt{1 - \rho_{XZ}^2}(\frac{1}{4} - X^2)\mathbf{I} - E(Y|X)E(Z|X)}{\sigma_{Y|X}\sigma_{Z|X}}$$

$$= 2\mathbf{I}.$$

Hence the conditional correlation $\rho_{YZ|X}$ does not depend on X, and is constant, say $\rho_{YZ|X} = \rho$. Moreover, this result does not depend on the copula f. Now we show that conditional correlation is equal to partial correlation. The partial correlation $\rho_{YZ;X}$ can be calculated in the following way

$$\rho_{ZY;X} = \frac{\rho_{ZY} - \rho_{XY}\rho_{XZ}}{\sqrt{(1 - \rho_{XY}^2)(1 - \rho_{XZ}^2)}}.$$

We also get

$$\rho = \rho_{YZ|X} = \frac{E(YZ|X) - E(Y|X)E(Z|X)}{\sigma_{Y|X}\sigma_{Z|X}} = \frac{E(YZ|X) - \rho_{XY}\rho_{XZ}X^2}{\sigma_{Y|X}\sigma_{Z|X}}.$$

Hence

$$E(YZ|X) = \rho\sigma_{Y|X}\sigma_{Z|X} + \rho_{XY}\rho_{XZ}X^2.$$

Since

$$\rho_{ZY} = \frac{E(E(YZ|X))}{\sigma_X^2},$$

then

$$\rho_{YZ;X} = \frac{\rho E(\sigma_{Y|X}\sigma_{Z|X})}{\sigma_X^2\sqrt{(1 - \rho_{XY}^2)(1 - \rho_{XZ}^2)}}.$$

Since by Proposition 3.18

$$\sigma_{Y|X} = \sqrt{\frac{1}{2}(1 - \rho_{XY}^2)\left(\frac{1}{4} - X^2\right)}, \quad \sigma_{Z|X} = \sqrt{\frac{1}{2}(1 - \rho_{XZ}^2)\left(\frac{1}{4} - X^2\right)};$$

we have

$$\rho_{YZ;X} = \frac{\rho E\left(\sqrt{\frac{1}{2}(1 - \rho_{XY}^2)(\frac{1}{4} - X^2)}\sqrt{\frac{1}{2}(1 - \rho_{XY}^2)(\frac{1}{4} - X^2)}\right)}{\sigma_X^2\sqrt{(1 - \rho_{XY}^2)(1 - \rho_{XZ}^2)}}$$

$$= \frac{\rho\frac{1}{2}E(\frac{1}{4} - X^2)}{\sigma_X^2} = \frac{\rho\frac{1}{2}(\frac{1}{4} - \frac{1}{12})}{\frac{1}{12}} = \rho. \quad \square$$

Proposition 3.21 Let X and Y be random variables with an Archimedean copula C with the generating function φ. Then the Kendall's tau for X and Y is given by

$$\tau(X, Y) = 1 + 4 \int_0^1 \frac{\varphi(t)}{\varphi'(t)} dt.$$

Proof.
By Proposition 3.13 we get

$$\tau = 4 \int_{[0,1]^2} C(u, v) dC(u, v) - 1 = 4 \int_{[0,1]^2} C(u, v) c(u, v) \, ds \, dt - 1. \quad (3.14)$$

Noticing that $C(u, v) = 0$ for all (u, v) such that $\varphi(u) + \varphi(v) = \varphi(0)$ and using the formula for $c(u, v)$, we can rewrite the integral in (3.14) as follows

$$\int_{[0,1]^2} C(u, v) c(u, v) \, du \, dv = - \int_{\varphi(u)+\varphi(v)<\varphi(0)} C(u, v) \frac{\varphi''(C)\varphi'(u)\varphi'(v)}{[\varphi'(C)]^3} \, du \, dv.$$

Taking transformation

$$s = C(u, v) = \varphi^{-1}[\varphi(u) + \varphi(v)] \quad t = v$$

with Jacobian $\frac{\partial(s,t)}{\partial(u,v)} = -\frac{\varphi'(u)}{\varphi'(C)}$ and $s \leq t$ we get

$$- \int_0^1 \int_s^1 s \frac{\varphi''(s)}{\varphi'(s)} \varphi'(t) \, dt \, ds.$$

Hence

$$\int_0^1 \int_s^1 s \frac{\varphi''(s)\varphi(s)}{\varphi'(s)} \, ds$$

Integrating by parts, the above integral becomes

$$\int_0^1 \frac{\varphi(s)}{\varphi'(s)} \, ds + \frac{1}{2}.$$

Substitution to (3.14) concludes the proof. $\square$

Proposition 3.25 Pearson (1904) Let (X, Y) have a joint normal distribution then

$$\rho(X, Y) = 2 \sin\left(\frac{\pi}{6} \rho_r(X, Y)\right).$$

Proof.
A density function of the standard normal vector (X, Y) with correlation ρ is

$$f(x, y) = \frac{1}{2\pi \sqrt{1 - \rho^2}} \exp\left[-\frac{1}{2(1 - \rho^2)} \left(x^2 - 2\rho xy + y^2\right)\right].$$

We measured ranks as deviations from their means, $\frac{1}{2}$

$$\xi = \int_{\mathbb{R}} \int_{-\infty}^{x} f \, dv \, dy = \frac{1}{\sqrt{2\pi}} \int_{0}^{x} e^{-\frac{v^2}{2}} \, dv.$$

$$\eta = \int_{\mathbb{R}} \int_{-\infty}^{y} f \, dw \, dx = \frac{1}{\sqrt{2\pi}} \int_{0}^{y} e^{-\frac{w^2}{2}} \, dw.$$

Then the rank correlation can be calculated

$$\rho_r = 12 \int_{\mathbb{R}^2} \xi \eta f \, dx \, dy$$

Differentiating $\log f$ with respect to ρ, we can show that

$$\frac{\partial^2 f}{\partial x \partial y} = \frac{\partial f}{\partial \rho}.$$

Since

$$\frac{\partial \rho_r}{\partial \rho} = 12 \int_{\mathbb{R}^2} \xi \eta \frac{\partial f}{\partial \rho} \, dx \, dy;$$

the above yields

$$\frac{\partial \rho_r}{\partial \rho} = 12 \int_{\mathbb{R}^2} \xi \eta \frac{\partial^2 f}{\partial x \partial y} \, dx \, dy.$$

By a partial integration with respect to x, we obtain

$$\frac{\partial \rho_r}{\partial \rho} = -12 \int_{\mathbb{R}^2} \eta \frac{\partial \xi}{\partial x} \frac{\partial f}{\partial y} \, dx \, dy.$$

Again by a partial integration with respect to y, we get

$$\frac{\partial \rho_r}{\partial \rho} = 12 \int_{\mathbb{R}^2} \frac{\partial \eta}{\partial y} \frac{\partial \xi}{\partial x} f \, dx \, dy.$$

Since

$$\frac{\partial \eta}{\partial y} = \frac{1}{\sqrt{2\pi}} e^{-\frac{y^2}{2}}$$

$$\frac{\partial \xi}{\partial x} = \frac{1}{\sqrt{2\pi}} e^{-\frac{x^2}{2}},$$

we find

$$\frac{\partial \rho_r}{\partial \rho} = \frac{12}{4\pi^2 \sqrt{1 - \rho^2}} \int_{\mathbb{R}^2} \exp\left[-\frac{x^2 - 2\rho x y + y^2}{2(1 - \rho^2)}\right] \exp\left[-\frac{x^2 + y^2}{2}\right] dx \, dy$$

$$= \frac{12}{4\pi^2 \sqrt{1 - \rho^2}} \int_{\mathbb{R}^2} \exp\left[-\frac{(2 - \rho^2)x^2 - 2\rho x y + (2 - \rho^2)y^2}{2(1 - \rho^2)}\right] dx \, dy,$$

which can be written in the following form

$$\frac{\partial \rho_r}{\partial \rho} = \frac{12}{4\pi^2 \sqrt{1-\rho^2}} \int_{\mathbb{R}^2} \exp\left[-\frac{1}{2}\left(\left[\sqrt{\frac{4-\rho^2}{2-\rho^2}}x\right]^2 + \left[\frac{\sqrt{2-\rho^2}y - \frac{\rho}{\sqrt{2-\rho^2}}x}{\sqrt{1-\rho^2}}\right]^2 \right) \right] dx\, dy.$$

Applying substitution

$$s = \sqrt{\frac{4-\rho^2}{2-\rho^2}}x$$

$$t = \frac{\sqrt{2-\rho^2}y - \frac{\rho}{\sqrt{2-\rho^2}}x}{\sqrt{1-\rho^2}}$$

for which the Jacobian is

$$\sqrt{\frac{1-\rho^2}{4-\rho^2}},$$

we obtain

$$\frac{\partial \rho_r}{\partial \rho} = \frac{6}{\pi\sqrt{4-\rho^2}}.$$

Since ρ_r and ρ vanish together, we get

$$\rho_r = \frac{6}{\pi} \arcsin\left(\frac{\rho}{2}\right)$$

and from the above

$$\rho = 2\sin\left(\frac{\pi}{6}\rho_r\right). \quad \square$$

Proposition 3.26 Let (X, Y) be random vectors with joint normal distribution, then

$$\tau(X, Y) = 2\arcsin\left(\rho(X, Y)\right)/\pi.$$

Proof.
Let $F(x, y)$ denote the cdf and $f(x, y)$ pdf of a normal distribution with correlation ρ:

$$f(x, y) = \frac{1}{2\pi\sqrt{1-\rho^2}} \exp\left[-\frac{1}{2(1-\rho^2)}(x^2 - 2\rho xy + y^2) \right];$$

$$F(x, y) = \frac{1}{2\pi\sqrt{1-\rho^2}} \int_{-\infty}^{x} \int_{-\infty}^{y} \exp\left[-\frac{1}{2(1-\rho^2)}(x^2 - 2\rho xy + y^2) \right] dy\, dx.$$

As in the proof of Proposition 3.13, we can show that

$$\tau(X, Y) = 4 \int_{\mathbb{R}^2} F(x, y) dF(x, y) - 1.$$

Differentiating τ with respect to ρ, we get

$$\frac{\partial \tau}{\partial \rho} = 4 \int_{\mathbb{R}^2} F(x, y) \frac{\partial f(x, y)}{\partial \rho} + f(x, y) \frac{\partial F(x, y)}{\partial \rho} \, dx \, dy.$$

As in Proposition 3.13; $\frac{\partial f(x,y)}{\partial \rho} = \frac{\partial^2 f(x,y)}{\partial x \partial y}$ so that

$$\frac{\partial \tau}{\partial \rho} = 4 \int_{\mathbb{R}^2} F(x, y) \frac{\partial^2 f(x, y)}{\partial x \partial y} + f^2(x, y) \, dx \, dy.$$

Integrating the first summand of the preceding equation by parts with respect to x and y, we obtain

$$\frac{\partial \tau}{\partial \rho} = 8 \int_{\mathbb{R}^2} f^2(x, y) \, dx \, dy.$$

Hence

$$\frac{\partial \tau}{\partial \rho} = \frac{8}{2\pi \sqrt{1 - \rho^2}} \int_{\mathbb{R}^2} \frac{1}{2\pi \sqrt{1 - \rho^2}} \exp\left[-\frac{2[x^2 - 2\rho xy + y^2]}{2(1 - \rho^2)}\right] dx \, dy.$$

Substituting $s = x/\sqrt{2}$ and $t = y/\sqrt{2}$ for which the Jacobian is equal to $\frac{1}{2}$, we get

$$\frac{\partial \tau}{\partial \rho} = \frac{8}{2\pi \sqrt{1 - \rho^2}} \frac{1}{2} = \frac{2}{\pi \sqrt{1 - \rho^2}}.$$

Since τ and ρ vanish together, we get

$$\tau = \frac{2}{\pi} \arcsin \rho,$$

which concludes the proof. $\square$

Proposition 3.29 (Kendall and Stuart (1961)) If X_1, X_2, X_3 follow a joint normal distribution, then $\rho(X_1|X_3, X_2|X_3) = \rho_{X_1 X_2; X_3}$.

Proof.
Without the loss of generality, we assume that X_1, X_2 and X_3 are standardized. We exclude the singular case as well. From (3.12)

$$f(x_1, x_2, x_3) = \frac{1}{(2\pi)^3 |C|} \exp\left\{-\frac{1}{2|C|} \sum_{i,j=1}^{3} C_{ij} x_i x_j\right\},$$

where C_{ij} is the cofactor of the (i, j)th element in the symmetric correlation determinant

$$|C| = \begin{vmatrix} 1 & \rho_{12} & \rho_{13} \\ & 1 & \rho_{23} \\ & & 1 \end{vmatrix}.$$

$\frac{C_{ij}}{|C|} = C^{ij}$ is the element of the reciprocal of C.

The c.f. of the distribution is

$$\phi(t_1, t_2, t_3) = \exp\left\{ -\frac{1}{2} \sum_{i,j=1}^{3} \rho_{ij} t_i t_j \right\}.$$

The conditional distribution g of X_1 and X_2 given X_3 is proportional to $g(x_1, x_2|x_3)$

$$\propto \exp\left\{ \tfrac{1}{2}(C^{11} x_1^2 + 2C^{12} x_1 x_2 + C^{22} x_2^2 + 2C^{23} x_2 x_3) \right\} \tag{3.15}$$

$$\propto \exp\left\{ \tfrac{1}{2}(C^{11}(x_1 - \xi_1)^2 + 2C^{12}(x_1 - \xi_1)(x_2 - \xi_2) + C^{22}(x_2 - \xi_2)^2) \right\}, \tag{3.16}$$

where

$$C^{11}\xi_1 + C^{12}\xi_2 = -C^{13} x_3,$$
$$C^{12}\xi_1 + C^{22}\xi_2 = -C^{23} x_3.$$

From (3.16), we see that given X_3, X_1 and X_2 are bivariate normally distributed with correlation coefficient

$$\rho_{12|3} = -\frac{C^{12}}{\sqrt{C^{11} C^{22}}}.$$

Cancelling the factor $|C|$, we have

$$\rho_{12|3} = -\frac{C_{12}}{\sqrt{C_{11} C_{22}}}$$

$$= \frac{\rho_{12} - \rho_{13}\rho_{23}}{\sqrt{(1 - \rho_{13}^2)(1 - \rho_{23}^2)}},$$

which by formula (3.3) is equal to the partial correlation $\rho_{12;3}$. $\square$

4

High-dimensional Dependence Modelling

4.1 Introduction

This chapter is the heart of this book. It addresses the question:

How should we encode information about the joint distribution of a large number of random variables in such a way that we, and our computer, can calculate and communicate efficiently?

How large is large? Current uncertainty analysis performed on personal computers may easily involve a few hundred variables. The set of variables of interest, or endpoints, may also be large. The input variables may not be assumed to be independent nor may they be assumed to fit any given parametric class.

Decision makers and problem owners are becoming more sophisticated in reasoning with uncertainty. They will increasingly be making demands on uncertainty modelling that analysts with the traditional toolbox will simply be unable to meet. This chapter is an attempt to anticipate these demands. The approach advanced here involves splitting the modelling effort into two separate tasks:

- Model the one-dimensional marginal distributions and

- Model the dependence structure.

The one-dimensional marginals are usually the most important and easiest to assess, either from data or from expert judgment. We want to focus our resources on this task. The dependence structure is in principle *much* more complex. With n uncertain variables, there are $2^n - 1$ distinct non-empty subsets of variables, each of which could have a distinct and important dependence structure. In most problems, this will not be the case, and most dependencies will be of secondary importance. In large problems, we will typically try to capture the most important dependence relations and leave others unspecified. Thus, we require methods for specifying dependence structures that can deal with arbitrary continuous invertible marginal distributions and with incomplete specifications. Moreover, we want sampling routines that will give us *exactly* what we specify, up to sampling error, with minimal additional information.

For most statisticians, the starting point for dependence modelling is the joint normal distribution. Why not just transform everything to a joint normal and use familiar techniques from the theory of the normal distribution? The limitations of the normal transform method encountered in the next section motivate the use of copula-tree and copula-vine methods, treated in succeeding sections. Issues arising with regard to positive definiteness and incomplete correlation matrices are the subject of a separate section.

4.2 Joint normal transform

The joint normal transform method for dependence modelling may be characterized as follows: Transform the marginals to normals, induce a dependence structure and transform back. This technique was implemented in Ghosh and Henderson (2002); Iman and Helton (1985). We present this procedure in the following steps:

1. Variables $X_1, X_2, \ldots, X_n$ are assigned continuous invertible univariate distribution functions $F_1, F_2, \ldots, F_n$, and the rank correlation matrix $\mathcal{R}_r$ is specified;

2. A sample $(y_1, y_2, \ldots, y_n)$ is drawn from a joint normal with standard normal margins and the rank correlation matrix $\mathcal{R}_r$;

3. If Φ denotes the standard normal distribution function, then a sample from a distribution with margins $F_1, F_2, \ldots, F_n$ and the rank correlation matrix $\mathcal{R}_r$ is calculated as:

$$\left(F_1^{-1}(\Phi(y_1)), F_2^{-1}(\Phi(y_2)), \ldots, F_n^{-1}(\Phi(y_n)) \right).$$

Since the transformations $F_i^{-1}(\Phi(y_i))$ are strictly increasing, $(X_1, \ldots, X_n)$ have the rank correlation matrix $\mathcal{R}_r$.

The first step is already slippery. We show in Section 4.4.6 that every three-dimensional correlation matrix (i.e. symmetric positive definite matrix with 1's on the main diagonal) is a rank correlation matrix, but for higher dimensions, the set of rank correlation matrices has not been characterized. Hence, we do not know whether the assignments in step 1 are consistent.

The second step also poses a problem: How de we create a joint normal distribution with a specified rank correlation matrix $\mathcal{R}_r$? Using the standard theory of normal distribution, we can find a linear combination of independent standard normal variables to realize a given product moment correlation matrix $\mathcal{R}$. We take the vector $Z^T = (Z_1, Z_2, \ldots, Z_n)$ of independent standard normal variables and apply the linear transformation $Y = LZ$; $\mathcal{R} = LL^T$, where L is a lower triangular matrix (the Cholesky decomposition of $\mathcal{R}$). In general, however, product moment and rank correlations are not equal, so at this point we have two options. We can either

a. ignore the difference between product moment and rank correlation matrices of the joint normal or

b. using the relation between product moment and rank correlation for the normal distribution (Proposition 3.25), calculate the product moment correlation matrix from

$$\rho(i, j) = 2 \sin\left(\frac{\pi}{6}\rho_r(i, j)\right), \tag{4.1}$$

where $\rho(i, j)$ and $\rho_r(i, j)$ denote the (i, j)th cell of $\mathcal{R}$ and $\mathcal{R}_r$ respectively.

We do know that every correlation matrix of dimension three is a rank correlation matrix (see Section 4.4.6). However, we also know that the joint normal distribution cannot realize all these rank correlation matrices. Consider the following example:

Example 4.1 *Let*

$$A = \begin{bmatrix} 1 & 0.7 & 0.7 \\ 0.7 & 1 & 0 \\ 0.7 & 0 & 1 \end{bmatrix}.$$

We can easily check that A is positive definite. However, matrix B, given by

$$B(i, j) = 2 \sin\left(\frac{\pi}{6}A(i, j)\right) \text{ for } i, j = 1, \ldots, 4,$$

that is,

$$B = \begin{bmatrix} 1 & 0.7167 & 0.7167 \\ 0.7167 & 1 & 0 \\ 0.7167 & 0 & 1 \end{bmatrix}$$

is not positive definite.

This should be taken into account in any procedure in which the rank correlation structure is induced by transforming distributions to standard normals and generating a dependence structure using the linear properties of the joint normal. From the preceding example, we can see that it is not always possible to find a product moment correlation matrix generating a given rank correlation matrix via relation (4.1). In fact, the probability that a randomly chosen correlation matrix stays positive definite after transformation (4.1) changes rapidly to zero with matrix dimension.

A generalization of the preceding example, the NORmal To Anything (NORTA) method (Carion and Nelsen (1997)), uses a normal transformation to generate random vectors with a prescribed correlation matrix. The key idea is to find a function d_{ij} that relates the correlation between variables X_i and X_j, say $\rho_X(i, j)$, to the correlation of the corresponding normal variables Y_i and Y_j, say $\rho_Y(i, j)$. It is shown in Carion and Nelsen (1997) that under certain mild conditions,

$$d_{ij}(\rho_Y(i, j)) = \rho_X(i, j) \tag{4.2}$$

is a non-decreasing, continuous function on $[-1, 1]$. This is used in a numerical procedure that finds a solution for (4.2) if this solution exists. When $X_i, i = 1, \ldots, n$, is uniform, the function d_{ij} is known as in (4.1). The NORTA method, however, experiences the same problems noted in the example in the preceding text, that is, nonexistence of a solution in (4.2). It is shown in Ghosh and Henderson (2002) that the probability of NORTA infeasibility rapidly increases with matrix dimension. A procedure is introduced to sample uniformly from the set of correlation matrices and determine empirically the relationship between matrix dimension and probability of NORTA infeasibility. Figure 4.1 shows how the probability of NORTA infeasibility grows with matrix dimension.

The preceding issues are of a rather theoretical nature and pale before a much more serious practical concern: This method requires that the entire rank correlation matrix $\mathcal{R}_r$ be specified. Consider a problem with 100 variables. After assessing

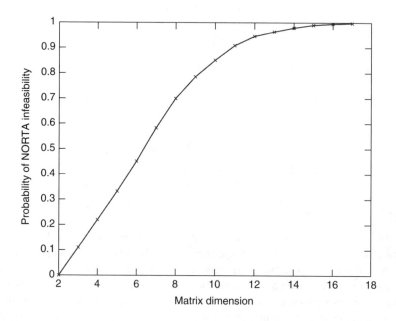

Figure 4.1 Dependence of the probability of NORTA infeasibility on matrix dimension.

the one-dimensional marginals, we turn to the problem of assessing dependencies. There are 4950 distinct correlations in the correlation matrix. It is highly unlikely that the 4950 correlations estimated by experts, or measured with error, will form a positive definite matrix. We then have the problem of altering the estimated or measured matrix to get something to which the normal transform method may be applied. The iterative proportional fitting (IPF) algorithm can be used when it is important to preserve zeros in the inverse correlation matrix (see Chapters 5 and 9); another procedure is presented at the end of this chapter.

Instead of specifying a (rank) correlation matrix, we can determine a joint normal distribution in different ways. It can be shown that the lower triangular matrix in the Cholesky decomposition is determined by a set of partial regression coefficients (see Section 3.3) together with conditional variances that are algebraically independent (Shachter and Kenley (1989)). For the joint normal distribution, the conditional variances are constant. Hence, if we can find experts who can assess partial regression coefficients for the normal transforms of our original variables, together with conditional variances of these transformed variables, then we can sidestep the problem of positive definiteness. The assessment task placed on these experts is not straightforward, to say the least.

Another possibility is to leave the rank correlation matrix only partially specified. In our 100×100 example, only a small number of the 4950 correlations may be deemed important enough to be assessed. The analyst who wishes to work with a partially specified correlation matrix is confronted with the so-called matrix completion problem: Can a partially specified matrix be extended to a positive definite matrix? This problem is hard. Given a partially specified correlation matrix, some procedure must be invoked to generate a positive definite matrix. The procedure of Iman et al. (1981) is illustrated as follows: If the following correlations are specified:

$$
\begin{bmatrix}
1 & 0.9 & \square \\
0.9 & 1 & 0.9 \\
\square & 0.9 & 1
\end{bmatrix}, \tag{4.3}
$$

where $\square$ denotes an unspecified entry, then this matrix is completed by taking the value of the unspecified entry as close to zero as possible.

Anticipating the developments further in this chapter, we can also represent the dependence structure of a joint normal distribution via a regular vine, which may be seen as an alternative to a correlation matrix. Since partial and conditional correlations are equal for the joint normal distribution, we can quantify the vine with conditional or partial correlations, or indeed with conditional rank correlations (see Proposition 3.25 and Section 4.4.5). According to Theorem 4.4, the partial correlations on a regular vine are algebraically independent and in one-to-one correspondence with correlation matrices. Hence this specification is always consistent. The completion problem is treated in Section 4.5.3 and in Kurowicka and Cooke (2003) using vines with partial correlations.

4.3   Dependence trees

A tree is an acyclic undirected graph; when trees are used to describe dependence structures in high-dimensional distributions, these are called *dependence trees*.

4.3.1   Trees

Definition 4.1 (Tree)[1] $\mathcal{T} = (N, E)$ *is a* **tree** *with nodes* $N = \{1, 2, \ldots, n\}$ *and edges E, where E is a subset of unordered pairs of N with no cycle; that is, there does not exist a sequence* $a_1, \ldots, a_k$ $(k > 2)$ *of elements of N such that*

$$\{a_1, a_2\} \in E, \ldots, \{a_{k-1}, a_k\} \in E, \{a_k, a_1\} \in E.$$

The **degree** *of node* $a_i \in N$ *is* $\#\{a_j \in N \mid \{a_i, a_j\} \in E\}$; *that is, the number of edges attached to* a_i.

In Figure 4.2 we can see two graphs. The graph on the left is a tree on six variables and the graph on the right is an undirected graph with a cycle.

4.3.2   Dependence trees with copulae

Edges of trees can be used to specify bivariate dependencies. If a bivariate distribution $H(x, y)$ has continuous marginal distribution functions F_X and F_Y, then there exists a unique copula $C(U_1, U_2)$, such that $H(X, Y) = C(F_X(X), F_Y(Y)) = C(U_1, U_2)$; if the distribution functions are invertible, then we can write $X \sim F_X^{-1}(U_1), Y \sim F_Y^{-1}(U_2)$. Figure 4.3 shows a non-invertible distribution function

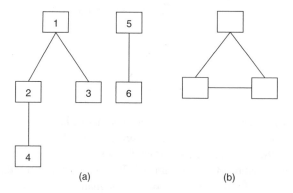

(a)   (b)

Figure 4.2   A tree with six nodes (a) and undirected graph with a cycle (b).

---

[1]Some authors additionally require that all nodes are connected: For any $a, b \in N$, there exists a sequence $c_2, \ldots, c_{k-1}$ of elements of N such that

$$\{a, c_2\} \in E, \{c_2, c_3\} \in E, \ldots, \{c_{k-1}, b\} \in E.$$

A tree, as defined here, is then called a forest of trees. We term a tree in which all nodes are connected as a *connected tree*.

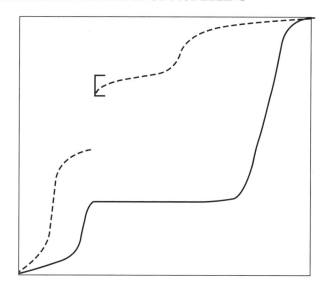

Figure 4.3 The dotted distribution function is not continuous, and the solid distribution function is not invertible.

and one that is non-continuous and non-invertible. Being continuous is not sufficient to guarantee the existence of a density; the Cantor distribution is continuous but is concentrated on a set of Lebesgue measure zero (Tucker (1967), p. 20) and cannot be written as an integral of a density. When we say that a distribution has a density, we mean that it can be written as an integral of a function that is positive on its domain.

For general bivariate tree specifications, we follow the original formulation from Cooke (1997a) in terms of distributions with given margins, and we specialize this for copula-trees. This requires distribution functions that are continuous and invertible on their domain. The following definitions formalize the concept of dependence trees.

Definition 4.2 (Bivariate- and Copula-tree specification) $(F, \mathcal{T}, B)$ *is a* **bivariate tree specification** *if*

1. $F = (F_1, \ldots, F_n)$ *is a vector of one-dimensional distribution functions for random vector* $(X_1, \ldots, X_n)$ *such that* $X_i \neq X_j$ *for* $i \neq j$.

2. $\mathcal{T}$ *is a tree on n elements, with nodes* $N = \{1, 2, \ldots, n\}$ *and edges* E.

3. $B = \{B_{ij} \mid \{i, j\} \in E$ *and* B_{ij} *is a non-empty subset of the set of bivariate distributions with margins* $F_i, F_j\}$.

A bivariate tree specification $(F, \mathcal{T}, B)$ *is a* **copula-tree** *if* $(F_1, \ldots, F_n)$ *are continuous invertible and if*

4. $B = \{C_{ij} \mid \{i, j\} \in E$ *and* C_{ij} *is the copula for* $(X_i, X_j)\}$.

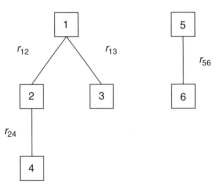

Figure 4.4 A tree with six nodes, with rank correlations assigned to the edges.

Bivariate tree specifications are always consistent. Copula-trees are special cases of bivariate trees in which the bivariate constraints B_{ij} contain one bivariate distribution specified by the copula C_{ij}. To be useful, we must have a convenient way of specifying the copulae. Rank correlation, that is, the correlation of the copula, together with a family of copula indexed by correlation, is the natural choice. We specify a copula-tree by assigning one-dimensional distribution functions to the nodes and rank correlations in $[-1, 1]$ to the edges.

In Figure 4.4 rank correlations are assigned to the edges of the tree in Figure 4.2.

Definition 4.3 (Tree dependence) 1. *A multivariate probability distribution G on $\mathbb{R}^n$ satisfies, or realizes, a bivariate tree specification (F, T, B) if the marginal distributions G_i of G equal F_i ($1 \leq i \leq n$) and if for $\{i, j\} \in E$ the bivariate distributions G_{ij} of G are elements of B_{ij}.*

2. *G has a tree dependence order M for T if $\{i, k_1\}, \ldots, \{k_m, j\} \in E$ implies that X_i and X_j are conditionally independent given any M of k_ℓ, $1 < \ell < m$, and X_i and X_j are independent if there is no path from i to j.*

3. *G has Markov tree dependence for T whenever disjunct subsets a and b of variables are separated by subset c of variables in T (i.e. every path from a to b intersects c), in which case the variables in a and b are conditionally independent given the variables in c.*

The Markov property is stronger than tree dependence of order M (see Exercise 4.3). The following theorem shows that bivariate tree specifications have Markov realizations. It is restricted to distributions with bivariate densities.

Theorem 4.1 *Let (F, T, B) be an n-dimensional bivariate tree specification that specifies the marginal densities f_i, $1 \leq i \leq n$, and the bivariate densities f_{ij}, $\{i, j\} \in E$. Then, there is a unique density g on $\mathbb{R}^n$, with margins $f_1, \ldots, f_n$, and bivariate* *margins f_{ij} for $\{i, j\} \in E$, such that g has Markov tree dependence for T. The density g is given by*

$$g(x_1, \ldots, x_n) = \frac{\prod_{\{i,j\} \in E} f_{ij}(x_i, x_j)}{\prod_{i \in N} (f_i(x_i))^{deg(i)-1}}, \tag{4.4}$$

where $deg(i)$ denotes the degree of node i and $f_i(x_i) > 0$, $i = 1, \ldots, n$. With copula densities c_{ij}, $\{i, j\} \in E$, this becomes

$$g(x_1, \ldots, x_n) = f_1(x_1) \ldots f_n(x_n) \prod_{\{i,j\} \in E} c_{ij}(F_i(x_i), F_j(x_j)). \tag{4.5}$$

Example 4.2 *The density function $g_{1\ldots6}$, with marginal densities $g_1, \ldots, g_6$ satisfying the rank correlation specification on the tree in Figure 4.4, where c_{ij} is density of a copula family indexed by correlation r_{ij}, is the following:*

$$g_{1\ldots6}(x_1, \ldots, x_6) = g_1(x_1) \ldots g_6(x_6) c_{12}(F_1(x_1), F_2(x_2)) c_{13}(F_1(x_1), F_3(x_3))$$

$$c_{24}(F_2(x_2), F_4(x_4)) c_{56}(F_5(x_5), F_6(x_6)).$$

Meeuwissen and Cooke (1994) and Cooke (1997a) showed that the minimum information realization of a bivariate tree specification is Markov and that the minimum information distribution with given marginals and rank correlations on a tree is the Markov realization of a copula-tree with the minimum information copula. This result easily follows from more general results on vines (Corollary 4.1).

Summarizing, the copula-tree method represents high-dimensional distributions by specifying the following elements:

1. Continuous invertible marginal distributions assigned to $X_1, \ldots, X_n$ nodes of the tree;

2. A set of copula assigned to the edges of the tree.

A copula-tree may have many realizations; the Markov realization makes non-adjacent variables on a path conditionally independent, given any set of variables separating them on the path. For the Markov realization of the tree in Figure 4.2 with six variables, variables 1 and 5 would be sampled independently. Variables 2 and 3 would then be sampled independently, conditional on the value of variable 1. Variable 6 would be sampled conditional on the value of variable 5. Finally, variable 4 would be sampled conditional on the value of variable 2. The sampling procedure for trees is described in Chapter 6. Note that for the Markov realization of Figure 4.4, variables 2 and 3 are conditionally independent given 1. It would be possible to enrich the copula-tree with additional information on the conditional dependence between 2 and 3 given 1. This is in fact a basic idea for vines. Also note that only correlations r_{12}, r_{13}, r_{24} and r_{56} are specified. Other correlations are determined by the realization and by the choice of copula. In a tree on

n variables only, $n - 1$ correlations out of $\binom{n}{2}$ in the correlation matrix can be specified.

4.3.3 Example: Investment

The following example is from a UNICORN project at the end of this chapter, in which we step through the details. We invest \$1000 for 5 years; the yearly interest (Vi; $i = 1, \ldots, 5$) is uniformly distributed on the interval $[0.05, 0.15]$, and successive yearly interests have a rank correlation of 0.70. This does not determine the joint distribution, and we explore the effects of different dependence structures. The first possibility is the tree given in Figure 4.5:

With this structure, the correlation between $V1$ and $V5$ is much less than that between $V1$ and $V2$.

The second possibility is to correlate all yearly interests to a latent variable, with rank correlation 0.84 (Figure 4.6). This dependence structure is symmetric, and all interests are correlated at 0.70.

Suppose that we now learn that the interest for the first year, $V1$, is in the upper 5% of its distribution. Figure 4.7 shows the result of conditionalizing on these highest values for the preceding two trees. Each broken line corresponds to one sample and intersects the vertical lines at the sample values for the corresponding variables.

Note that the first tree has a more diffuse distribution for $V5$, conditional on high values for $V1$, than the second tree. The return after 5 years (leftmost variable) is also higher with the second tree.

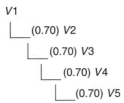

Figure 4.5 Tree for Investment.

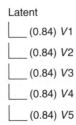

Figure 4.6 Tree for Investment with latent variable.

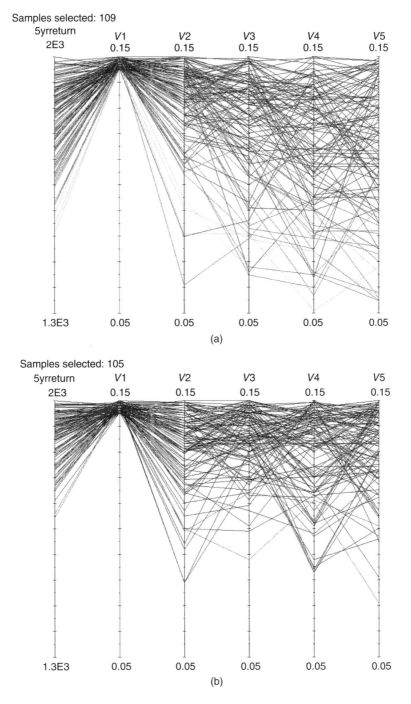

Figure 4.7 Cobweb plots conditional on the 100 highest values for the two trees; plots for Fig. 4.5 (a) and Fig. 4.6 (b)

4.4   Dependence vines

A graphical model called vines was introduced in Cooke (1997a) and studied in
Bedford and Cooke (2001b, 2002); Kurowicka and Cooke (2003). A *vine* on *n*
variables is a nested set of trees, where the edges of the tree *j* are the nodes of
the tree *j* + 1, and each tree has the maximum number of edges. The 'higher-level
trees' enable us to encode conditional constraints. When used to specify dependence
structures in high-dimensional distributions, the trees are called dependence vines.

4.4.1   Vines

Figure 4.8 shows examples of a tree and a vine on three variables. In the vine, a
conditional constraint on *Y* and *Z* given *X* can be imposed.

A *regular* vine on *n* variables is one in which two edges in tree *j* are joined
by an edge in tree *j* + 1 only if these edges share a common node. Figure 4.9
shows a regular vine and a non-regular vine on four variables. The nested trees
are distinguished by the line style of the edges; tree 1 has solid lines, tree 2 has
dashed lines and tree 3 has solid, bend lines.

In the definition below, we introduce the regular vine in a more formal fashion.

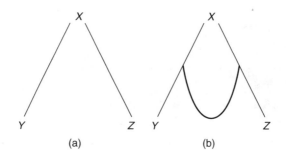

Figure 4.8   A tree (a) and a vine (b) on three elements.

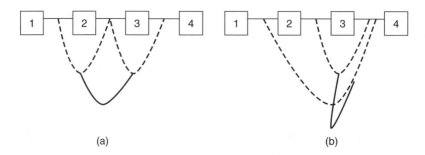

Figure 4.9   A regular vine (a) and a non-regular vine (b) on four variables.

Definition 4.4 (Vine, regular vine) $\mathcal{V}$ *is a* **vine** *on n elements if*

1. $\mathcal{V} = (T_1, \ldots, T_{n-1})$.

2. T_1 *is a connected tree with nodes* $N_1 = \{1, \ldots, n\}$ *and edges* E_1; *for* $i = 2, \ldots, n-1$, T_i *is a connected tree with nodes* $N_i = E_{i-1}$.

$\mathcal{V}$ *is a* **regular vine** *on n elements if additionally*

3. *(proximity) For* $i = 2, \ldots, n-1$, *if* $\{a, b\} \in E_i$, *then* $\#a \triangle b = 2$, *where* $\triangle$ *denotes the symmetric difference. In other words, if a and b are nodes of* T_i *connected by an edge in* T_i, *where* $a = \{a_1, a_2\}$, $b = \{b_1, b_2\}$, *then exactly one of the* a_i *equals one of the* b_i.

In Figure 4.10, we can see that different vines can be obtained even when the first tree in both cases is the same.

In the case of the regular vine in Figure 4.9, the representation is unique, given the first tree, since each node in the first tree of this vine has a degree of at most 2. There are two families of vines whose members are fixed simply by fixing the initial ordering of the variables. These are the D-vine and, the canonical (C-) vine, characterized by minimal and maximal degrees of nodes in the trees[2].

Definition 4.5 (D-vine, C-vine) *A regular vine is called a*

- **D-vine** *if each node in* T_1 *has a degree of at most 2.*

- **Canonical** *or* **C-vine** *if each tree* T_i *has a unique node of degree* $n - i$. *The node with maximal degree in* T_1 *is the* **root.**

The following definitions provide the vocabulary for studying vines.

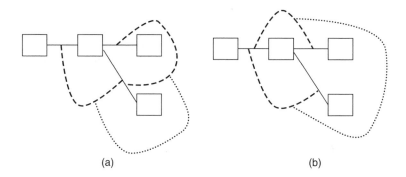

(a)   (b)

Figure 4.10 Two different regular vines on four variables.

---

[2]D-vines were originally called 'drawable vines'; canonical vines owe their deferential name to the fact that they are the most natural for sampling. The suggestion that 'D-vine' is a sacrilegious pun is thoroughly unfounded.

Definition 4.6 (m-child; m-descendent) *If node e is an element of node f, we say that e is an **m-child** of f; similarly, if e is reachable from f via the membership relation: $e \in e_1 \in \ldots \in f$, we say that e is an **m-descendent** of f.*

For each edge of a vine, we define *constraint, conditioned* and *conditioning* sets of this edge as follows: The variables reachable from a given edge are called the constraint set of that edge. When two edges are joined by an edge of the next tree, the intersection of the respective constraint sets form the conditioning set, and the symmetric difference of the constraint sets is the conditioned set.

Definition 4.7 (Conditioning, conditioned and constraint sets)

1. *For $j \in E_i, i \leq n - 1$, the subset $U_j(k)$ of $E_{i-k} = N_{i-k+1}$, defined by $U_j(k) = \{e \mid \exists\ e_{i-(k-1)} \in e_{i-(k-2)} \in \ldots \in j,\ e \in e_{i-(k-1)}\}$, is called the **k-fold union** $U_j^* = U_j(i)$ is the **complete union** of j, that is, the subset of $\{1, \ldots, n\}$ consisting of m-descendants of j.*
   If $a \in N_1$, then $U_a^* = \emptyset$.
   $U_j(1) = \{j_1, j_2\} = j$.
   By definition we write $U_j(0) = \{j\}$.

2. *The **constraint set** associated with $e \in E_i$ is U_e^*.*

3. *For $i = 1, \ldots, n - 1, e \in E_i, e = \{j, k\}$, the **conditioning** set associated with e is*

$$D_e = U_j^* \cap U_k^*$$

   *and the **conditioned set** associated with e is*

$$\{C_{e,j}, C_{e,k}\} = U_j^* \triangle U_k^* = \{U_j^* \setminus D_e,\ U_k^* \setminus D_e\}.$$

   The order of node e is $\#D_e$.

Note that for $e \in E_1$, the conditioning set is empty.
For $e \in E_i, i \leq n - 1, e = \{j, k\}$ we have $U_e^* = U_j^* \cup U_k^*$.

Figures 4.11 and 4.12 show D- and C-vines on five variables, with the conditioned and conditioning sets for each edge. We use the D-vine in Figure 4.11 to illustrate Definition 4.7. We get

$$T_1 = (N_1, E_1),\ N_1 = \{1, 2, \ldots, 5\},$$

$$E_1 = \{\{1, 2\}; \{2, 3\}; \{3, 4\}; \{4, 5\}\};$$

$$T_2 = (N_2, E_2),\ N_2 = E_1,$$

$$E_2 = \{\{\{1, 2\}, \{2, 3\}\}; \{\{2, 3\}, \{3, 4\}\}; \{\{3, 4\}, \{4, 5\}\}\};$$

$$\vdots$$

The complete union of $j = \{1, 2\}$ is $U_j^* = \{1, 2\}$, and for $k = \{2, 3\}$, $U_k^* = \{2, 3\}$. Hence, the conditioning set of the edge $e = \{\{1, 2\}, \{2, 3\}\}$ in T_2 is

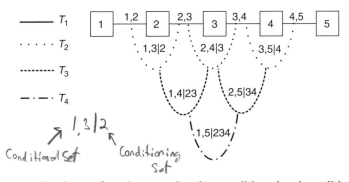

Figure 4.11 A D-vine on five elements showing conditioned and conditioning sets.

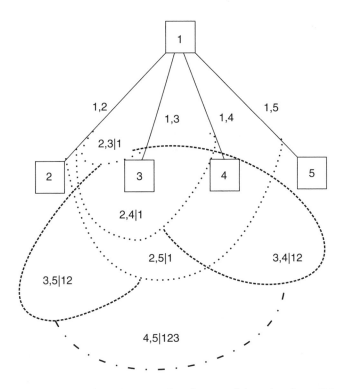

Figure 4.12 A C-vine on five elements showing conditioned and conditioning sets.

$D_e = U_j^* \cap U_k^* = \{1, 2\} \cap \{2, 3\} = \{2\}$. The conditioned set consists of $C_{e,j} = U_j^* \setminus D_e = \{1, 2\} \setminus \{2\} = \{1\}$ and $C_{e,k} = U_k^* \setminus D_e = \{2, 3\} \setminus \{2\} = \{3\}$. The dotted edge of T_2 between $\{1, 2\}$ and $\{2, 3\}$ in Figure 4.11 is denoted as $1, 3|2$, which gives the elements of the conditioned sets $\{1\}, \{3\}$ before '|' and of the conditioning set $\{2\}$ after '|'.

For regular vines, the structure of the constraint set is particularly simple, as shown by the following lemmas (Cooke (1997a); Kurowicka and Cooke (2003)).

Lemma 4.1 *Let $\mathcal{V}$ be a regular vine on n elements, and let $j \in E_i$. Then*

$$\#U_j(k) = 2\#U_j(k-1) - \#U_j(k-2); k = 2, 3, \ldots i.$$

Lemma 4.2 *Let $\mathcal{V}$ be a regular vine on n elements, and let $j \in E_i$. Then*

$$\#U_j(k) = k + 1; \ k = 0, 1, \ldots, i.$$

The regularity condition ensures that the symmetric difference of the constraint sets always consists of two variables.

Lemma 4.3 *If $\mathcal{V}$ is a regular vine on n elements, then for all $i = 1, \ldots n - 1$ and all $e \in E_i$, the conditioned set associated with e is a doubleton; $\#U_e^* = i + 1$, and $\#D_e = i - 1$.*

Lemma 4.4 *Let $\mathcal{V}$ be a regular vine, and suppose for $j, k \in E_i, U_j^* = U_k^*$, then $j = k$.*

Since there are $\binom{n}{2}$ nodes in a regular vine, the following lemma shows that every pair of variables occurs exactly once as the conditioned set of some node.

Lemma 4.5 *If the conditioned sets of nodes i, j in a regular vine are equal, then $i = j$.*

Lemma 4.6 *For any node M of order $k > 0$ in a regular vine, if variable i is a member of the conditioned set of M, then i is a member of the conditioned set of exactly one of the m-children of M and the conditioning set of an m-child of M is a subset of the conditioning set of M.*

4.4.2   Bivariate- and copula-vine specifications

As with trees, we first define bivariate-vine dependence for regular vines and then specialize to copula-vines. If we wish to impose only bivariate and conditional bivariate constraints on a joint distribution, then we evidently restrict our attention to regular vines. For dependence on general vines, see Bedford and Cooke (2002).

Definition 4.8 (Bivariate- and Copula-vine specification) $(F, \mathcal{V}, B)$ *is   a* **bivariate-vine specification** *if*

1. *$F = (F_1, \ldots, F_n)$ is a vector of distribution functions for random vectors $(X_1, \ldots, X_n)$ such that $X_i \neq X_j$ for $i \neq j$.*

2. *$\mathcal{V}$ is a regular vine on n elements.*

3. $B = \{B_{jk} \mid e(j,k) \in \bigcup_{i=1}^{n-1} E_i$, where $e(j,k)$ is the unique edge with conditioned set $\{j,k\}$, and B_{jk} is a non-empty subset of the set of margins on $\{X_j, X_k\}$ conditional on $D_{e(j,k)}\}$;

A bivariate-vine specification $(F, \mathcal{V}, B)$ is a **copula-vine** if $(F_1, \ldots, F_n)$ are continuous invertible and,

4. $B = \{C_{jk} \mid e(j,k) \in \bigcup_{i=1}^{n-1} E_i$, where $e(j,k)$ is the unique edge with conditioned set $\{j,k\}$, and C_{jk} is a copula for $\{X_j, X_k\}$ conditional on $D_{e(j,k)}\}$;

A distribution realizes a vine specification if the one-dimensional marginals agree with F and the conditional distributions (X_j, X_k) conditioned on $D_{e(j,k)}$ are in B_{jk}. A convenient way to specify the copulae is to choose a family of copulae indexed by correlation and assign a conditional rank correlation to each edge. In general, conditional correlations may depend on the value of the conditioning variables. For the normal vines introduced in Section 4.4.5 however, the conditional rank correlations must be constant.[3] Copula-vine specifications are always consistent. The proof of this fact is given by the sampling routine (see Chapter 6). The constant conditional rank correlations may be chosen arbitrarily in the interval $[-1, 1]$.

Just as for the tree-copula dependence structure, we can express a regular vine distribution in terms of its density (Bedford and Cooke (2001b)).

Theorem 4.2 Let $\mathcal{V} = (T_1, \ldots, T_{n-1})$ be a regular vine on n elements. For each edge $e(j,k) \in T_i$, $i = 1, \ldots, n-1$ with conditioned set $\{j,k\}$ and conditioning set D_e, let the conditional copula and copula density be $C_{jk|D_e}$ and $c_{jk|D_e}$ respectively. Let the marginal distributions F_i with densities f_i, $i = 1, \ldots, n$ be given. Then, the vine-dependent distribution is uniquely determined and has a density given by

$$f_{1\ldots n} = \left(\prod_{i=2}^{n-1} \prod_{e(j,k)\in E_i} c_{jk|D_e}(F_{j|D_e}, F_{k|D_e}) \right) \cdot \frac{\prod_{\{j,k\}\in E_1} f_{jk}}{\prod_{j\in N_1} f_j^{deg(j)-1}} \qquad (4.6)$$

$$= f_1 \ldots f_n \prod_{i=1}^{n-1} \prod_{e(j,k)\in E_i} c_{jk|D_e}(F_{j|D_e}, F_{k|D_e}). \qquad (4.7)$$

Example 4.3 The density function $f_{1\ldots4}$ of the distribution satisfying a copula-vine specification of the D-vine on four variables with the marginal densities $f_1, \ldots, f_4$ is the following:

$$f_{1\ldots4} = f_1 \ldots f_4 c_{12}(F_1, F_2)c_{23}(F_2, F_3)c_{34}(F_3, F_4)c_{13|2}(F_{1|2}, F_{3|2})$$

$$c_{24|3}(F_{2|3}, F_{4|3})c_{14|23}(F_{1|23}, F_{4|23}).$$

The existence of the unique minimum information distribution with respect to the product of margins for regular vines can be shown:

---

[3]Currently UNICORN is also restricted to constant conditional correlations.

Theorem 4.3 *Let g be an n-dimensional density satisfying the bivariate-vine speci-fication $(F, \mathcal{V}, B)$ with density g and one-dimensional marginal densities $g_1, \ldots, g_n$; then*

$$I\left(g \mid \prod_{i=1}^{n} g_i\right) = \sum_{i=1}^{n-1} \sum_{e(j,k) \in E_i} E_{D_e} I(g_{C_{e,j}, C_{e,k}|D_e} \mid g_{C_{e,j}|D_e} \cdot g_{C_{e,k}|D_e}). \quad (4.8)$$

If $g_{\{C_{e,j}, C_{e,k}\}|D_e}$ is the unique density satisfying $(F, \mathcal{V}, B)$, which minimizes

$$I(g_{C_{e,j}, C_{e,k} D_e} \mid g_{C_{e,j} D_e} \cdot g_{C_{e,k}|D_e}); \quad i = 1, \ldots, n-1; e = \{j, k\} \in E_i,$$

then g is the unique density satisfying $(F, \mathcal{V}, B)$ and minimizing

$$I\left(g \mid \prod_{i=1}^{n} g_i\right).$$

The quantity on the left-hand side of (4.8) is often called the *mutual information* of g. Theorem 4.3 says that the mutual information can be written as the sum over the nodes of a regular vine of expected mutual information of conditional bivariate densities, with expectation taken over the conditioning sets of the nodes.

We note that for the (conditional) independent copula, the variables $\{C_{e,j}, C_{e,k}\}$ are conditionally independent given D_e; the corresponding term in the sum of (4.8) is zero, and the corresponding term in the density in (4.6), $c_{jk|D_e}(F_{j|D_e}, F_{k|D_e})$, is the independent copula, that is, unity. If all constraints B_{jk} corresponding to edges in trees $T_2 \ldots T_{n-1}$ contain the independent copula, then the minimum information will be realized by the (conditional) independent copula. The density then has the form (4.4) and we have:

Corollary 4.1 *The minimum information realization of a copula-tree specification, relative to the independent distribution with the same one-dimensional marginals, is Markov.*

We will often require that the copulae realizing the conditional correlations have the *zero independence* property: Zero (conditional) rank correlation entails (conditional) independence. The diagonal band (DB), Frank's and minimum infor-mation copulae have the zero independence property, with which it is easy to deal with unspecified correlations. If a (conditional) rank correlation for a regular vine is unspecified, then setting this correlation equal to zero makes the condi-tioned variables conditionally independent. This is always consistent and yields, by Theorem 4.3, the minimum information joint distribution, given the specified rank correlation and associated copula.

4.4.3 Example: Investment continued

We return to the investment example and reflect that if the interest is high in year 2, it is unlikely to be high in both years 1 and 3. We can capture this with a D-vine with

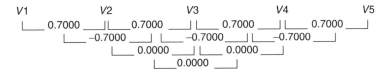

Figure 4.13 D-vine for investment; Frank's copula.

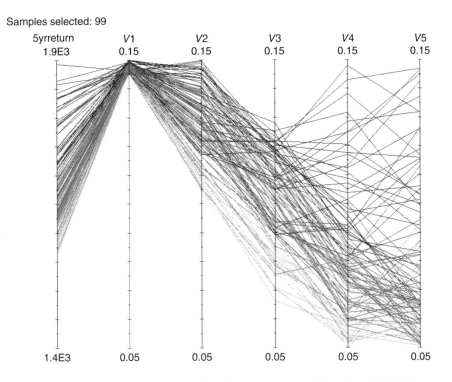

Figure 4.14 Cobweb plot for D-vine investment, conditional on high $V1$.

rank correlation -0.70 between Vi and $Vi + 2$, conditional on $Vi + 1, i = 1, 2, 3$ (Figure 4.13).

Other conditional rank correlations are zero; with Frank's copula this implies that $V1$ and $V4$ are conditionally independent given $V2$ and $V3$. The effect of the negative conditional rank correlation is obvious in Figure 4.14.

4.4.4 Partial correlation vines

Recall that the partial correlation $\rho_{12;3,\dots,n}$ can be interpreted as the correlation between the orthogonal projections of X_1 and X_2 on the plane orthogonal to the linear space spanned by $X_3, \dots, X_n$. The edges of a regular vine may also be associated with partial correlations, with values chosen arbitrarily in the interval $(-1, 1)$ in the following way:

For $i = 1, \ldots, n - 1$, to $e \in E_i$, with conditioned and conditioning variables $\{j, k\}$ and D_e respectively, we associate

$$\rho_{j,k;D_e}.$$

This is not a vine specification in the sense of Definition 4.8, as the one-dimensional marginals are not specified; indeed, a given set of marginals may not be consistent with a given set of partial correlations (see, however, the next section). Even if they are consistent, this would not determine the joint distribution and not lead to a sampling routine. Nonetheless, because of the convenience of calculating with partial correlations, and their approximate equality to conditional correlations and conditional rank correlations, it is convenient to assign partial correlations to the edges of a regular vine and to call the result a *partial correlation vine*.

In Theorem 4.4 Bedford and Cooke (2002) shows that each such partial correlation vine specification uniquely determines the correlation matrix, and every full-rank correlation matrix can be obtained in this way. In other words, a regular vine provides a bijective mapping from $(-1, 1)^{\binom{n}{2}}$ into the set of positive definite matrices with 1's on the diagonal.

Theorem 4.4 *For any regular vine on n elements there is a one-to-one correspondence between the set of $n \times n$ full-rank correlation matrices and the set of partial correlation specifications for the vine.*

All assignments of the numbers between -1 and 1 to the edges of a partial correlation regular vine are consistent in the sense that there is a joint distribution realizing these partial correlations, and all correlation matrices can be obtained this way.

It can be verified that the correlation between ith and jth variables can be computed from the sub-vine generated by the constraint set of the edge, the conditioned set of which is $\{i, j\}$ using the recursive formulae (3.3) and the following lemma.

Lemma 4.7 *If $z, x, y \in (-1, 1)$, then also $w \in (-1, 1)$, where*

$$w = z\sqrt{(1 - x^2)(1 - y^2)} + xy.$$

A regular vine may thus be seen as a way of picking out partial correlations, which uniquely determine the correlation matrix and which are algebraically independent. The partial correlations in a partial correlation vine need not satisfy any algebraic constraint like positive definiteness. The 'completion problem' for partial correlation vines is therefore trivial: An incomplete specification of a partial correlation vine may be extended to a complete specification by assigning arbitrary numbers in the $(-1, 1)$ interval to the unspecified edges in the vine.

Partial correlation vines have another important property that plays a central role in model inference in the next chapter. The product of 1 minus the square partial correlations equals the determinant of the correlation matrix.

Theorem 4.5 *Let D be the determinant of the correlation matrix of variables $1, \ldots, n$, with $D > 0$. For any partial correlation vine,*

$$D = \prod_{i=1}^{n-1} \prod_{e \in E_i} (1 - \rho_{j,k;D_e}^2),$$

where $\{j, k\}$ and D_e are conditioned and conditioning sets of e.

4.4.5 Normal vines

A normal vine arises when $(Y_1, \ldots, Y_n)$ have a joint normal distribution, and the edges of a regular vine on n nodes are assigned the partial correlations of this distribution. By Proposition 3.29, the partial and conditional correlations are equal for the joint normal distribution. With Proposition 3.25 these conditional correlations may be transformed to conditional rank correlations.

This offers a convenient way of realizing any rank correlation specification on a regular vine. Suppose that $X_1, \ldots, X_n$ are assigned arbitrary continuous invertible distribution functions and that rank correlations on a regular vine $\mathcal{V}$ are specified. We can realize this structure in the following way:

- Create a partial correlation vine $\mathcal{V}'$ by assigning to each edge e in $\mathcal{V}$ the partial correlation $\rho_{i,j;D_e} = 2\sin(\frac{\pi}{6}r_{i,j|D_e})$, where $\{i, j\}$ and D_e are the conditioned and conditioning sets of e, and $r_{i,j|D_e}$ is the conditional rank correlation assigned to e in $\mathcal{V}$.

- Compute the unique correlation matrix R determined by the partial correlation vine $\mathcal{V}'$, and sample a joint normal distribution $(Y_1, \ldots, Y_n)$ with standard normal margins and correlation R.

- Write $x_i = F_i^{-1}\Phi(y_i)$, where F_i is the cumulative distribution function of X_i, $i = 1 \ldots n$, Φ is the cumulative distribution function of the standard normal distribution and $y_1, \ldots, y_n$ is a sample of $Y_1, \ldots, Y_n$.

This procedure realizes $X_1, \ldots, X_n$ with the stipulated conditional rank correlations. We avoid the problems encountered in Section 4.2 because we do not specify a rank correlation *matrix*, but rather a rank correlation vine. This sampling method is also fast.

4.4.6 Relationship between conditional rank and partial correlations on a regular vine

For the vine-copula method, we would like a copula for which the relation between conditional rank correlations and partial correlations is known. We now examine a few copulae from this perspective.

Elliptical copula   The elliptical copula is a copula for which partial and constant conditional product moment correlations are equal. More precisely, it is known that when (X, Y) and (X, Z) are joined by elliptical copulae (Section 3.4), and when the conditional copula of (Y, Z) given X does not depend on X, then the conditional correlation of (Y, Z) given X is equal to the partial correlation (Kurowicka and Cooke (2001) and Proposition 3.19). Hence, we can find a relationship between partial and conditional rank correlations using techniques as in the proof of Proposition 3.19. We could also incorporate the sampling algorithm for a C-vine from Chapter 6, and then

$$\rho_{23;1} = \frac{12 \int_{I^3} x_2 x_3 du_1 du_2 du_3 - r_{12} r_{13}}{\sqrt{(1 - r_{12}^2)(1 - r_{13}^2)}}, \tag{4.9}$$

where $I = [-\frac{1}{2}, \frac{1}{2}]$. Calculating the integral given earlier (with x_2 and x_3 given by the sampling algorithm (6.2), inverse conditional distributions of the elliptical copula and simplifying), we get

$$\rho_{23;1} = 2 \int_{I^2} \sin(\pi u_2) \sin\left(\pi \left[\sqrt{1 - r_{23|1}^2}\sqrt{\frac{1}{4} - u_2^2}\sin(\pi u_3) + r_{23|1} u_2\right]\right), \tag{4.10}$$

where the integration is with respect to u_2 and u_3. Notice that $\rho_{23;1}$ does not depend on r_{12}, r_{13}; it depends only on $r_{23|1}$. This is very specific for the elliptical copula. We denote the relationship (4.10) as

$$\rho_{23;1} = \psi(r_{23|1}). \tag{4.11}$$

Now we can easily show that if $r_{23|1} = 1$, then

$$\rho_{23;1} = 2 \int_{I^2} \sin(\pi u_2) \sin(\pi u_2) du_2 du_3 = 1,$$

and if $r_{23|1} = -1$,

$$\rho_{23;1} = 2 \int_{I^2} \sin(\pi u_2) \sin(-\pi u_2) du_2 du_3 = -1.$$

From the preceding result, Hoeffding's theorem (Hoeffding (1940), Theorem 3.1) and Theorem 4.4 we get that there exists trivariate uniform distribution realizing any correlation structure.

Example 4.4 *Construct a trivariate distribution with the following rank correlation structure:*

$$A = \begin{bmatrix} 1 & 0.7 & 0.7 \\ 0.7 & 1 & 0 \\ 0.7 & 0 & 1 \end{bmatrix}. \tag{4.12}$$

The copula-vine with elliptical copulae provides a very convenient way of constructing a distribution with rank correlation matrix (4.12). We have $\rho_{12} = \rho_{13} = 0.7$ and $\rho_{23} = 0$. The partial correlation $\rho_{23;1}$ can be calculated from (3.3) as -0.96. From (4.10) we find conditional rank correlation $r_{23|1} = -0.9635$. Using the sampling algorithm for the C-vine (see Chapter 6), we can sample a distribution with the rank correlation matrix (4.12) very efficiently. Notice that this rank correlation matrix could not be realized by the joint normal transformation method (see Example 4.1).

Unfortunately, the preceding result cannot be generalized to higher dimensions. The supplement discusses two fourvariate correlation matrices; the first can be realized with the elliptical copulae but not by the joint normal method, while the second cannot be realized with elliptical copulae.

Other copulae   Using techniques presented in the previous subsection, we can find a relationship between partial and conditional rank correlations for other copulae. If in (4.9) we use x_2, x_3 when calculating the integral, with the inverse conditional distributions of, for example, the DB or the Frank's copula, then the relationship between the partial correlation and a parameter of the copula that corresponds to $r_{23|1}$ can be established. However, in contrast to the elliptical copula, $\rho_{23;1}$ will also depend on r_{12} and r_{13}.

Interestingly, the correlation matrix in Example 4.4 cannot be realized with the vine method with the DB copula. We may use formula (4.9) to search for $r_{23|1}$ that corresponds to the partial correlation -0.96, and we find that $r_{23|1} = -1$ yields partial correlation equal only to -0.9403.

At present, there are no analytical results concerning the relationship between partial and constant conditional rank correlations for the DB or the Frank's copula. It would be interesting to know whether there exists a copula for which this relation is known. We could also consider incorporating non-constant conditional rank correlations assigned to the edges of a regular vine and see how they relate to partial correlations.

Predicting correlation matrices – simulation results   Having specified a joint uniform distribution with the copula-vine method, we would like to predict the resulting correlation matrix. In general, we cannot calculate the correlation matrix, but if we *pretend* that the conditional rank correlations satisfy the recursive relations for partial correlations, we can predict the correlation matrix with some error. In this section, we estimate this error for the elliptical DB and Frank's copulae with simulation. The procedure is as follows:

1. Sample a correlation matrix of size n with 'onion' method of Ghosh and Henderson (2002).

2. Find the partial correlation specification on the C-vine on n variables.

3. Pretend that conditional rank and partial correlation specifications are equal.

4. Draw 10,000 samples of the n-dimensional distribution described by the rank correlation specification on the C-vine with the elliptical (E), the DB and the Frank's (F) copula (in the sampling procedure, the same independent samples are used for all copulae).

5. Calculate correlation matrices form these samples and compare with the target correlation matrix (as a measure of difference between matrices, we take the sum of absolute differences between elements of the target and sampled matrix);

The preceding procedure was repeated 500 times for each dimension. Error is defined per matrix as the sum, over all cells, of the absolute difference between the target and the sampled matrix. In Table 4.1 we record the maximum absolute error over the 500 matrices, the average absolute error over the 500 matrices and percentage of the 500 simulated matrices for which the given copula led to the minimum error. To compare these results with the joint normal method, we sampled 500 random correlation matrices with the onion method and used these matrices as product moment correlation matrices of the joint normal. We drew 10,000 samples from the joint normal and calculated the difference between the correlation and the rank correlation matrices from these samples. Because we sampled from normal distributions instead of uniform distributions, the results per matrix could not be compared with the vine-copula matrices, but the maximum and average error can be compared.

Notice that for three- and four-dimensional matrices, the elliptical copula gives the smallest error. Five-, six- and seven-dimensional matrices are best approximated

Table 4.1   Performance of elliptical, diagonal band and Frank's copulae (E, DB and F) in the vine-copula method. Simulation results for joint normal method (N).

Copula/dim		3	4	5	6	7	8	9	10
E	%min error	60.6	53.6	28.6	8.4	1	0	0	0
	max error	0.11	0.21	0.45	0.70	1.09	1.52	2.02	2.81
	average error	0.03	0.10	0.22	0.41	0.70	1.05	1.50	1.99
DB	%min error	25.4	33.6	53.4	62.6	61	46.4	27.4	9.4
	max error	0.13	0.23	0.37	0.56	0.77	1.00	1.24	1.57
	average error	0.04	0.10	0.19	0.32	0.47	0.65	0.87	1.04
F	%min error	14	12.8	18	29	38	53.6	72.6	90.6
	max error	0.27	0.29	0.42	0.67	0.82	1.05	1.28	1.48
	average error	0.05	0.13	0.22	0.35	0.48	0.64	0.83	0.94
N	max error	0.20	0.36	0.57	0.64	0.87	1.00	1.20	1.55
	average error	0.08	0.16	0.26	0.38	0.52	0.69	0.86	1.07

Table 4.2   Average error per cell for vine-copula method with E, DB and F copulae
and for joint normal method (N).

Copula/dim	3	4	5	6	7	8	9	10
E	0.006	0.008	0.011	0.014	0.017	0.019	0.021	0.022
DB	0.007	0.009	0.010	0.011	0.011	0.012	0.012	0.012
F	0.009	0.011	0.0110	0.012	0.012	0.012	0.012	0.011
N	0.013	0.013	0.013	0.013	0.012	0.012	0.012	0.012

by the vine-diagonal band method, and then Frank's copula starts to produce the
smallest error. Of course, the average error over 500 matrices is affected by the fact
that the number of cells over which we sum increases with dimension. Table 4.1
shows average error, divided by the number of off-diagonal cells.

The average error per cell for the vine-copula method and joint normal method
is presented in Table 4.2. Notice that the average error per cell using the normal
method is always greater than the average error per cell for the vine method with
the diagonal band and Frank's copula. However, the difference decreases with
matrix dimension.

From these numerical experiments, we may conclude that Frank's copula is
a good choice if we wish to predict the correlation matrix from a rank correlation vine.

4.5   Vines and positive definiteness

The copula-vine method does not work with correlation matrices; it specifies a joint
distribution by choosing a copula family and assigning conditional rank correlations
to the edges of a regular vine. Regular vines can however be used in problems that
arise if we work with correlation matrices, as in the normal transform method. In
Theorem 4.4 a one-to-one correspondence between correlation matrices and partial
correlation specifications on a regular vine is shown. This relationship can be used
to check positive definiteness, to repair violations of positive definiteness and to
attack the completion problem.

4.5.1   Checking positive definiteness

If A is an $n \times n$ symmetric matrix with positive numbers on the main diagonal,
we may transform A to a matrix $\overline{A} = DAD$, where elements of D are as follows:

$$d_{ij} = \begin{cases} \frac{1}{\sqrt{a_{ii}}} & \text{if } i = j \\ 0 & \text{otherwise.} \end{cases}$$

Thus,

$$\overline{a_{ij}} = \frac{a_{ij}}{\sqrt{a_{ii}a_{jj}}}. \tag{4.13}$$

$\overline{A}$ has 1's on the main diagonal. Since it is known that $\overline{A}$ is positive definite if and only if all principle sub-matrices[4] are positive definite, we can restrict our further considerations to the matrices, which after transformation (4.13) have all $\overline{a_{ij}} \in (-1, 1)$, where $i \neq j$.

Definition 4.9 (proto-correlation matrix) *A symmetric matrix with all off-diagonal elements in the interval (−1, 1) and with 1's on the main diagonal is called a* **proto-correlation matrix**.

It is well known that A is positive definite $(A \succ 0)$ if and only if $\overline{A}$ is positive definite.

In order to check positive definiteness of the matrix $\overline{A}$, we will use the partial correlation specification for the C-vine. Because of the one-to-one correspondence between partial correlation specifications on a regular vine and positive definite matrices given in Theorem 4.4, it is enough to check whether all partial correlations from the partial correlation specification on the C-vine are in the interval $(-1, 1)$ to decide whether $\overline{A}$ is positive definite.

Example 4.5 *Let us consider the matrix*

$$A = \begin{bmatrix} 25 & 12 & -7 & 0.5 & 18 \\ 12 & 9 & -1.8 & 1.2 & 6 \\ -7 & -1.8 & 4 & 0.4 & -6.4 \\ 0.5 & 1.2 & 0.4 & 1 & -0.4 \\ 18 & 6 & -6.4 & -0.4 & 16 \end{bmatrix}$$

and transform A to proto-correlation matrix using formula (4.13). Then we get

$$\overline{A} = \begin{bmatrix} 1 & 0.8 & -0.7 & 0.1 & 0.9 \\ 0.8 & 1 & -0.3 & 0.4 & 0.5 \\ -0.7 & -0.3 & 1 & 0.2 & -0.8 \\ 0.1 & 0.4 & 0.2 & 1 & -0.1 \\ 0.9 & 0.5 & -0.8 & -0.1 & 1 \end{bmatrix}.$$

Since

$$\begin{bmatrix} \rho_{23;1}, & \rho_{24;1}, & \rho_{25;1} \end{bmatrix} = \begin{bmatrix} 0.6068, & 0.5360, & -0.8412 \end{bmatrix},$$

$$\begin{bmatrix} \rho_{34;12}, & \rho_{35;12}, \end{bmatrix} = \begin{bmatrix} 0.0816, & -0.0830, \end{bmatrix} \text{ and}$$

$$\begin{bmatrix} \rho_{45;123} \end{bmatrix} = \begin{bmatrix} 0.0351 \end{bmatrix}$$

are all between (−1, 1), it follows that, $\overline{A}$ and A are positive definite.

---

[4]A principle sub-matrix of an $n \times n$ matrix A is the matrix obtained by removing from A the rows and columns indexed by subset of $\{1, \ldots, n\}$.

In general, for an $n \times n$ correlation matrix we must calculate

$$\sum_{k=1}^{n-2} \binom{n-k}{2} = \frac{(n-2)(n-1)n}{6} < \frac{n^3}{6}$$

partial correlations using formula (3.3). The complexity of this algorithm, however, is not smaller than that for other known procedures such as, for example, Cholesky decomposition. This algorithm is implemented in UNICORN to check positive definiteness of a specified matrix.

4.5.2 Repairing violations of positive definiteness

In physical applications, it often happens that correlations are estimated by noisy procedures. It may thus arise that a measured correlation matrix is not positive definite. If we want to use this matrix, we must change it to get a positive definite matrix, which is as close as possible to the measured matrix.

Partial correlation specifications on a C-vine can be used to alter a non-positive definite matrix A so as to obtain a positive definite matrix B. If the matrix is not positive definite, then there exists at least one element in the partial correlation specification of the C-vine that is not in the interval $(-1, 1)$. We will find the first such element, change the value and recalculate partial correlations on the vine using the following algorithm:

for $1 \leq s \leq n - 2$, $j = s + 2, s + 3, \ldots, n$,

$$\rho_{s+1,j;12\ldots s} \notin (-1, 1) \rightarrow \rho_{s+1,j;12\ldots s} := V\left(\rho_{s+1,j;12\ldots s}\right),$$

where $V\left(\rho_{s+1,j;12\ldots s}\right) \in (-1, 1)$ is the altered value of $\rho_{s+1,j;12\ldots s}$.
Recalculate partial correlations of lower order as follows:

$$V(\rho_{s+1,j;1\ldots t-1}) = V(\rho_{s+1,j;1\ldots t})\sqrt{(1 - \rho_{t,s+1;1\ldots t-1}^2)(1 - \rho_{s+1,j;1\ldots t-1}^2)}$$
$$+ \rho_{t,s+1;1\ldots t-1}\rho_{s+1,j;1\ldots t-1}, \tag{4.14}$$

where $t = s, s - 1, \ldots, 1$.

Theorem 4.6 *The following statements hold:*

a. *All recalculated partial correlations are in the interval $(-1, 1)$.*

b. *Changing the value of the partial correlation on the vine leads to changing only one correlation in the matrix and does not affect the correlations that were already changed.*

c. *There is a linear relationship between the altered value of partial correlation and the correlation with the same conditioned set in the proto-correlation matrix.*

d. *This method always produces a positive definite matrix.*

From the statement (c) of Theorem 4.6, we can obtain the following result.

Corollary 4.2 *If*

$$|\rho_{s+1,j;12...s} - V(\rho_{s+1,j;12...s})^{(1)}| < |\rho_{s+1,j;12...s} - V(\rho_{s+1,j;12...s})^{(2)}|,$$

then

$$|\rho_{s+1,j} - \rho_{s+1,j}^{(1)}| < |\rho_{s+1,j} - \rho_{s+1,j}^{(2)}|,$$

where $V(\rho_{s+1,j;12...s})^{(1)}$ and $V(\rho_{s+1,j;12...s})^{(2)}$ are two different choices of $V(\rho_{s+1,j;12...s})$ in (4.14).

Let us consider the following example:

Example 4.6 *Let*

$$A = \begin{bmatrix} 1 & -0.6 & -0.8 & 0.5 & 0.9 \\ -0.6 & 1 & 0.6 & -0.4 & -0.4 \\ -0.8 & 0.6 & 1 & 0.1 & -0.5 \\ 0.5 & -0.4 & 0.1 & 1 & 0.7 \\ 0.9 & -0.4 & -0.5 & 0.7 & 1 \end{bmatrix}.$$

We get $\rho_{34;12} = 1.0420$; hence A is not positive definite.

Since $\rho_{34;12} > 1$, we will change its value to $V(\rho_{34;12}) = 0.9$ and recalculate the lower-order correlations

$$V(\rho_{34;1}) = V(\rho_{34;12})\sqrt{(1 - \rho_{23;1}^2)(1 - \rho_{24;1}^2)} + \rho_{23;1}\rho_{24;1}$$

and

$$V(\rho_{34}) = V(\rho_{34;1})\sqrt{(1 - \rho_{13}^2)(1 - \rho_{14}^2)} + \rho_{13}\rho_{14}.$$

We find $V(\rho_{34;1}) = 0.9623$ and, finally, the new value in the proto-correlation matrix $V(\rho_{34}) = 0.0293$. Next, we will apply the same algorithm to verify that this altered matrix is positive definite. We obtained the matrix

$$B = \begin{bmatrix} 1 & -0.6 & -0.8 & 0.5 & 0.9 \\ -0.6 & 1 & 0.6 & -0.4 & -0.4 \\ -0.8 & 0.6 & 1 & 0.0293 & -0.5 \\ 0.5 & -0.4 & 0.0293 & 1 & 0.7 \\ 0.9 & -0.4 & -0.5 & 0.7 & 1 \end{bmatrix},$$

which is positive definite. Note that only cell (3, 4) is altered.

Remark 4.1 *In Example 4.6, if we could choose a new value for the correlation $\rho_{34;12}$, that is, $V(\rho_{34;12})$, as 0.99, then the altered value of correlation ρ_{34} is 0.0741. If we change $\rho_{34;12}$ to 0.999, then we calculate ρ_{34} to be 0.0786, but in this case we obtain the value of $\rho_{45;123}$ to be equal to -1.6224. If we change this value to -0.999 we calculate $\rho_{45} = 0.7052$.*

Remark 4.2 *Note that the choice of vine has a significant effect on the resulting altered matrix. The C-vine favours entries in the first row. These values are not changed. The further we go from the first row, the greater the changes are. Hence, when fixing a matrix using the C-vine, one should rearrange variables to have the most reliable entries in the first row. Alternatively, different regular vines can be used.*

4.5.3 The completion problem

Let A be an $n \times n$ partially specified proto-correlation matrix such that the unspecified cells are given by index pairs

$$(i_k, j_k), (j_k, i_k), \quad k = 1, \ldots, K. \tag{4.15}$$

We must fill these unspecified elements such that the resulting matrix $B = [b_{ij}]_{i,j=1,\ldots,n}$ is positive definite. Thus, we must find a vector $(x_1, \ldots, x_k)$ such that

$$b_{i_k, j_k} = b_{j_k, i_k} = x_k, \quad k = 1, \ldots, K,$$

$$b_{ij} = a_{ij}, \qquad \text{otherwise}$$

and B is positive definite.

We could approach this problem by trying to find a projection of A on the set of positive definite matrices. However, as we have already discussed in Chapter 3, the constraint of positive definiteness is quite strong. The set of positive definite matrices is not simple. Thus, algorithms that search elements of this set are complicated.

The partial correlations specified on a regular vine are algebraically independent, and they uniquely determine the correlation matrix. Thus, the partial correlation vine can be seen as an algebraically independent parametrization of the set correlation matrices.

We thus formulate the completion problem as the following optimization problem. Let A be a partially specified proto-correlation matrix for n variables, let $\mathcal{V}$ be a regular vine on n variables, let x be a vector of partial correlations assigned to the edges of $\mathcal{V}$ and let $B(x)$ be the correlation matrix calculated from $\mathcal{V}$ with x. We then minimize

$$\sum |A_{ij} - B(x)_{ij}|,$$

where the sum is over the specified cells of A. If the sum is zero, then A is completable. Notice that the set of vectors x that we must search is simply $(-1, 1)^{\binom{n}{2}}$.

For some special cases of partially specified proto-correlation matrices, the completion problem is trivial. If specified cells of A correspond to the edge set of a tree, then A is always completable, and we could find all completions of A by taking a regular vine with the first tree equal to specified cells in A and assigning freely the values in $(-1, 1)$ to all partial correlations of order greater then zero. If these partial correlations are chosen to be equal to zero, then using

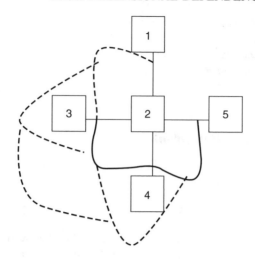

Figure 4.15 A regular vine for Example 4.7, with dashed edges corresponding to partial correlations that can be chosen freely.

Theorem 4.5 one can easily see that the resulting completion matrix will have maximum determinant.

If A corresponds to more complicated graph, then a smart choice of a vine can simplify the search procedure significantly or even allow the description of all possible completions for A. In the Supplement, preliminary results of the completion problem for special cases of graphs are shown.

Example 4.7 *Let us consider the following partially specified matrix*

$$A = \begin{bmatrix} 1 & \rho_{12} & \square & \square & \square \\ \rho_{12} & 1 & \rho_{23} & \rho_{24} & \rho_{25} \\ \square & \rho_{23} & 1 & \rho_{34} & \square \\ \square & \rho_{24} & \rho_{34} & 1 & \rho_{45} \\ \square & \rho_{25} & \square & \rho_{45} & 1 \end{bmatrix},$$

where $\square$ denotes unspecified entry in A.

By choosing a D-vine for this case, one can get all the correlations in the first tree $1 - 2 - 3 - 4 - 5$. The partial correlation $\rho_{24;3}$ can be also calculated from ρ_{23}, ρ_{34} and ρ_{24}. One must search for values of the remaining correlations: $\rho_{13;2}$, $\rho_{35;4}$, $\rho_{14;24}$, $\rho_{25;34}$ and $\rho_{15;234}$. Notice that they cannot be chosen freely, as they have to agree with correlation ρ_{25}. In this case, we could however work with the regular vine in Figure 4.15 and then $\rho_{34;2}$ and $\rho_{45;2}$ can be calculated and $\rho_{13;2}$, $\rho_{14;23}$, $\rho_{35;24}$ and $\rho_{15;234}$ can be chosen freely. Hence, the whole set of completions can be described.

In Kurowicka and Cooke (2003), the partial correlation specification on the C-vine was used to find the completion of some types of partially specified matrices.

4.6   Conclusions

This chapter addresses the problem of representing high-dimensional distributions with dependence. The joint normal transform method involves specifying a complete rank correlation matrix and yields approximate results.

The copula-tree method in Section 4.3 allows effective specification of a joint distribution. A sampling procedure for the distribution specified by the tree structure is presented in Chapter 6. A tree on n variables, however, allows specification of only $n - 1$ rank correlations. This is a significant limitation.

The copula-vine method presented in this chapter generalizes the copula-tree approach. It uses conditional dependence to construct multidimensional distributions from two-dimensional and conditional two-dimensional distributions of uniform variables. This approach yields a sampling algorithm that is fast and accurate, giving us exactly what we specify up to sampling error (see Chapter 6). Moreover, the rank correlations in copula-vines are algebraically independent. Hence, every rank correlation specification on a regular vine is consistent and can be realized.

Section 4.4.6 shows that the elliptical copula in vines ensures the existence of a trivariate joint uniform with an arbitrary correlation structure. From the results of Section 4.4.6, we conclude that there are correlation matrices of size four that cannot be realized with the vine-elliptical copula method, but this method realizes more than joint normal method does (Example 4.8).

Many questions remain. We do not know as to which n-dimensional correlation matrices can be realized by a joint uniform distribution, that is, which of the matrices are rank correlation matrices. We would like to determine the correlation matrices that are rank correlation matrices and how this set relates to the set of matrices that can be realized by copula-vine method. The examples in this chapter show that the choice of copula affects the set of realizable rank correlation matrices. Generalizations of the copula-vine method could be contemplated; in particular, we could consider non-regular vines and non-constant conditional rank correlations.

4.7   Unicorn projects

Project 4.1 Investment
In this project, we step through the construction of the tree and vine dependence structures discussed in this chapter. Create a case called 'invest_tree' with one constant variable 'start', the value of which is 1000. This is the initial capital. Add variables $V1, \ldots, V5$ for the yearly interest, each with a uniform distribution on [0.05, 0.15]. Go to the Formula panel and enter the formula called '5yrReturn':

$$start * (1 + V1) * (1 + V2) * (1 + V3) * (1 + V4) * (1 + V5).$$

Now go to the Dependence panel, click on Add New and select Dependence tree. The variables appear in the left window. Click on V1; it moves from the left to the right window. Select V1, and while it is highlighted, click on V2. V2 appears on the right, attached to V1, with correlation 0.00. Select V2 and click on V3, and so on, until the right window's tree looks like Figure 4.5. Use Ctrl click to select all the correlations 0.00 in the tree and go to the Rank Correlation input box and enter 0.70. This value now appears at all positions in the tree. In the copula list box, select Frank's copula.

Go to the Run panel; you are advised to save your structure; do so. Choose 20 runs (2000 samples) and check the save input and save output boxes. The default for Repeat Whole Simulation is '1 time', do not change this. Click on Run. When the simulation is finished, you can generate and view the report. You may check the items you wish to see in the report, and select the output format. Returning to the Run panel; click on Display Graphics. You see a cobweb plot with all input and output variables and 200 samples. In the Variables check box, unselect 'start', as this is a constant. In the cobweb plot, use the mouse to drag the variable '5yrReturn' to the left most position. The Plot Samples slider allows you to choose the number of samples shown. Move the slider all the way to the right to see all 2000 samples. The cobweb plot colour codes the leftmost variable. Selecting Options Colour... you can change the number of colours and the colours themselves. Experiment with this feature. You can also choose the variables scale; with the Natural scale, you see the minimum and maximum values for each variable below and above the corresponding vertical lines. With the mouse right click and hold, you can select the values of V1 on which to conditionalize. Starting at the top, select the highest 100 values of V1. It may be difficult to select exactly 100 values owing to granularity in the graphics. By releasing the right click on the mouse, you see the conditional cobweb plot of Figure 4.7. From the file menu you can export this picture to a bitmap or jpg file.

To create the symmetric dependence structure, go back to the variables panel and add a uniform [0, 1] variable 'LATENT'. Go to the dependence panel. With the pointer in the right window, right click. The Tree options box appears; select Remove Entire dependence structure. All variables return to the left window. Now from Dependence/Add new, select Dependence tree. Attach variables V1, ..., V5 to LATENT and assign them the rank correlation 0.84. Proceed as before to produce the cobweb plot in Figure 4.7.

To create the D-vine dependence structure, go back to the Dependence panel and Remove Entire Dependence Structure. Now from the Dependence menu, Add New and select D-vine. Notice that the minimum information copula is now unavailable; since its distribution function is not available in closed form, it cannot be used in sampling a vine. A D-vine structure is specified by specifying an order of the variables. Click on V1, ..., V5. The first tree appears in the right window. Assign rank correlations 0.70 to all edges in this first tree. The higher trees are created by hitting the 'Transcend' button. The higher trees can only be created when the first tree is finished. In the second tree, choose rank correlations −0.70 (to see the rank correlation matrix approximating the vine structure, check the 'Matrix View' box). The result should look like Figure 4.13. Simulating and displaying graphics works the same way as with trees.

Project 4.2 Mission Completion times

This project is a stylized version of an application that illustrates some of UNICORN's handy built-in functions. A system of communications satellites has a nominal mission time of 15 years, but the actual mission time is uncertain; it could be as low as 8 years and as high as 18 years, depending on obsolescence and market growth, and on how well or badly the satellites function. At least 15 satellites are required for the system to function according to specifications, and we are launching 20 satellites, each with a median life of 10 years. The input data are shown in Figure 4.16. Note that the descriptive fields have been filled to identify the variables. We created a dummy variable 'dummy' to construct symmetric dependencies. The 'missiontime' has a beta distribution on the interval [8, 18], with parameters $\alpha = 4$ and $\beta = 2$. Although there are evident dependencies in this problem, we first treat the satellites and mission time as independent. We are interested in the number of years for which the system is OK, that is, at least 15 of the satellites are working. We are also interested in the total number of functional satellite years up to mission completion. The following user-defined functions (UDFs) allow us to accomplish this (the UDF name proceeds the colon):

1. *t: vary{1,missiontime + 1,1},*

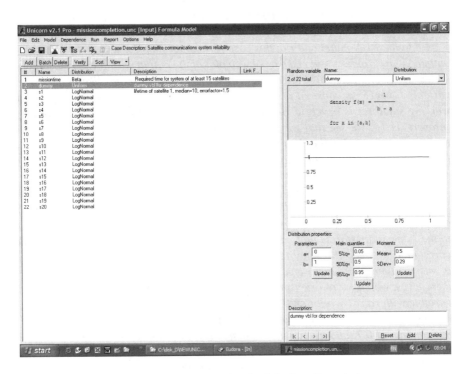

Figure 4.16 Input variables for Mission Completion.

2. *over15:*
   i15{t,s1,s2,s3,s4,s5,s6,s7,s8,s9,s10,s11,s12,s13,s14,s15,s16,s17,s18,s19,
   s20,≫},

3. *nrovert:*
   i#{t,s1,s2,s3,s4,s5,s6,s7,s8,s9,s10,s11,s12,s13,s14,s15,s16,s17,s18,s19,
   s20,≫},

4. *totunityrs: sum(nrovert),*

5. *totsysok: sum(over15).*

The first UDF is a variable (the name 'vary' is obligatory) running from 1 up to, and not *including, missiontime + 1, with step size 1. In other words, 'vary' runs from 1 to missiontime, inclusive of missiontime. UDF 2 is a generalized indicator, which returns 1 if the first and last arguments contain at least 15 of the middle arguments, otherwise it returns 0. The leftmost argument ≫ denotes infinity. In other words, 'over15' is 1 if and only if at least 15 of the satellites live at least up to time t and is zero otherwise. UDF 3, 'nrovert' is a generalized indicator, which counts the number of intermediate variables between the left- and rightmost arguments, inclusive. UDF 4 sums the number of systems alive after t, for t running from one to mission time with step size 1. UDF 5 counts the number of years for which the system is OK, that is, at least 15 of the satellites are working.*

Go to the RUN panel and run this case with 2000 samples, saving the input and output. Go to graphics and make a cobweb plot of the output variables. By dragging the variables and choosing colours, you should get the plot to look like Plate 2. The variable 'over15' is constant at zero because at the last *value of t, the number of working satellites is never 15 or over. Indeed, we see that the maximum value of 'nrovert' is 14. We see that we get at most 10 years of system performance. With the mouse, select the value 10 for the variable 'totsysOK;' the number of samples for which 'totsysOK' is equal to 10 is 37. The maximum value of the total number of satellite years in service up to mission completion is 232.*

There are dependencies in this problem. Most notably, mission time is allowed to depend on how well the satellites function. Also, the satellites have been designed as identical and manufactured by the same company. In service they will be exposed to the same solar storms that can influence life. Experts agree that there is a rank correlation between the satellites' life and the mission time of 0.7. Independently of mission time, there is a weak rank correlation, of 0.09, of each pair of satellites themselves, which we realize (approximately) by rank correlating them to a dummy variable with rank correlation 0.3. This suggests a C-vine with missiontime as root and dummy as the second variable. Satellites S1...S20 are rank correlated with missiontime at 0.7, and given *mission time, they are rank correlated to the dummy at 0.3. Higher-order correlations are zero.*

The results are shown in Plate 3. Note that the maximal number of years of 'totsysOK' is now 12, and 362 samples yield 10 years of system functioning. On the other hand, 'nrovert' has its maximal value equal to 9. This means that when t = missiontime, the number of systems still alive is much less than that in the independent case shown in the preceding text. Why? This is because the relatively strong correlation between missiontime and satellite life induces a positive correlation of about 0.64 between satellites. Hence, the satellites' deaths tend to be less spread out. Since the mean is 10 years, we find that 15 survive less often up to 18 years. You can find that the total number of satellite years now has a maximal value of 270 rather than 232.

4.8 Exercises

Ex 4.1 *Let X and Y be uniform on [0, 1] and Z = X + Y. Calculate Var(Z) when X, Y are:*

1. *independent,*

2. *completely positively correlated,*

3. *completely negatively correlated.*

Ex 4.2 *Let us consider a tree on three variables X, Y and Z, which are uniform on $(-\frac{1}{2}, \frac{1}{2})$. Let $(X, Y), (X, Z)$ be joined by the elliptical copula with correlation r_{XY} and r_{XZ}, respectively. Y and Z are conditionally independent given X. Show that the correlation r_{YZ} is equal to the product of r_{XY} and r_{XZ}.*

Ex 4.3 *Let X_i be the outcome (heads or tails) on the i-th toss of a fair coin, $i = 1, \ldots, 3$. Let $X_p = 1$ if the number heads on $X_1, \ldots, X_3$ is even, and $= 0$ otherwise. Show that the tree*

$$X_1 \text{ --- } X_p \text{ --- } X_2 \text{ --- } X_3$$

has tree dependence of order 1 but not order 2 and is not Markov. Show that the tree

$$X_1 \text{ --- } X_p \text{ --- } X_2$$
$$\lfloor X_3$$

has tree dependence of order m, for all m, but is not Markov.

Ex 4.4 *Determine whether the correlation matrix in Example 4.1 can be realized with the vine-Frank's copula method (this exercise requires numerical integration).*

Ex 4.5 *Use the techniques from Section 4.4.6 to show that if variables X, Y and Z are uniform on $(-\frac{1}{2}, \frac{1}{2})$, with (X, Y) and (X, Z) joined by the elliptical copula with rank correlations r_{XY} and r_{XZ}, respectively, and the conditional rank correlation is as follows:*

$$r_{YZ|X} = \begin{cases} -1 & X \leq 0 \\ 1 & X > 0 \end{cases}$$

then

$$r_{YZ} = r_{XY} r_{XZ}.$$

Ex 4.6 *Solve the completion problem for the matrix in (4.3) using the partial correlation vine method.*

4.9 Supplement

4.9.1 Proofs

Theorem 4.1 (Meeuwissen and Cooke (1994)) Let (F, T, B) be an n-dimensional bivariate tree specification that specifies the marginal densities f_i, $1 \le i \le n$, and the bivariate densities f_{ij}, $\{i, j\} \in E$, the set of edges of T. Then there is a unique density g on $\mathbb{R}^n$ with marginals $f_1, \ldots, f_n$ and bivariate marginals f_{ij} for $\{i, j\} \in E$ such that g has Markov tree dependence described by T. The density g is given by

$$g(x_1, \ldots, x_n) = \frac{\prod_{\{i,j\}\in E} f_{ij}(x_i, x_j)}{\prod_{i \in N}(f_i(x_i))^{deg(i)-1}}. \qquad (4.16)$$

With copula densities c_{ij}, $\{i, j\} \in E$, this becomes

$$g(x_1, \ldots, x_n) = f_1(x_1) \ldots f_n(x_n) \prod_{\{i,j\}\in E} c_{ij}(F_i(x_i), F_j(x_j)). \qquad (4.17)$$

Proof.
The proof is by induction on the number of variables n. If $n = 2$, the theorem is trivial. Without loss of the generality, we may assume that the tree is connected. Fix $i \in N$ with a degree of at least 2 and let D_i denote the set of neighbours of i: $D_i = \{j \mid j \in N, \{i, j\} \in E\}$. Now, for each $j \in D_i$, consider the subtrees T_j with set of nodes

$$N_j = \{i, j\} \cup \{k \mid k \in N, \text{there is a path from } k \text{ to } j \text{ that does not include } i\}$$

and set of edges

$$E_j = \{\{k, l\} \mid \{k, l\} \in E, k, l \in N_j\}.$$

Thus, $\bigcup_{j \in D_i} E_j = E$ and for all $j, k \in D_i$: $E_j \cap E_k = \emptyset$ and $N_j \cap N_k = \{i\}$ because T is a tree. Further, let $deg_j(k)$ denote degree of the node k in T_j; that is, $deg_j(i) = 1$ and $deg_j(k) = deg(k)$ for all other $k \in N_j$. Now, let g_j be the unique distribution satisfying the theorem for the subtree T_j for all $j \in D_i$. By induction, we have for $N_j = \{l_1, l_2, \ldots, l_{n(j)}\}$

$$g_j(x_{l_1}, x_{l_2}, \ldots, x_{l_{n(j)}}) = \frac{\prod_{\{l_i,l_j\}\in E_j} f_{l_i,l_j}(x_{l_i}, x_{l_j})}{\prod_{h \in N_j}(f_{l_h}(x_{l_h}))^{deg_j(l_h)-1}}.$$

This expression is equivalent to

$$g_j(x_{l_1}, x_{l_2}, \ldots, x_{l_{n(j)}}) = f_{i,j}(x_i, x_j) \frac{\prod_{\{l_i, l_j\} \in E_j \setminus \{i,j\}} f_{l_i, l_j}(x_{l_i}, x_{l_j})}{\prod_{h \in N_j \setminus \{i\}} (f_{l_h}(x_{l_h}))^{deg_j(l_h) - 1}}.$$

Denote the conditional density of g_j given X_i as $g_{j|i} = g_j / f_i$. Since the sets $N_j \setminus \{i\}$, $j \in D_i$ are disjoint, we may set

$$g(x_1, x_2, \ldots, x_n) = f_i(x_i) \prod_{j \in D_i} g_{j|i}(x_{l_1}, x_{l_2}, \ldots, x_{l_{n(j)}}),$$

which is equal to (4.16).

To verify that g has the Markov tree dependence property, let X_a and X_b be disjunct subsets of variables separated by variables in set X_c. To prove that X_a and X_b are conditionally independent given X_c, it suffices to factorize the marginal density $g(x_a, x_b, x_c)$ as

$$g(x_a, x_b, x_c) = H(x_a, x_c) J(x_b, x_c) \tag{4.18}$$

for some functions H, J (Whittaker (1990)). Such a factorization is immediately obtained from (4.16), integrating out irrelevant variables.

To verify that g is unique, let $\widetilde{g}$ be another density satisfying the theorem; let x_1 be associated with a node with degree 1 and let $\{1, 2\}$ be the edge attached to node 1. Let $I = N \setminus \{1, 2\}$. Since g and $\widetilde{g}$ both have the Markov property,

$$g(x_1, \ldots, x_n), = g(x_1, x_I | x_2) g(x_2)$$
$$= g(x_I | x_2) g(x_1 | x_2) g(x_2),$$

with a similar equation for $\widetilde{g}$. By the induction hypotheses and by the equality of the first and second dimensional marginals, it follows that

$$g(x_1, \ldots, x_n) = \widetilde{g}(x_1, \ldots, x_n).$$

Replacing each term f_{ij} in (4.16) of Theorem 4.1 by

$$c_{ij}(F_i, F_j) f_i f_j$$

yields 4.17. $\square$

Lemma 4.1 Let $\mathcal{V}$ be a regular vine on n elements and let $j \in E_i$. Then

$$\#U_j(k) = 2\#U_j(k-1) - \#U_j(k-2); k = 2, 3, \ldots. \tag{4.19}$$

Proof.
For $e \in U_j(k-1)$, write $e = \{e_1, e_2\}$ and consider the lexigraphical ordering of the names e_c, $c = 1, 2$. There are $2\#U_j(k-1)$ names in this ordering. $\#U_j(k)$ is the

number of names in the ordering, diminished by the number of names that refer to
an element that is already named earlier in the ordering. By regularity, for every
element in $U_j(k-2)$, there is exactly one name in the lexigraphical ordering that
denotes an element previously named in the ordering. Hence (4.19) holds. $\square$

Lemma 4.2 Let $\mathcal{V}$ be a regular vine on n elements and let $j \in E_i$. Then

$$\#U_j(k) = k + 1; \; k = 0, 1, \ldots, i.$$

Proof.
The statement clearly holds for $k = 0$, $k = 1$. By the proximity property, it follows
immediately that it holds for $k = 2$. Suppose that the results holds up to $k - 1$. Then
$\#U_j(k-1) = k$. By Lemma 4.1,

$$\#U_j(k) = 2\#U_j(k-1) - \#U_j(k-2).$$

With the induction hypothesis we conclude

$$\#U_j(k) = 2k - (k-1) = k + 1. \quad \square$$

Lemma 4.3 If $\mathcal{V}$ is a regular vine on n elements then for all $i = 1, \ldots n - 1$, and
all $e \in E_i$, the conditioned set C_e associated with e is a doubleton, $\#U_e^* = i + 1$,
and $\#D_e = i - 1$.

Proof.
Let $e \in E_i$ and $e = \{j, k\}$. By Lemma 4.2 $\#U_e^* = i + 1$. Let $D_e = U_j^* \cap U_k^*$ and
$C_e = U_j^* \triangle U_k^*$. It suffices to show that $\#C_e = 2$. We get

$$i + 1 = \#D_e + \#C_e \tag{4.20}$$

and

$$2i = \#U_j^* + \#U_k^* = \#C_e + 2\#D_e. \tag{4.21}$$

Divide (4.21) by 2 and subtract from (4.20); then

$$\#C_e = 2.$$

Hence $\#(U_j^* \setminus D_e) = 1$, $\#(U_k^* \setminus D_e) = 1$ and $\#D_e = i - 1$. $\square$

Lemma 4.4 Let $\mathcal{V}$ be a regular vine, and suppose for $j, k \in E_i$, $U_j^* = U_k^*$, then
$j = k$.

Proof.
We claim that $U_j(x+1) = U_k(x+1)$ implies $U_j(x) = U_k(x)$. In any tree, the
number of edges between y vertices is less than or equal to $y - 1$. $\#U_j(x+1) =$
$x + 2$ and $U_j(x+1) \subseteq N_{i-x}$, so in tree T_{i-x} the number of edges between the
nodes in $U_j(x+1)$ is less than or equal to $x + 1$. $\#U_j(x) = x + 1 = \#U_k(x)$, so

both these sets must consist of the $x + 1$ possible edges between the nodes of T_{i-x} that are in $U_j(x + 1) = U_k(x + 1)$. Hence, $U_j(x) = U_k(x)$. Since $U_j^* = U_k^*$, that is, $U_j(i) = U_k(i)$, repeated application of this result produces $U_j(1) = U_k(1)$, that is, $j = k$. □

Lemma 4.5 If the conditioned sets of edges i, j in a regular vine are equal, then $i = j$.

Proof.
Suppose i and j have the same conditioned sets. By Lemma 4.3, the conditioned set is doubleton, say $\{a, b\}, a \in N, b \in N$. Let D_i and D_j be the conditioning sets of edges i and j, respectively. Then, in the tree T_1, there is a path from a to b through the nodes in D_i and also a path from a to b through the nodes in D_j. If $D_i \neq D_j$, then there must be a cycle in the edges E_1, but this is impossible since T_1 is a tree. It follows that $D_i = D_j$, and from Lemma 4.4, it follows that $i = j$. □

Lemma 4.6 For any node M of order $k > 0$ in a regular vine, if variable i is a member of the conditioned set of M, then i is a member of the conditioned set of exactly one of the m-children of M, and the conditioning set of an m-child of M is a subset of the conditioning set of M.

Proof.
If the conditioning set of an m-child of M is vacuous, the proposition is trivially true; we therefore assume $k > 1$. Let $M = \{A, B\}$, where A, B are nodes of order $k - 1$. By regularity we may write $A = \{A1, D\}, B = \{B1, D\}$, where $A1, B1, D$ are nodes of order $k - 2$. $U_M^* = U_A^* \cup U_B^*$. By assumption, $i \in U_A^* \triangle U_B^*$. Suppose $i \in U_A^*$, then $i \notin U_B^*$. $U_A^* = U_{A1}^* \cup U_D^*$, and since $U_D^* \subseteq U_B^*$ and $i \notin U_B^*$, we have $i \notin U_D^*$. It follows that $i \in U_{A1}^* \triangle U_D^*$; that is, i is in the conditioned set of A. Since the conditioning set of A is $U_{A1}^* \cap U_D^* \subseteq U_B^*$, we have $U_{A1}^* \cap U_D^* \subseteq U_A^* \cap U_B^*$; that is, the conditioning set of A is a subset of the conditioning set of M. □

Proposition 4.1 *Let $\mathcal{V}$ be a regular vine on n elements and i an integer, $1 \leq i < n - 1$. Given a node m in tree T_i, there are exactly $deg(m) - 1$ edges in T_{i+1} around m, where an edge in T_{i+1} is around m if both its elements contain m.*

Proof. Without loss of generality we may assume that $i = 1$. First we show that $deg(m) - 1$ is the maximal number of edges in T_2 around m. Clearly, there are $deg(m)$ edges joining m to other nodes. These are the nodes in T_2 that will be around m when joined by edges. Because T_2 has to be a tree, there can be no cycles of edges, so there are at most $deg(m) - 1$ different edges in T_2 around m.

We now show that there are exactly $deg(m) - 1$ edges in T_2 around m. Note first that an edge in T_2 can only be around one node of T_1, as otherwise there would be a cycle in T_1. If some node m of T_1 has less than $deg(m) - 1$ edges in

T_2 around it, then we can count the total number of edges in T_2 as

$$\sum_{j \in N_1} \# \text{ edges around } j < \sum_{j \in N_1} (deg(j) - 1)$$

$$= \left(\sum_{j \in N_1} deg(j) \right) - n$$

$$= 2(n - 1) - n = n - 2.$$

This contradicts the fact that T_2 has $n - 2$ edges. $\square$

Theorem 4.2 Let $\mathcal{V} = (T_1, \ldots, T_{n-1})$ be a regular vine on n elements. For each edge $e(j, k) \in T_i$, $i = 1, \ldots, n - 1$ with conditioned set $\{j, k\}$ and conditioning set D_e, let the conditional copula and copula density be $C_{jk|D_e}$ and $c_{jk|D_e}$ respectively. Let the marginal distributions F_i with densities f_i, $i = 1, \ldots, n$ be given. Then the vine-dependent distribution is uniquely determined and has a density given by

$$f_{1 \ldots n} = \left(\prod_{i=2}^{n-1} \prod_{e(j,k) \in E_i} c_{jk|D_e}(F_{j|D_e}, F_{k|D_e}) \right) \cdot \frac{\prod_{\{j,k\} \in E_1} f_{jk}}{\prod_{j \in N_1} f_j^{deg(j)-1}} \qquad (4.6)$$

$$= f_1 \ldots f_n \prod_{i=1}^{n-1} \prod_{e(j,k) \in E_i} c_{jk|D_e}(F_{j|D_e}, F_{k|D_e}). \qquad (4.7)$$

Proof. The first statement is proved by reverse induction on the level of the tree in the vine. We claim that for every $2 \le M \le n - 1$,

$$f_{1 \ldots n} = \left(\prod_{i=M}^{n-1} \prod_{e(j,k) \in E_i} c_{jk|D_e}(F_{j|D_e}, F_{k|D_e}) \right) \cdot \frac{\prod_{e \in E_{M-1}} f_{U_e^*}}{\prod_{e \in N_{M-1}} f_{U_e^*}^{deg(e)-1}}.$$

The inductive claim holds for $M = n - 1$, which can be seen as follows: For the one edge in T_{n-1}, say $e = \{e_1, e_2\}$ with $U_{e_1}^* = \{j\} \cup D_e$ and $U_{e_2}^* = \{k\} \cup D_e$, we have

$$f_{1 \ldots n} = f_{jk|D_e} f_{D_e},$$

$$= c_{jk|D_e}(F_{j|D_e}, F_{k|D_e}) \cdot f_{j|D_e} f_{k|D_e} f_{D_e},$$

$$= c_{jk|D_e}(F_{j|D_e}, F_{k|D_e}) \frac{f_{U_{e_1}^*} f_{U_{e_2}^*}}{f_{D_e}}.$$

Since T_{n-2} is a tree with two edges and three nodes, one of the nodes, say m, has to have degree 2, and the edge e of T_{n-1} must be around m. Hence, $D_e = U_m^*$ and the claim is demonstrated.

For the inductive step assume that the formula holds for M. We show that it holds for $M - 1$. To see this, apply first the same decomposition as the one in the preceding text for the marginal distribution corresponding to each edge of T_M. For $e \in E_{M-1}$, there are nodes in N_{M-1}, or equivalently, edges in E_{M-2} that we call e_1 and e_2, such that $e = \{e_1, e_2\}$. The decomposition immediately gives all the claimed copula density terms, but the remaining term built from marginal densities of f is of the form:

$$\left(\prod_{e \in E_{M-1}} \frac{f_{U_{e_1}^*} f_{U_{e_2}^*}}{f_{D_e}} \right) \frac{1}{\prod_{m \in N_{M-1}} f_{U_m^*}^{deg(m)-1}}.$$

In order to show that this reduces to the formula claimed for the induction step, we have to show two things: (1) The extra multiplicity of $f_{U_{e_1}^*}$ terms arising because a node of N_{M-1} is cancelled by $\prod_{m \in N_{M-1}} f_{U_m^*}^{deg(m)-1}$; and (2)

$$\prod_{e \in E_{M-1}} f_{D_e} = \prod_{m \in N_{M-2}} f_{U_m^*}^{deg(m)-1}.$$

The requirement (1) is clear, since the degree of a node is just a number of edges attached to it. Hence, the multiplicity of a term f_{e_i} in the denominator is $deg(e_i)$, so that after cancellation we retain each term exactly once. For (2) note that if $e \in E_{M-1}$ then D_e equals U_m^* for some $m \in N_{M-1}$ and, furthermore, that e is around m. The claim follows immediately from Proposition 4.1, which completes the proof of the first statement.

The second statement is proved by replacing each term f_{jk} in (4.6) by $c_{jk}(F_j, F_k) f_j f_k$. $\square$

Lemma 4.7 If $z, x, y \in (-1, 1)$, then even $w \in (-1, 1)$, where

$$w = z\sqrt{(1 - x^2)(1 - y^2)} + xy.$$

Proof.
We substitute $x = \cos \alpha$, $y = \cos \beta$, and use

$$1 - \cos^2 \alpha = \sin^2 \alpha;$$

$$\cos \alpha \cos \beta = \frac{\cos(\alpha - \beta) + \cos(\alpha + \beta)}{2};$$

$$\sin \alpha \sin \beta = \frac{\cos(\alpha - \beta) - \cos(\alpha + \beta)}{2};$$

and find

$$z \left| \frac{\cos(\alpha - \beta) - \cos(\alpha + \beta)}{2} \right| + \frac{\cos(\alpha - \beta) + \cos(\alpha + \beta)}{2} = w.$$

Write this as

$$z \left| \frac{a - b}{2} \right| + \frac{a + b}{2} = w,$$

where $a, b \in (-1, 1)$. As the left-hand side is linear in z, its extreme values must occur when $z = 1$ or -1. It is easy to check that in these cases $w \in (-1, 1)$. □

Let g be a density on $\mathbb{R}^n$ for which all marginal and conditional marginal densities satisfy the absolute continuity conditions implicit in the relative information integrals. $g_{1,\dots,k}$ denotes the marginal over $x_1, \dots, x_k$ and $g_{1,\dots,k-1|k,\dots,n}$ denotes the marginal over $x_1, \dots, x_{k-1}$ conditional on $x_k, \dots, x_n$. $E_{1,\dots,k}$ denotes expectation taken over $x_1, \dots, x_k$.

Lemma 4.8

1.

$$I\left(g \mid \prod_{i=1}^{n} g_i\right) = I\left(g_{k,\dots,n} \mid \prod_{i=k}^{n} g_i\right) + E_{k,\dots,n} I\left(g_{1,\dots,k-1 \mid k,\dots,n} \mid \prod_{i=1}^{k-1} g_i\right).$$

2.

$$I\left(g \mid \prod_{i=1}^{n} g_i\right) = \sum_{j=1}^{n-1} E_{1,\dots,j} I(g_{j+1|1,\dots,j} \mid g_{j+1}).$$

3.

$$E_{2,\dots,n} I(g_{1|2,\dots,n} \mid g_1) + E_{1,\dots,n-1} I(g_{n|1,\dots,n-1} \mid g_n) =$$
$$E_{2,\dots,n-1} \left(I(g_{1,n|2,\dots,n-1} \mid g_{1|2,\dots,n-1} g_{n|2,\dots,n-1}) + I(g_{1,n|2,\dots,n-1} \mid g_1 g_n)\right).$$

4.

$$2I\left(g \mid \prod_{i=1}^{n} g_i\right) = I\left(g_{2,\dots,n} \mid \prod_{i=2}^{n} g_i\right) + I\left(g_{1,\dots,n-1} \mid \prod_{i=1}^{n-1} g_i\right) +$$
$$+ E_{2,\dots,n-1} I(g_{1,n|2,\dots,n-1} \mid g_{1|2,\dots,n-1} g_{n|2,\dots,n-1}) + I(g \mid g_1 g_n g_{2,\dots,n-1}).$$

Proof.
We indicate the main steps, leaving the computational details to the reader.

1. For g on the left-hand side fill in $g = g_{1,\dots,k-1|k,\dots,n} g_{k,\dots,n}$.

2. This follows from the above by iteration.

3. The integrals on the left-hand side can be combined, and the logarithm under the integral has the argument:

$$\frac{g g}{g_{2,\dots,n} g_{1,\dots,n-1} g_1 g_n}.$$

This can be re-written as

$$\frac{g_{1,n|2,\dots,n-1}}{g_{1|2,\dots,n-1} g_{n|2,\dots,n-1}} \frac{g_{1,n|2,\dots,n-1}}{g_1 g_n}.$$

By writing the log of this product as the sum of logarithms of its terms, the result on the right-hand side is obtained.

4. This follows from the previous statement by noting

$$E_{2,...n-1}I(g_{1,n|2,...n-1} \mid g_1 g_n) = I(g \mid g_1 g_n g_{2,...n-1}). \quad \square$$

Theorem 4.3 Let g be an n-dimensional density satisfying the bivariate-vine specification $(F, \mathcal{V}; B)$ with density g and one-dimensional marginal densities $g_1, \ldots, g_n$, then

$$I\left(g \mid \prod_{i=1}^{n} g_i\right) = \sum_{i=1}^{n-1} \sum_{e(j,k) \in E_i} E_{D_e} I(g_{C_{e,j}, C_{e,k}|D_e} \mid g_{C_{e,j}|D_e} \cdot g_{C_{e,k}|D_e}). \quad (4.22)$$

If $g_{C_{e,j}, C_{e,k}|D_e}$ is the unique density satisfying $(F, \mathcal{V}; B)$, which minimizes

$$I(g_{C_{e,j}, C_{e,k}|D_e} \mid g_{C_{e,j}|D_e} \cdot g_{C_{e,k}|D_e}); \quad i = 1, \ldots, n-1; e(j,k) \in E_i,$$

then g is the unique density satisfying $(F, \mathcal{V}, B)$ and minimizing

$$I\left(g \mid \prod_{i=1}^{n} g_i\right).$$

Proof.
The theorem is proved by induction on n. E_{n-1} has one element, say $e(1, 2)$, and we may assume that $C_{e,1} = x_1$, $C_{e,2} = x_n$. We define a vine specification $(F^1, \mathcal{V}^1, B^1)$ on $\{x_2, \ldots, x_n\}$:

$$F^1 = F_2, \ldots, F_n;$$
$$N_i^1 = N_i^1 \backslash x_1;$$
$$E_i^1 = E_i \backslash \{j, k\} \text{ if } j = x_1 \text{ or } k = x_1;$$
$$B_{e(j,k)}^1 = B_{e(j,k)} \text{ if } \{C_{e,j}, C_{e,k}, D_e\} \subset \{x_2, \ldots, x_n\}.$$

We define vine specifications $(F^n, \mathcal{V}^n, B^n)$ on $\{x_1, \ldots, x_{n-1}\}$ and $(F^{1,n}, \mathcal{V}^{1,n}, B^{1,n})$ on $\{x_2, \ldots, x_{n-1}\}$ in the same manner. From the definition of regularity, it follows immediately that $\mathcal{V}^1$, $\mathcal{V}^n$ and $\mathcal{V}^{1,n}$ are regular. $g_{2,...,n}$, $g_{1,...,n-1}$ and $g_{2,...,n-1}$ satisfy the conditions of the theorem for these specifications. In other words,

$$I\left(g_{1,...,n-1} \mid \prod_{i=1}^{n-1} g_i\right)$$

is minimal for densities satisfying B^n, and

$$I\left(g_{2,...,n} \mid \prod_{i=2}^{n} g_i\right)$$

is minimal for densities satisfying B^1.

We now claim that

$$I\left(g \mid \prod_{i=1}^{n} g_i\right) = \sum_{i=1}^{n-1} \sum_{e(j,k) \in E_i} E_{D_e} I(g_{C_{e,j}, C_{e,k} \mid D_e} \mid g_{C_{e,j} \mid D_e} \cdot g_{C_{e,k} \mid D_e}). \quad (4.23)$$

The claim is proved by applying Lemma 4.8(4); the last term in the preceding sum is the expectation in Lemma 4.8(4). By applying the induction hypothesis to the vine specification $(F^{1,n}, \mathcal{V}^{1,n}, B^{1,n})$, we note that the terms in the expansion of $I(g_{2,\dots n-1} \mid \prod_{i=2}^{n-1} g_i)$ are exactly those that are counted twice in the expansion of

$$I\left(g_{2,\dots,n} \mid \prod_{i=2}^{n} g_i\right) + I\left(g_{1,\dots,n-1} \mid \prod_{i=1}^{n-1} g_i\right),$$

from which the claim follows. Since g minimizes each information term in equation (4.23), it also minimizes each expectation, and the theorem is proved. □

Theorem 4.4 For any regular vine on n elements, there is a one-to-one correspondence between the set of $n \times n$ full-rank correlation matrices and the set of partial correlation specifications for the vine.

Proof.
Clearly, the correlations determine the partial correlations via the recursive relations (3.3).

(1) We first show that the correlations $\rho_{ij} = \rho(X_i, X_j)$ can be calculated from the partial correlations specified by the vine. The theorem is proved by induction on the number of elements n. The base case $(n = 2)$ is trivial. Assume that the theorem holds for $i = 2, \dots, n - 1$. For a regular vine over n elements, the tree T_{n-1} has one edge, say $e = \{j, k\}$. By Lemma 4.3, $\#D_e = n - 2$. By re-indexing the variables $X_1, \dots X_n$ if necessary, we may assume that

$$C_{e,j} = U_j^* \setminus D_e = X_1,$$

$$C_{e,k} = U_k^* \setminus D_e = X_n,$$

$$U_j^* = \{1, \dots, n - 1\}$$

$$U_k^* = \{2, \dots, n\}$$

$$D_e = \{2, \dots, n - 1\}.$$

The correlations over U_j^* and U_k^* are determined by the induction step. The correlation ρ_{1n} remains to be determined. The left-hand side of

$$\rho_{1n;2\dots n-1} = \frac{\rho_{1n;3\dots n-1} - \rho_{12;3\dots n-1}\rho_{2n;3\dots n-1}}{\sqrt{1 - \rho_{12;3\dots n-1}^2}\sqrt{1 - \rho_{2n;3\dots n-1}^2}}$$

is determined by the vine specification. The terms

$$\rho_{12;3\dots n-1}, \quad \rho_{2n;3\dots n-1}$$

are determined by the induction hypothesis. It follows from Lemma 4.7 that we can solve the preceding equation for $\rho_{1n;3\ldots n-1}$, and write

$$\rho_{1n;3\ldots n-1} = \frac{\rho_{1n;4\ldots n-1} - \rho_{13;4\ldots n-1}\rho_{3n;4\ldots n-1}}{\sqrt{1 - \rho_{13;4\ldots n-1}^2}\sqrt{1 - \rho_{3n;4\ldots n-1}^2}}$$

Proceeding in this manner, we eventually find

$$\rho_{1n;n-1} = \frac{\rho_{1n} - \rho_{1n-1}\rho_{nn-1}}{\sqrt{1 - \rho_{1n-1}^2}\sqrt{1 - \rho_{nn-1}^2}}.$$

This equation may now be solved for ρ_{1n}. This shows that if two distributions have the same partial correlations on a regular vine, then they have the same correlation matrix.

(2) To go the other way, we show that for any regular vine and any assignment of partial correlations to its edges with values in $(-1, 1)$, there is a joint distribution with these partial correlations. This is proved by induction on the number n of variables. As in (1), we may assume that the variables are indexed so that $\rho_{1,n\,;\,2,\ldots,n-1}$ is the partial correlation of the tree T_{n-1}. By the induction hypothesis, there are variables $X_1, \ldots, X_{n-1}$ and $\widetilde{X}_2, \ldots, \widetilde{X}_n$ realizing the partial correlations of the sub-vines on U_j^* and U_k^*. Without loss of generality, we may assume that these are (possibly correlated) standard normal variables and that $(X_2, \ldots, X_{n-1}) = (\widetilde{X}_2, \ldots, \widetilde{X}_{n-1})$. We may further assume that

$$X_1 = A_1 W_1 + \sum_{j=2}^{n-1} A_j X_j, \tag{4.24}$$

$$X_n = B_n W_n + \sum_{j=2}^{n-1} B_j X_j, \tag{4.25}$$

where W_1, W_n are any standard normal variables independent of $(X_2, \ldots, X_{n-1})$. Indeed, the coefficients A_i are obtained from the *linear least squares predictor* $\widehat{X}_1$ of X_1 from $W_1, X_2, \ldots, X_{n-1}$ (Whittaker (1990)):

$$\widehat{X}_1 = Cov(X_1, (W_1, X_2, \ldots, X_{n-1}))Var(W_1, X_2, \ldots, X_{n-1})^{-1}$$
$$(W_1, X_2, \ldots, X_{n-1})^T,$$

and similarly, for X_n, $X_1, \ldots, X_n$ are joint normal. Since for the joint normal distribution partial and conditional correlation are equal (Proposition 3.29), we have

$$\rho_{1,n\,;\,2,\ldots,n-1} = \rho_{1,n\,|\,2,\ldots,n-1} = \rho(A_1 W_1, B_n W_n) = \rho(W_1, W_n), \tag{4.26}$$

and we may choose W_1, W_n to have the required correlation. $\quad\square$

For the next theorem, we reproduce the relevant formulae for multiple correlation $R_{1\{2\ldots n\}}$ for convenience.

$$1 - R_{1\{2,\ldots,n\}}^2 = (1 - \rho_{1,n}^2)(1 - \rho_{1,n-1;n}^2) \cdots (1 - \rho_{1,2;3\ldots n}^2). \qquad (4.27)$$

$R_{1\{2,\ldots,n\}}$ is invariant under permutation of $\{2,\ldots,n\}$ and

$$D = \left(1 - R_{1\{2,\ldots,n\}}^2\right)\left(1 - R_{2\{3,\ldots,n\}}^2\right) \cdots \left(1 - R_{n-1\{n\}}^2\right), \qquad (4.28)$$

where $R_{n-1\{n\}} = \rho_{n-1,n}$.

Theorem 4.5

Let D be the determinant of the correlation matrix of variables $1,\ldots,n$; with $D > 0$. For any partial correlation vine;

$$D = \prod_{i=1}^{n-1} \prod_{e \in E_i} (1 - \rho_{j,k;D_e}^2), \qquad (4.29)$$

where $\{j, k\}$ and D_e are the conditioned and conditioning set of e.

Proof. Re-indexing if necessary, let $\{1, 2|3,\ldots,n\}$ denote the constraint of the single node of the topmost tree T_{n-1}. Collect all m-descendents of this node containing variable 1. By Lemma 4.6; 1 occurs only in the conditioned sets of the m-descendent nodes, and the conditioning set of an m-child is a subset of the conditioning set of its m-parent. By Lemma 4.5 variable 1 occurs exactly once with every other variable in the conditioned set of some node. By re-indexing $\{2,\ldots,n\}$ if necessary, we may write the constraints of the m-descendents containing 1 of the top node as

$$\{1, 2|3,\ldots,n\}, \ \{1, 3|4,\ldots,n\}, \ \ldots \{1, n-1|n\}, \ \{1, n\}.$$

The partial correlations associated with these m-descendent nodes are

$$\rho_{1,2;3,\ldots,n}, \ \rho_{1,3;4,\ldots,n}, \ \cdots \rho_{1,n-1;n}, \ \rho_{1,n}$$

and are exactly the terms occurring in (4.27); hence, we may replace the terms in the product on the right-hand side of (4.29) containing these partial correlations by $1 - R_{1\{2,\ldots,n\}}^2$. Note that (4.27) is invariant under permutation of $\{2,\ldots,n\}$. Remove variable 1 and nodes containing 1; these are just the nodes whose constraints are given in the preceding text. We obtain the sub-vine over variables $\{2,\ldots,n\}$. By Lemma 4.6; 2 is in the conditioned set of the top node of this sub-vine. We apply the same argument, re-indexing $\{3,\ldots,n\}$ if necessary. With this re-indexing, we may replace the product of terms in (4.29)

$$(1 - \rho_{2,3;4,\ldots,n}^2), \ (1 - \rho_{2,4;5,\ldots,n}^2), \ \ldots (1 - \rho_{2,n}^2)$$

by $1 - R_{2\{3,\ldots,n\}}^2$. Proceeding in this way we obtain (4.29). $\square$

4.9.2 Results for Section 4.4.6

We now show how the relationship between partial and conditional correlations on a vine can be established for a fourvariate distribution obtained by the vine-elliptical copula method. For a given 4×4 correlation matrix, using the recursive formula (3.3), a partial correlation specification on the C-vine on four variables can be obtained

$$
\begin{array}{ccc}
\rho_{12}, & \rho_{13}, & \rho_{14}, \\
 & \rho_{23;1}, & \rho_{24;1}, \\
 & & \rho_{34;12}.
\end{array}
\tag{4.30}
$$

We must find a rank correlation specification on the C-vine that corresponds to the partial correlation specification (4.30):

$$
\begin{array}{ccc}
r_{12}, & r_{13}, & r_{14}, \\
 & r_{23|1}, & r_{24|1}, \\
 & & r_{34|12}.
\end{array}
\tag{4.31}
$$

Clearly, $r_{1i} = \rho_{1i}$, $i = 2, 3, 4$. The $r_{23|1}, r_{24|1}$ that correspond to $\rho_{23;1}, \rho_{24;1}$, respectively, can be found from (4.11), hence $r_{23|1} = \psi^{-1}(\rho_{23;1})$ and $r_{24|1} = \psi^{-1}(\rho_{24;1})$. Now we must find $r_{34|12}$ that corresponds to $\rho_{34;12}$. Using the sampling procedure for the C-vine (see Chapter 6), the correlation ρ_{34} can be calculated as

$$
\rho_{34} = 12 \int_{I^4} x_3 x_4 \, du_1 \, du_2 \, du_3 \, du_4.
$$

By simplifying and using partial correlation formula (3.3), we get

$$
\rho_{34;1} = 2 \int_{I^3} g(r_{23|1}, u_2, u_3) \cdot g(r_{24|1}, u_2, g(r_{34|12}, u_3, u_4)),
\tag{4.32}
$$

where the integration is with respect to u_2, u_3 and u_4 and

$$
g(r, u, v) = \sin\left[\pi(\sqrt{1 - r^2}\sqrt{\frac{1}{4} - u^2} \sin(\pi v) + ru) \right].
$$

Hence, $\rho_{34;1}$ depends on $r_{23|1}, r_{24|1}$ that are already chosen and $r_{34|12}$ that we are looking for. Denoting this relationship by $\rho_{34;1} = \Phi(r_{34|12}, r_{23|1}, r_{24|1})$ and using the partial correlation formula (3.3), the relationship between $\rho_{34;12}$ and $r_{34;12}$ can be denoted as

$$
\rho_{34;12} = \frac{\Phi(r_{34|12}, r_{23|1}, r_{24|1}) - \rho_{23;1}\rho_{24;1}}{\sqrt{(1 - \rho_{23;1}^2)(1 - \rho_{24;1}^2)}}.
\tag{4.33}
$$

This cannot be solved analytically, but using numerical integration, we can search for $r_{34|12}$ that corresponds to the given $\rho_{34;12}$.

Notice that for

$$\rho_{34;12} = \frac{\rho_{34;1} - \rho_{23;1}\rho_{24;1}}{\sqrt{(1 - \rho_{23;1}^2)(1 - \rho_{24;1}^2)}},$$

where $\rho_{34;1}$ is given by (4.32) and $\rho_{23;1}$, given by (4.10) and $\rho_{24;1}$ calculated as $\rho_{24;1} = \frac{\rho_{24} - \rho_{12}\rho_{14}}{\sqrt{1 - \rho_{12}^2}\sqrt{1 - \rho_{14}^2}}$. If $r_{34|12} = 1$, then

$$\rho_{34;12} =$$

$$\frac{2\int_{I^2} g(r_{23|1}, u_2, u_3)g(r_{24|1}, u_2, u_3) - 4\int_{I^2} \sin(\pi u_2)g(r_{23|1}, u_2, u_3)\int_{I^2} \sin(\pi u_2)g(r_{24|1}, u_2, u_3)}{\sqrt{(1 - (2\int_{I^2} \sin(\pi u_2)g(r_{23|1}, u_2, u_3))^2)(1 - (2\int_{I^2} \sin(\pi u_2)g(r_{24|1}, u_2, u_3))^2))}},$$

where all integrals are with respect to u_2, u_3. This, in general, is not equal to 1, but we can show the following:

Proposition 4.2 *If $r_{34|12} = 1$ and $r_{23|1} = r_{24|1}$, then $\rho_{34;12} = 1$.*

Proof. It suffices to show that

$$2\int_{I^2} g(r_{23|1}, u_2, u_3)^2 \, du_2 \, du_3 = 1.$$

We get

$$2\int_{I^2} g(r_{23|1}, u_2, u_3)^2 \, du_2 \, du_3$$

$$= 2\int_{I^2} \sin^2(\pi[\sqrt{1 - r_{23|1}^2}\sqrt{\tfrac{1}{4} - u_2^2}\sin(\pi u_3) + r_{23|1}u_2]) \, du_2 \, du_3$$

$$= \int_{I^2} 1 - \cos(2\pi[\sqrt{1 - r_{23|1}^2}\sqrt{\tfrac{1}{4} - u_2^2}\sin(\pi u_3) + r_{23|1}u_2]) \, du_2 \, du_3.$$

Using the formula for cosine of a sum of two angles and noticing that $\int_{I^2} \sin(2\pi r_{23|1}u_2) \, du_2 \, du_3 = 0$, we obtain

$$1 - \int_{I^2} \cos(2\pi\sqrt{1 - r_{23|1}^2}\sqrt{\tfrac{1}{4} - u_2^2}\sin(\pi u_3)) \cos(2\pi r_{23|1}u_2) \, du_2 \, du_3.$$

Integrating the preceding integral by parts and simplifying, we get

$$1 - \frac{1}{2\pi r_{23|1}} \int_{I^2} \sin(2\pi\sqrt{1 - r_{23|1}^2}\sqrt{\tfrac{1}{4} - u_2^2}\sin(\pi u_3)) \sin(2\pi r_{23|1}u_2)$$

$$\frac{u_2}{\sqrt{\tfrac{1}{4} - u_2^2}} \, du_2 \, du_3.$$

Since the integrand in preceding integral is such that $f(u_2, u_3) = -f(-u_2, -u_3)$ and $f(-u_2, u_3) = -f(u_2, -u_3)$, we find $\rho_{34;12} = 1$. $\square$

4.9.3 Example of fourvariate correlation matrices

The following example shows how the rank correlation specification on the C-vine on four variables can be found to realize a given correlation structure.

Example 4.8 *Let us consider a matrix*

$$A = \begin{bmatrix} 1.0000 & -0.3609 & 0.3764 & -0.3254 \\ -0.3609 & 1.0000 & 0.6519 & -0.3604 \\ 0.3764 & 0.6519 & 1.0000 & -0.2919 \\ -0.3254 & -0.3604 & -0.2919 & 1.0000 \end{bmatrix}.$$

The partial correlation specification on the C-vine is

$$\begin{array}{llll} \rho_{12}, & \rho_{13}, & \rho_{14}, & -0.3609, \quad 0.3764, \quad -0.3254, \\ & \rho_{23;1}, & \rho_{24;1}, & = & 0.9117, \quad -0.5419, \\ & & \rho_{34;12}, & & 0.8707. \end{array}$$

The corresponding rank correlation specification is the following:

$$\begin{array}{llll} r_{12}, & r_{13}, & r_{14}, & -0.3609, \quad 0.3764, \quad -0.3254, \\ & r_{23|1}, & r_{24|1}, & = & 0.9170, \quad -0.5557, \\ & & r_{34|12}, & & 0.9392, \end{array}$$

where $r_{23|1}, r_{24|1}$ are found by applying (4.11) to $\rho_{23;1}$ and $\rho_{24;1}$ respectively and $r_{34|12}$ from (4.33).

The sampling procedure (6.2) with the elliptical copula and preceding rank correlations gives us a distribution with rank correlation matrix A up to sampling and numerical errors.

The matrix A in preceding example was chosen such that after transformation (4.1), it becomes non-positive definite. Hence, A is not a rank correlation matrix of joint normal distribution but can be realized with the vine-elliptical copula method.

The next example shows a four-dimensional correlation matrix that cannot be realized with the vine-elliptical copula method.

Example 4.9 *Let us consider a matrix*

$$A = \begin{bmatrix} 1.0000 & 0.8000 & 0.6000 & -0.3000 \\ 0.8000 & 1.0000 & 0.2400 & -0.6979 \\ 0.6000 & 0.2400 & 1.0000 & 0.5178 \\ -0.3000 & -0.6979 & 0.5178 & 1.0000 \end{bmatrix}.$$

The partial correlation specification on the C-vine is

$$\begin{array}{llll} \rho_{12}, & \rho_{13}, & \rho_{14}, & 0.8, \quad 0.6, \quad -0.3, \\ & \rho_{23;1}, & \rho_{24;1}, & = & -0.5, \quad -0.8, \\ & & \rho_{34;12}, & & 0.99. \end{array}$$

We can find that $r_{23|1} = -0.5137$, and $r_{24|1} = -0.8101$, which correspond to $\rho_{23;1} = -0.5$, and $\rho_{24;1} = -0.8$, respectively. However, assigning $r_{34|12} = 1$ yields $\rho_{34;12}$ that is equal only to 0.9892. The preceding matrix is also not a rank correlation matrix for joint normal.

4.9.4   Results for Section 4.5.2

Theorem 4.6 The following statements hold:

a. All recalculated partial correlations are in the interval $(-1, 1)$.

b. Changing the value of the partial correlation on the vine leads to changing only one correlation in the matrix and does not affect correlations that were already changed.

c. There is a linear relationship between the altered value of the partial correlation and the correlation with the same conditioned set in the proto-correlation matrix.

d. This method always produces a positive definite matrix.

Proof.

a. This condition follows directly from Lemma 4.7.

b. Observe that changing the value of the correlation $\rho_{s+1,j;12...s}$ in the algorithm (4.14) leads to the recalculation of correlations of lower order but only with the same indices before ';', that is, $s + 1, j$.

c. Since $\rho_{s+1,j;12...t-1}$ is linear in $\rho_{s+1,j;12...t}$ for all $t = s, s - 1, \ldots, 1$, the linear relationship between $\rho_{s+1,j}$ and $\rho_{s+1,j;12...s}$ follows by substitution.

d. Applying the algorithm (4.14) whenever a partial correlation is found outside the interval $(-1, 1)$, we eventually obtain that all partial correlations in the partial correlation specification on the vine are in $(-1, 1)$; that is, the altered matrix is positive definite. $\square$

5

Other Graphical Models

5.1 Introduction

This chapter discusses other graphical models, in particular, Bayesian belief nets
and independence graphs. These models are not intended for, and are not suited for,
generic uncertainty analysis, and our treatment will not be exhaustive. Some results
published in the literature will be mentioned without proof. The most significant
point of comparison with the vine-copula approach of the previous chapter is the
problem of inferring a model from data. This provides the occasion for elaborating
a theory of vine inference. We will refer to specific instances of sampling regular
vines, but the full development of sampling procedures must wait for Chapter 6.

5.2 Bayesian belief nets

Bayesian belief nets (bbn's) are directed acyclic graphs (DAGs). They provide a
compact representation of high-dimensional uncertainty distribution over a set of
variables $(X_1, \ldots, X_n)$ (Cowell et al. (1999); Jensen (1996, 2001); Pearl (1988)).
A bbn encodes the probability density or mass function (whichever is appropriate)
on $(X_1, \ldots, X_n)$ by specifying a set of conditional independence statements in a
form of acyclic directed graph and a set of probability functions.

Given any ordering of the variables, the joint density, or mass function can be
written as:

$$f(x_1, x_2, \ldots, x_n) = f(x_1) \prod_{i=2}^{n} f(x_i | x_1 \ldots x_{i-1}). \qquad (5.1)$$

Note that specifying this joint mass/density involves specifying values of an
n-dimensional function. The directed graph of a bbn induces a (generally

non-unique) ordering, and stipulates that each variable is conditionally independent of all predecessors in the ordering given its direct predecessors, that is its parents[1].

Each variable is associated with a conditional probability function of the variable given its parents in the graph, $f(X_i|X_{pa(i)})$, $i = 1, \ldots, n$. The conditional independence statements encoded in the graph allow us to express the joint probability as

$$f(x_1, x_2, \ldots, x_n) = \prod_{i=1}^{n} f(x_i|x_{pa(i)}),$$

where $f(x_i|x_{pa(i)}) = f(x_i)$ if $pa(i) = \emptyset$. If k is the maximal number of parents of any node in the graph, we now have only to specify functions of dimension not greater than k. This is the simplification achieved by representing the density/mass function via a bbn.

5.2.1 Discrete bbn's

In Figure 5.1 a simple example of bbn on 4 variables is shown. This graph tells us that variables 1, 2 and 3 are independent and the distribution of 4 is the assessed conditional on 1, 2 and 3.

Assuming that all variables take only two values, say 'true' and 'false' then to specify a joint distribution given by this structure the marginal distributions of 1, 2 and 3

1	True	False
	0.5	0.5

2	True	False
	0.8	0.2

3	True	False
	0.1	0.9

and the conditional distribution of 4 given 1, 2, 3 as given below are required.

1	True				False			
2	True		False		True		False	
3	True	False	True	False	True	False	True	False
True	0.6	0.5	1	0.1	0.7	0.6	0.3	1
False	0.4	0.5	0	0.9	0.3	0.4	0.7	0

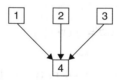

Figure 5.1 Bayesian belief net on 4 variables.

[1]Note that any variables X and Y are independent given X

Even for such a simple example, figuring out the right probabilities in the probability tables requires some work (e.g. statistical data or expert's opinions). For a large net with many dependencies and nodes taking more values, this is daunting. A project at the end of this chapter shows how to create conditional probability tables using vine dependence.

Applications involving high complexity in data-sparse environments, relying on expert judgment, place a premium on traceability, modelling flexibility and maintainability. These features stress the bbn methodology in current form at some of its weakest points which are as follows:

1. *Assessment burden/traceability*: Most serious is the very high assessment burden. If a given node X has K incoming 'influences', where each influence originates from a chance node with M possible outcomes, then the conditional distribution of X must be assessed for each of the M^K input influences. The excessive assessment burden invites rapid, informal and indefensible numerical input.

2. *Discretization/flexibility*: This assessment burden can only be reduced by grossly coarse-graining the outputs from nodes and/or introducing simplifying assumptions for the compounding of 'influences'. In practice, chance nodes are often restricted to two possible values. In many cases, it would be more natural to use continuous nodes, but these are currently insupportable unless the joint distribution is joint normal.

3. *Maintainability*: Whereas bbn's are very flexible with respect to recalculation and updating; they are *not* flexible with respect to changes in modelling. If we add one parent node, then we must redo all previous quantification for the children of this node. In a fluid modelling environment, this is a serious drawback. We should much prefer to be able to add a new node by adding *one number* for each child node, indicating influence, without redoing the previous quantification.

Such considerations motivate the development of a vine-based approach to bbn modelling.

Updating bbn's   The main use of bbn's is to update distributions given observations. If some variables have been observed, we want to infer the probabilities of other events, which have not yet been observed. Using Bayes Theorem, it is then possible to update the values of all the other probabilities in the bbn. Updating bbn's is complex, but with the algorithm proposed by Lauritzen and Spiegelhalter (1998), it is possible to perform fast updating in large bbn's.

5.2.2   Continuous bbn's

Discrete-normal continuous bbn's   Continuous bbn's (Pearl (1988); Shachter and Kenley (1989)) developed for joint normal variables interpret 'influence' of the parents on a child as partial regression coefficients when the child is regressed on the parents. They require means, conditional variances and partial regression coefficients, which can be specified in an algebraically independent manner. The partial regression coefficients together with the conditional standard deviations form a Cholesky decomposition of the covariance matrix.

More precisely, let $X = (X_1, \ldots, X_n)$ have a multivariate normal distribution with mean vector $\mu = (\mu_1, \ldots, \mu_n)$ and covariance matrix Σ. For normal bbn's, the conditional probability functions in (5.1) are of the form

$$f(X_i | X_{pa(i)}) \sim \mathcal{N}\left(\mu_i + \sum_{j \in pa(i)} b_{ij}(X_j - \mu_j); v_i\right) \qquad (5.2)$$

where $v = (v_1, \ldots, v_n)$ is a vector of conditional variances and b_{ij} are linear coefficients that can be thought of as partial regression coefficients

$$b_{ij} = b_{ij; pa(i) \setminus j}.$$

We show now that means, conditional variances and partial regression coefficients can be specified in an algebraically independent manner (Shachter and Kenley (1989)).

Theorem 5.1 Σ *is positive (semi-)definite if and only if* $v > 0$ (≥ 0). *Furthermore, the rank of* Σ *is equal to the number of nonzero elements in* v.

In 'Discrete normal' bbn's, continuous variables follow a multivariate normal distribution given the discrete variables. Discrete normal bbn's allow discrete parents of continuous nodes, but no discrete children of continuous nodes.

Continuous bbn's as above are much easier to construct than their discrete counterparts. The price, of course, is the restriction to the joint normal distribution and in the absence of data to experts who can estimate partial regression coefficients and conditional variances. With regard to maintainability the following should be noted: If a continuous parent node is added or removed, the child node must be requantified. This reflects the fact that partial regression coefficients depend on the set of regressors; adding or removing a regressor entails changing all partial regression coefficients. To circumvent the restriction to normality, one could transform a given distribution to joint normal (Rosenblat (1952)). This involves transforming conditional distributions $(X_k | X_1 \ldots X_{k-1})$ to normal with required mean and variance, for every value of the conditioning variables. This option is primarily of theoretical interest. Another idea is to use the theory of linear least squares predictors as applied to arbitrary joint distributions. Suppose $(X_1, \ldots, X_{k-1})$ are the ancestors of X_k in an ordered Bbn.[2] We could interpret the 'influence' of X_j on X_k as the partial regression of X_k on X_j given $1, \ldots, j-1, j+1, \ldots, k-1$.

---

[2]Y is an ancestor of X with respect to an ordering of the variables, which preserves the parent–child relations, that is, an ordering such that parents occur before their children in the ordering.

If j is not a parent of k, then j and k are conditionally independent given the parents of k; however, it is *not* generally true that the partial regression of k on j is zero (Proposition 3.12 and Kurowicka and Cooke (2000)). This means that the partial regression coefficients do not reflect the conditional independence structure of the bbn.

Non-parametric continuous bbn's   Let us associate nodes of a bbn with continuous univariate random variables and each arc with a (conditional) parent–child rank correlation according to the protocol given below. We specify nested sets of high-dimensional joint distributions using the vine-copula approach described in Chapter 4, where any copula with invertible conditional cumulative distribution functions may be used as long as the chosen copula represents (conditional) independence as zero (conditional) correlation. The conditional rank correlations (like the partial regression coefficients) are algebraically independent, and there are tested protocols for their use in structured expert judgment (discussed in Chapter 2).

We note that quantifying bbn's in this way requires assessing *all* (continuous, invertible) one-dimensional marginal distributions. On the other hand, the dependence structure is meaningful for any such quantification, and need not be revised if the univariate distributions are changed. In fact, when comparing different decisions or assessing the value of different observations, it is frequently sufficient to observe the effects on the quantiles of each node. For such comparisons, we do not need to assess the one-dimensional margins at all.

The theory presented here can be extended to include 'ordinal' variables; that is, variables, which can be written as monotone transforms of uniform variables, perhaps taking finitely many values. The dependence structure must be defined with respect to the uniform variates. Further, we consider here only the case where the conditional correlations associated with the nodes of vines are constant; however, the sampling algorithms discussed in Chapter 6 will work mutatis mutandis for conditional correlations depending on the values of the conditioning variables.

The vine-based approach is quite general and of course this comes at a price: These bbn's must be evaluated by Monte Carlo simulation. We assume throughout that all univariate distributions have been transformed to uniform distributions on $(0, 1)$. To determine which (conditional) correlations are necessary, we adopt the following protocol:

1. Construct a sampling order for the nodes, that is, an ordering such that all ancestors of node i appear before i in the ordering. A sampling order begins with a source node and ends with a sink node. Of course, the sampling order is not in general unique. Index the nodes according to the sampling order $1, \dots, n$.

2. Factorize the joint in the standard way following the sampling order. If the sampling order is $1, 2, \dots, n$, write:

$$P(1, \dots, n) = P(1)P(2|1)P(3|21) \dots P(n|n-1, n-2, \dots, 1).$$

3. Underscore those nodes in each condition, which are not parents of the conditioned variable and thus, are not necessary in sampling the conditioned variable. This uses some of the conditional independence relations in the belief net. Hence, if in sampling $2, \ldots, n$ variable 1 is not necessary (i.e. there is no influence from 1 to any other variable) then

$$P(1, \ldots, n) = P(1)P(2|\underline{1})P(3|2\underline{1}) \ldots P(n|n-1, n-2, \ldots, \underline{1}). \quad (5.3)$$

The underscored nodes could be omitted thereby yielding the familiar factorization of the bbn as a product of conditional probabilities, with each node conditionalized on its parents (for source nodes the set of parents is empty).

4. For each term, i with parents (non-underscored variables) $i_1 \ldots i_{p(i)}$ in (5.3), associate the arc $i_{p(i)-k} \longrightarrow i$ with the conditional rank correlation

$$r(i, i_{p(i)}) \; ; k = 0$$

$$r(i, i_{p(i)-k} | i_{p(i)}, \ldots i_{p(i)-k+1}); \; 1 \le k \le p(i) - 1, \quad (5.4)$$

where the assignment is vacuous if $\{i_1 \ldots i_{p(i)}\} = \emptyset$. Assigning conditional rank correlations for $i = 1, \ldots, n$, every arc in the bbn is assigned a conditional rank correlation between parent and child.

Let $\mathcal{D}^i$ denote a D-vine on i variables ordered $(i, i-1, \ldots, 1)$. The following theorem shows that these assignments uniquely determine the joint distribution and are algebraically independent:

Theorem 5.2 *Given*

1. *a directed acyclic graph (DAG) with n nodes specifying conditional independence relationships in a bbn,*

2. *the specification of conditional rank correlations (5.4), $i = 1, \ldots, n$ and*

3. *a copula realizing all correlations $[-1, 1]$ for which correlation 0 entails independence;*

the joint distribution is uniquely determined. This joint distribution satisfies the characteristic factorization (5.3) and the conditional rank correlations in (5.4) are algebraically independent.

Sampling procedures for regular vines are discussed in the next chapter. Suffice to say here that we can sample X_i using the sampling procedure for $\mathcal{D}^i$. When using vines to sample a continuous bbn, it is not in general possible to keep the same order of variables in successive D-vines. In other words, we will have to reorder the variables before constructing $\mathcal{D}^{i+1}$ and sampling X_{i+1}, and this will involve calculating some conditional distributions.

Example 5.1 *For the bbn in Figure 5.1 the factorization is* $P(1)P(2|\underline{1})$ $P(3|\underline{21})P(4|321)$. *Hence the preceding protocol would require assignment of (conditional) correlations* r_{43}, $r_{42|3}$ *and* $r_{41|23}$ *to the edges of that bbn. In this case, the order of variables in each* $\mathcal{D}^K$ *can be kept the same. Figure 5.2 shows* $\mathcal{D}^2, \mathcal{D}^3, \mathcal{D}^4$. *Hence this bbn can be represented as one D-vine in Figure 5.3.*

Example 5.2 *Consider the following bbn on 5 variables.*

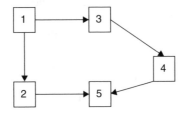

Choose the sampling order: 1, 2, 3, 4, 5.

The factorization is then: $P(1)P(2|1)P(3|\underline{21})P(4|3\underline{21})P(5|4\underline{321})$.

The following rank correlations have to be assessed:

$$r_{21}, r_{31}, r_{43}, r_{54}, r_{52|4}.$$

In this case $\mathcal{D}^4 = D(4, 3, 2, 1)$, but the order of variables in $\mathcal{D}^5$ must be $D(5, 4, 2, 3, 1)$. Hence this bbn cannot be represented as one vine. Rather, we must

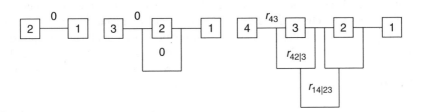

Figure 5.2 $\mathcal{D}^2, \mathcal{D}^3, \mathcal{D}^4$ for example 5.1.

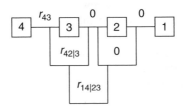

Figure 5.3 D-vine for example 5.1.

recalculate the conditional rank correlations in $\mathcal{D}^5$. A more detailed discussion is given in the next chapter.

Updating Continuous bbn's   The main use of bbn's is in situations that require updating on observations. After observing some events, we might want to infer the probabilities of other unobserved events. Alternatively, we might consider performing (expensive) observations and may wish to see how the results would impact our beliefs in other events of interest. Updating continuous bbn's requires resampling. We discuss this approach in Chapter 6. Updating with resampling is not so elegant as updating schemes for discrete or discrete-normal bbn's. For large models other options should be considered.

For updating continuous bbn's, a hybrid technique can be used that combines the virtues of continuous approach (influences as conditional correlations) with the fast updating for discrete bbn's (see Section 6.4.4).

Example: Flight crew alertness model   We show how continuous bbn's can be applied to quantify and update the flight crew alertness model adopted from the discrete model described in Roelen et al. (2003). In the original model, all chance nodes were discretized to take one of two values 'OK' or 'Not OK'. The names of nodes has been altered to indicate how, with greater realism, these can be modelled as continuous variables. Alertness is measured by performance on a simple tracking test programmed on a palmtop computer. Crew members did this test during breaks in-flight under various conditions. The results are scored on an increasing scale and can be modelled as a continuous variable.

In Figure 5.4 the flight crew alertness model is presented. Continuous distributions for each node must be gathered from existing data or expert judgment. The distribution functions are used to transform each variable to uniform on the interval (0,1). Required (conditional) rank correlations are found using the protocol described in Section 5.2.2. These can be assessed by experts in the way described in Chapter 2. In Figure 5.4, a (conditional) rank correlation is assigned to each arc of the bbn. These numbers are chosen to illustrate this approach and are based on in-house expert judgment.

The sampling procedure and the updating for this model are described in Chapter 6.

The main use of bbn's in decision support is updating on the basis of possible observations. Let us suppose that we know before the flight that the crew did not have enough sleep. Let us assume that the crew's hours of sleep correspond to the 25th percentile of hours of sleep distribution. We would like to know how this information will influence a distribution of the crew alertness. Without this assumption the crew alertness distribution would be uniform on (0,1). The distribution of crew alertness given that hours of sleep of the crew are equal to 25th percentile is shown in Figure 5.5. We can see that knowing that the flight crew did not have enough sleep reduces crew alertness (e.g. with probability 50% the crew

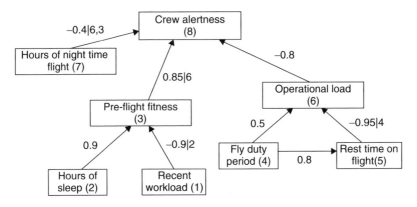

Figure 5.4 Flight crew alertness model.

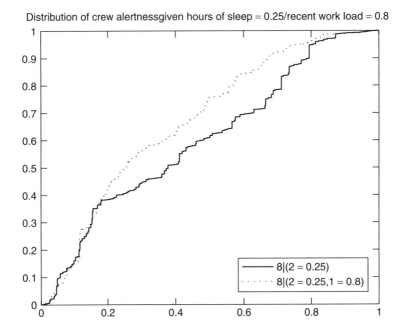

Figure 5.5 Distributions of crew alertness given hours of sleep equal to 25th percentile and recent workload equal to 80th percentile.

alertness is less than or equal to 35th percentile). If we assume additionally that the crew had significant recent workload (corresponding to the 80th percentile) then situation further deteriorates (now, with probability 50 25th percentile, see Figure 5.5).

Distribution of crew alertess given hours of sleep = 0.25 and flight duty period = 0.1 or 0.8

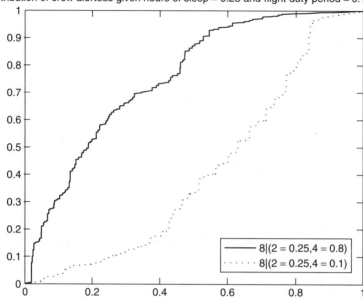

Figure 5.6   Distributions of crew alertness given hours of sleep equal to 25th percentile and fly duty period equal to 10th or 80th percentile.

Figure 5.6 shows distributions of the crew alertness in situations when the crew's hours of sleep are equal to 25th percentile and flight duty period is short (equal to 10th percentile) or long (equal to 80th percentile). We see that the short hours of sleep are not so important in case of a short flight but in case of a long flight the effect on crew alertness is alarming (e.g. with probability 50% alertness is less than or equal to 15th percentile of its unconditional distribution).

Is it possible to improve alertness of the flight crew in case the crew did not have much sleep and their recent flight duty period was long? We could compensate loss of the crew alertness in this situation by introducing a few policies. Firstly, we require that the number of night hours on the flight should be small (equal to 10th percentile). This improves the situation a bit (dotted line in Figure 5.7). Alternatively we could require having long resting time on a flight (equal to 90th percentile). This results in a significant improvement of the crew alertness distribution (see dashed line in Figure 5.7). Combining both these policies improves the result even more.

Notice that in comparing different polices it is not necessary to know the actual distributions of the variables. Our decisions can be based on quantile information. We might think of the transformation from quantiles to physical units of

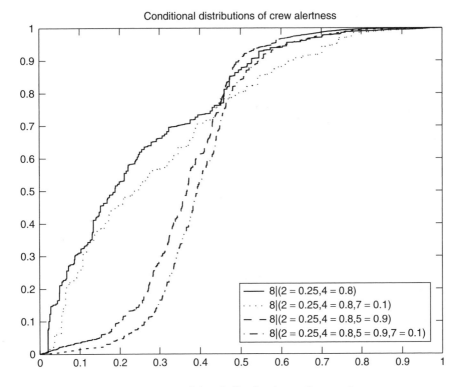

Figure 5.7 Four conditional distributions of crew alertness.

the variables as being absorbed into a monotonic utility function. Thus, conclusions based on quantiles will hold for all monotonic utility functions of the random variables.

If the individual nodes are discretized to two possible outcomes, we must assess 22 independent probabilities. If we wish to add a third possible outcome to each node, the number of independent probabilities to be assessed jumps to 104; for five outcomes the number is 736. On the other hand, the quantification with continuous nodes requires eight algebraically independent numbers. This demonstrates the reduction of assessment burden obtained by quantifying influence as conditional rank correlation.

5.3   Independence graphs

The conditional independence graph is a powerful way of representing conditional independence relationships between variables (Kiiveri and Speed (1984); Whittaker

(1990)). This graphical structure, however, does not specify a joint distribution. It gives only information about conditional independence statements.

An undirected graph is an independence graph, if there is no edge between two vertices whenever the pair of variables is independent given all remaining variables.

Definition 5.1 (Independence graph) *Let* $X = (X_1, X_2, \ldots, X_n)$. *The independence graph of* X *is the undirected graph* $G = (N, E)$, *where* $N = \{1, 2, \ldots, n\}$ *and* $\{i, j\} \notin E$ *if and only if* X_i *and* X_j *are conditionally independent given* $X_{N \setminus \{i,j\}}$.

If there is no edge between X_i and X_j in the independence graph, then X_i and X_j are conditionally independent given $X_{N \setminus \{i,j\}}$, which we denote $X_i \perp X_j | X_{N \setminus \{i,j\}}$. For independence graphs one can conclude that not all variables $X_{N \setminus \{i,j\}}$ are necessary to insure conditional independence of X_i and X_j. This set can be reduced to variables that separate X_i and X_j in the graph.

Theorem 5.1 (The separation theorem) *(Whittaker (1990)) If* X_a, X_b *and* X_c *are vectors containing disjoint subsets of variables from* X, *and if, in the independence graph of* X, *each vertex in* b *is separated from each vertex in* c *by the subset of* a, *then*

$$X_b \perp X_c | X_a.$$

5.4   Model inference

In situations when the data does not exist or is very sparse we must rely on expert judgment to define the graphical structure and assess required parameters. However, if the data is available we would like to extract a best fitting model from the data. *Model learning* or *model inference* is concerned with this problem. We discuss here briefly basic ideas behind model inference for bbn's, independence graphs and vines. To illustrate the methods that will be presented we use the following example treated in Callies et al. (2003).

Example 5.3 *Observations of local December mean temperatures at four European stations were taken from the World Monthly Station Climatology of the National Center for Atmospheric Research. Corresponding diagnostic 'forecasts' between 1990 and 1993 were produced by regressing local temperatures on monthly mean regional scale atmospheric sea level pressure distributions as represented by* $5^o \times 5^o$ *analysis Trenberth and Paolino (1980) at 60 grid points covering region* $40^o N$ *to* $64^o N$ *and* $20^o W$ *to* $25^o E$. *The regressions scheme was calibrated for 1960–1980. Prior to regression, both regional and local data were filtered by standard principle components analysis to reduce the number of degrees of freedom and to avoid overfitting. Only four degrees of freedom were retained.*

The following (8 × 8) sample correlation matrix contains all information about the interactions between observations, θ and the corresponding forecasts, F, at the four stations Geneva (G), Innsbruck (I), Budapest (B) and Copenhagen (K).

$$S = \begin{matrix} \theta^K & \theta^G & \theta^I & \theta^B & F^K & F^G & F^I & F^B \end{matrix}$$

$$S = \begin{bmatrix} 1 & .35 & .50 & .49 & .68 & .38 & .50 & .59 \\ & 1 & .79 & .69 & .12 & .64 & .62 & .49 \\ & & 1 & .72 & .18 & .61 & .58 & .43 \\ & & & 1 & .05 & .46 & .47 & .43 \\ & & & & 1 & .33 & .51 & .71 \\ & & & & & 1 & .97 & .77 \\ & & & & & & 1 & .90 \\ & & & & & & & 1 \end{bmatrix} \quad (5.5)$$

5.4.1 Inference for bbn's

There is a rich literature concerning learning bbn's, an introduction to which is found in Cowell et al. (1999); Hackerman (1998) together with references for more extensive treatment of this problem. The present brief discussion merely gives the flavour. It is well known that probabilistic influence is non-directional; if random variable X influences random variable Y in the sense that $P(Y|X) \neq P(Y)$, then Y also influences X. Hence the directionality in a bbn cannot in general[3] be inferred from data with purely statistical methods. It necessarily reflects some external structure, for example, causal or temporal ordering. We find it convenient to think of the directionality in a bbn as specifying an information flow in a sampling algorithm. Commercial interest in these algorithms may result in incomplete descriptions in the open literature. Be that as it may, learning bbn's consists of two parts as follows:

1. Model selection,

2. Learning the conditional probabilities.

Inferring conditional probabilities, given the model structure, can proceed using a number of standard statistical techniques. We focus on model selection. The graphical probabilistic model learning is usually approached in one of two ways: the *search and scoring methods* and the *dependency analysis methods*. In the first approach, we start with a graph without any edges. We use a search method to add an edge to the current graph and then use some scoring function to decide whether the new structure is better then the old one. This procedure iterates until no new structure is better then the previous ones. The second approach tries to discover dependencies from the data and uses them to infer the structure. We sketch the algorithm explained in detail in Cheng et al. (1997). Recall that the *mutual information* of a bivariate distribution is the relative information of this distribution with respect to the product of the one-dimensional margins.

---

[3]The directionality can be inferred from data in some cases. The bbn in Figure 5.1 is a good example.

1. Calculate the mutual information of each pair of nodes. Store those that are greater than a certain small value ϵ and sort them in descending order. Create a draft tree, by connecting nodes with the highest mutual information.

2. Examine all pairs of nodes for which the mutual information is bigger than ϵ but are not directly connected in the draft. Check if there exists a cut set in the graph that makes them conditionally independent given variables in the cut set; if not, connect these nodes. A group of conditional independence tests is used here.

3. Apply groups of conditional independence tests to check if the new connections are really necessary. If not then, the edge is removed permanently.

4. Orient the edges of the graph by identifying triplets of nodes (X, Y, Z) such that X and Y and Y and Z are directly connected but X and Z are not directly connected. Only the structure $X \rightarrow Y \leftarrow Z$ corresponds to the preceding information. Y is called a *collider*, hence this algorithm finds all colliders in the graph and orients edges in the bbn accordingly. After this procedure, however, some edges may not be oriented.[4]

5.4.2   Inference for independence graphs

In Whittaker (1990) the problem of model inference is cast as a problem of identifying conditional independence. The joint distribution of variables $X_1, \ldots, X_n$ is assumed to be joint normal. The approach is sketched as follows:

1. Estimate the variance V by the sample variance matrix S.

2. Compute its inverse S^{-1}, rescale this inverse so that the diagonal entries are 1; the off-diagonal cell $\{i, j\}$ contains the negative of the partial correlation of i, j with respect to all remaining variables K.

3. Set any sufficiently small partial correlations in S^{-1} equal to zero; call the resulting matrix P^{*}.[5]

4. Find a positive definite matrix P 'as close as possible' to P^*; the zero's of P correspond to pairs of variables which are modelled as conditionally independent given all other variables.

---

[4]We were unable to find a Bayesian Belief Net programme, which has implemented a learning algorithm for continuous variables. Although several programs have algorithms, which could be applied to discretized versions of continuous distributions, the results were found to depend strongly on the method of discretization.

[5]The process of removing links corresponding to small partial correlations can be carried out sequentially. After each removal, the correlation matrix and partial correlations are updated. The decision to remove a link is based on the (a) 'edge exclusion-' and (b) 'edge inclusion-deviance'. These are entropy-based measures that reflect (a) the additional disturbance of the original distribution caused by removing an additional edge and (b) the reduction in disturbance achieved by restoring previously removed edge. The process stops when the minimum edge exclusion deviance is much bigger than the maximum edge inclusion deviance.

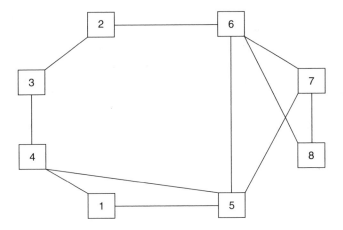

Figure 5.8 Independence graph from Whitakker method.

The procedure for finding the positive definite matrix P uses iterative proportional fitting, and relies heavily on the properties of the joint normal distribution. P and P^* have the same zeros. Goodness of fit tests are available to determine whether P^{-1} is 'close enough' to the original sample variance S.

The resulting structure is expressed as an independence graph. Starting with the saturated undirected graph over all variables, edges between variables are removed if the partial correlation of these variables, given the remaining variables, is zero. The remaining edges connect variables having an interaction in the inferred model.

In Figure 5.8, the independence graph for Example 5.3 is shown. Whittaker's method removed 17 of the 28 edges in the saturated graph.

5.4.3   Inference for vines

We shall approach model learning as a problem of inferring a regular vine from data with certain desirable properties. To this end, we shall associate nodes in a regular vine with partial correlations (see Chapter 3); partial correlation $\rho_{ij;K}$ may be associated with an edge in a regular vine if K is the conditioning set, and $\{i, j\}$ the conditioned set of the edge. The result of such an association is a partial correlation vine (see Chapter 4).

As shown in Chapter 4, a partial correlation vine fully characterizes the correlation structure of the joint distribution and the values of the partial correlations are algebraically independent. Unlike the values in a correlation matrix, the partial correlations in a regular vine need not satisfy an algebraic constraint like positive definiteness. Moreover, the partial correlation vine represents a factorization of the determinant of the correlation matrix. The determinant of the correlation matrix is a measure of linear dependence in a joint distribution. If all variables are independent, the determinant is 1, and if there is linear dependence between the variables, the determinant is zero. Intermediate values reflect intermediate dependence.

Following the approach of Whittaker, we would like to change partial corre-lations in the vine to zero while disturbing the determinant as little as possible. In this case, however, we can first choose the partial correlation vine, which best lends itself to this purpose. Moreover, by Theorem 4.4, any such change will be consistent. There is no need for an analogue to the iterative proportional fitting algorithm in the independence graph method.

As proven in Chapter 4, the product of 1 minus the square partial correlations on a regular vine equals the determinant of the correlation matrix. We can write

$$-\log(D) = -\sum_{\{i,j\}} \log(1 - \rho_{ij;K(ij)}^2) \tag{5.6}$$

$$= -\sum_{\{i,j\}} a_{ij;K(ij)}, \tag{5.7}$$

where D is the determinant of the correlation matrix. The terms $a_{ij;K(ij)}$ will depend on the regular vine, which we choose to represent the second order structure, how-ever, the sum of these terms must satisfy (5.7). We seek a partial correlation vine for which the terms $a_{ij;K(ij)}$ in (5.7) are 'as spread out' as possible. This concept is made precise with the notion of *majorization* (Marshall and Olkin (1979)).

Definition 5.2 *Let* $x, y \in \mathbb{R}^n$ *be such that* $\sum_{i=1}^n x_i = \sum_{i=1}^n y_i$; *then* x *majorizes* y *if for all* k; $k = 1, \ldots, n$

$$\sum_{j=1}^k x_{(j)} \le \sum_{j=1}^k y_{(j)}, \tag{5.8}$$

where $x_{(j)}$ *is the increasing arrangement of the components of* x, *and similarly for* y.

In view of (5.7), the model inference problem may be cast as the problem of finding a regular vine whose terms $a_{ij;K(ij)}$ are non-dominated in the sense of majorization. In that case, setting those partial correlations equal to zero whose square is smallest will change the determinant as little as possible. Finding non-dominated solutions may be difficult, but a necessary condition for non-dominance can be found by maximizing any Schur convex function.

Definition 5.3 *A function* $f : \mathbb{R}^k \to \mathbb{R}$ *is Schur convex if* $f(x) \ge f(y)$ *whenever* x *majorizes* y.

Schur convex functions have been studied extensively. A sufficient condition for Schur convexity is given by (Marshall and Olkin (1979)):

Proposition 5.1 *If* $f : \mathbb{R}^k \to \mathbb{R}$ *may be written as* $f(x) = \sum f_i(x_i)$ *with* f_i *convex, then* f *is Schur convex.*

The following strategy for model inference suggests itself:

1. Choose a Schur convex function $f : \mathbb{R}^{\frac{n(n-1)}{2}} \to \mathbb{R}$;

2. Find a partial correlation vine $\mathcal{V}$ whose vector $a_{ij;K(ij)}$ maximizes f;

3. Set the partial correlations in $\mathcal{V}$ equal to zero for which the terms $a_{ij;K(ij)}$ are smallest;

4. Using the sampling distribution for the determinant, verify that the change in the determinant is not statistically significant.

If the joint distribution is normal, then the sampling distribution of the determinant is given by (Rao (1973)):

Theorem 5.1 *Let $X = (X_1, \ldots, X_N)$ be samples of a n-dimensional normal vector with sample mean $\overline{X}$, variance V and normalized sample variance*

$$S = N^{-1} \sum_{i=1}^{N} (X_i - \overline{X})(X_i - \overline{X})^T,$$

then

$$\frac{N \det(S)}{\det(V)} \sim \prod_{i=1}^{N} T_i, \tag{5.9}$$

where $\{T_i\}$ are independent chi-square distributed variables with $N - i + 1$ degrees of freedom.

If the distribution is not joint normal, we could estimate the distribution of the determinant with the bootstrap. Evidently, this strategy depends on the choice of Schur convex function. Searching the set of all partial correlation vines is not easy; at present heuristic constraints are required.

The determinant D of the sample correlation matrix (5.5) is:

$$-\log(D) = 11.4406.$$

We wish to 'add independence'. This will produce a new correlation matrix whose determinant D^* will be larger, or equivalently, $-\log(D^*) < -\log(D)$. Roughly, we would like to add as much independence as possible, while keeping the increment in the determinant as small as possible. Application of the independence graph method (previous subsection) led to setting 17 of the 28 partial correlations in the scaled inverse covariance matrix equal to zero. We find in this case $-\log(D^*) = 10.7763$.

To compare this with the vine-based approach, we adopt a heuristic search on the basis of maximizing the Schur convex function

$$f(x) = \sum x_i ln(x_i) \tag{5.10}$$

The heuristic works as follows:

1. Choose an ordering of the variables;

2. Start with subvine consisting of variables 1 and 2 in the ordering. For $j = 3$ to n; find the subvine extending the current subvine by adjoining variable $j + 1$, so as to maximize (5.10). Store the vine obtained for $j = n$;

3. Go to 1;

4. Choose the optimal partial correlation vine maximizing (5.10) among all those stored.

In general, it is not feasible to search all permutations; heuristic search methods or Monte Carlo sampling must be used.

When we set the 17 smallest partial correlations in the optimal vine equal to zero, we find $-\log(D^*) = 11.0970$, which is closer to the sample value than that found by the independence graph method. In this sense, the vine method retains the same number of interactions as the independence graph method, while perturbing the sample distribution less. It must be noted, however, that both the independence graph and the regular vine entail more conditional independencies than are directly represented in the graphs.

Figure 5.9 shows the first tree in the optimal vine. Note that it is neither a canonical nor D-vine; node 2 has degree 4. Figure 5.10 shows the matrix of partial correlations, in which the 17 smallest partial correlations have been set equal to zero. Figure 5.11 shows the independence graph obtained by the method of Whittaker (1990). Figure 5.12 uses a similar format to show the 11 significant interactions from the vine method. We term this an 'interaction graph'; it is not an independence graph in so far as the partial correlations corresponding to edges may not be conditioned on all remaining variables.

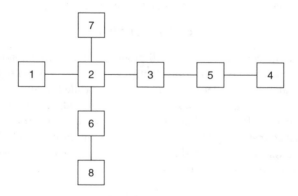

Figure 5.9   Tree 1 in optimal vine, node 2 has degree 4.

	2	3	4	5	6	7	8
1	12 = 00	13;2 = 0	14;235 = 0	15;32 = 0.70	16;235 = 0	17;235 = 0	18;2356 = 0
2		23 = 0.79	24;53 = 0	25;3 = 0	26 = 0.64	27 = 0.62	28;6 = 0
3			34;5 = 0.72	35 = 0	36;2 = 0	37;2 = 0	38;26 = 0
4				45 =0	46;2351 = 0	47;2351 = 0	48;12356 = 0.49
5					56;23 = 0	57;23 = 0.54	58;236 = 0.78
6						67;12345 = 0.98	68 = 0.77
7							78;123456 = 0.89

Figure 5.10 Partial correlations after setting 17 smallest equal to zero in optimal vine.

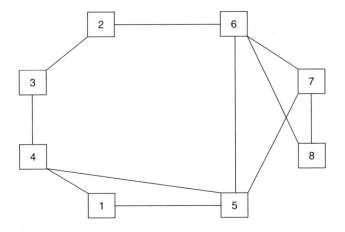

Figure 5.11 Independence graph from method of Whittaker.

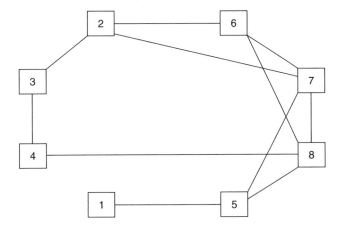

Figure 5.12 Interaction graph from optimal vine.

5.5 Conclusions

Independence graphs provide a powerful and elegant way of inferring conditional independence structure in joint normal data. This method has been adopted to vine inference. For the joint normal distribution, conditional and partial correlations are equal. The vine inference strategy makes a similar assumption. However, it does not require iterative proportional fitting and in this sense is less beholden to the joint normal distribution.

Model inference is a new and active area, and much remains to be done. In particular, methods for searching the set of regular vines and incorporating extra probabilistic information require further work. The vine inference could be followed by a copula inference procedure. In combination with the vine, this could be used to infer a joint distribution, which could be tested against the sample.

5.6 Unicorn projects

A chapter on other graphical models is a good place to introduce the min-cost-flow solver in UNICORN. Three illustrative projects are given illustrating an expanding set of features and applications.

UNICORN incorporates a network optimization program DUALNET developed at the TU Delft (Koster (1995)). DUALNET computes minimum cost flows on directed networks. The version of DUALNET adopted uses integer-valued costs and bounds. You may assign these (positive) random variables in UNICORN with continuous distributions, and the distributions will be converted to integer-valued distributions. Lower bounds, upper bounds and supplies must be non-negative; costs may be positive or negative. The min-cost solver computes the cheapest way of shipping the supplies to the demands, where the supplies, demands, costs and bounds may be random variables, or user-defined functions (UDF's) of random variables.

We refer to the UNICORN help file for a full discussion. All simulations are done with random seed 1 and 1 run (100 simulations).

Project 5.1 Min cost flow networks

This simple project simply introduces the capabilities. From the FILE menu choose New DUALNET Model. The graphic model builder is opened. You can specify supply nodes and demand nodes, and also one arc between nodes. A UNICORN variable file is created automatically, according to the following conventions:

- *each node is represented by an integer between 1 and 99, except for the 'last node', which is represented as 'x';*

- *each arc is associated with three input variables with names: 'iCj', 'iLj', 'iUj'. i and j are the names of the source and sink nodes respectively; (i.e. i is an integer between 1 and 99, or is 'x') iCj is the cost of shipping one unit from i*

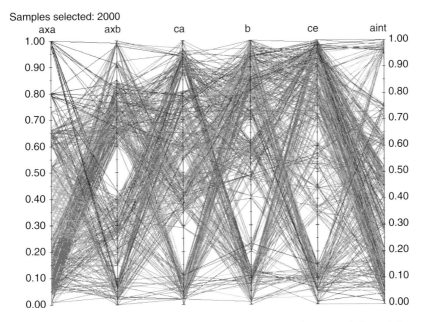

Plate 1 Cobweb plot for the transfer coefficients in the extended model.

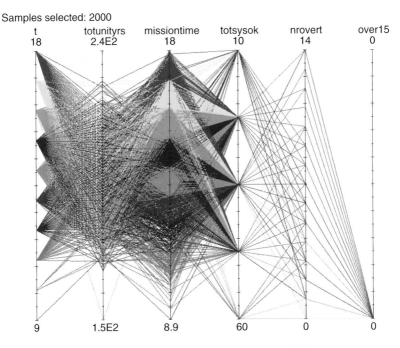

Plate 2 Cobweb plot for Mission Completion.

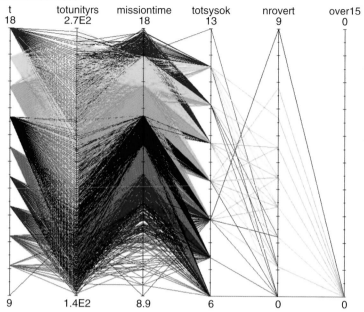

Plate 3 Cobweb plot for Mission Completion with dependence.

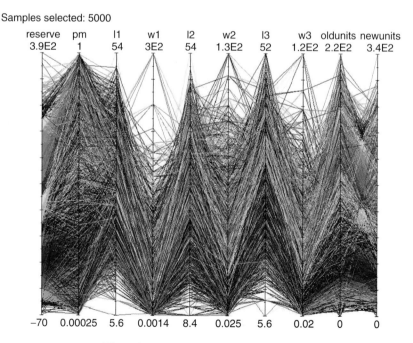

Plate 4 Unconditional cobweb plot.

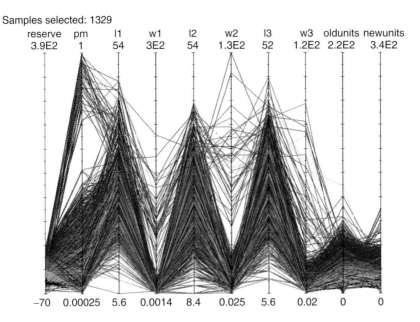

Samples selected: 1329

Plate 5 Conditional cobweb plot.

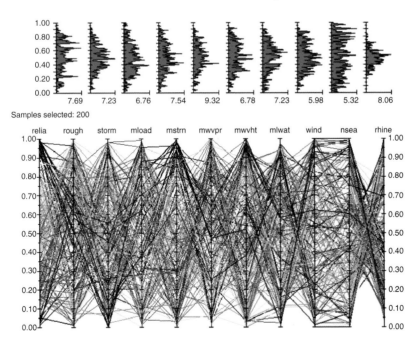

Samples selected: 200

Plate 6 Cobweb for dike ring.

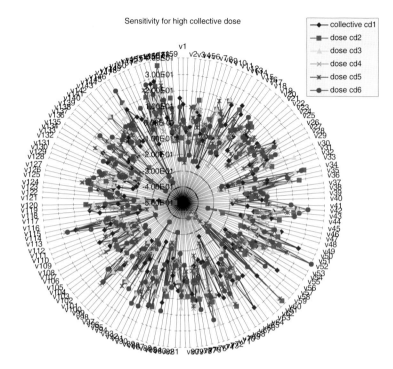

Plate 7 Radar plot; LPSM for collective dose.

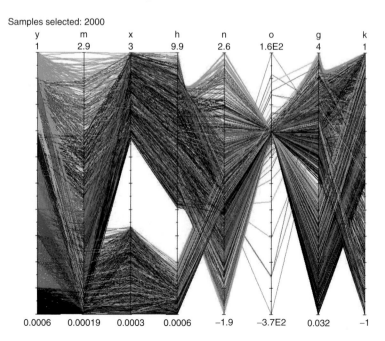

Plate 8 Cobweb for probdiss.

to j, iLj is the lower bound of the number of items shipped, iUj is the upper bound of items shipped. iLj and/or iUj are assigned default constant values 0 and 1 respectively; they may be deleted from the list of variables and these assignments will be maintained. Alternatively, they may be assigned arbitrary positive valued distributions.

- supply to node i is represented by an input variable named Si,

- demand to be extracted from node j is named Dj.

These conventions must be strictly followed, names cannot be changed. The graphical model builder always adds cost and flow variables from the lower index to the higher index, in the case of supply nodes, or from supply to demand. You may add reverse flows in the UNICORN variable file.

On each simulation, the supply must equal demand; therefore if supply and/or demand is uncertain, a special demand variable $Dx = \sum supplies - \sum demands$ is needed. Dx must be positive. Dx may be assigned an arbitrary distribution, and its value will be adjusted if necessary to insure that supply equals demand. Supplies and demands must be non-negative. These variables are created as random variables by the graphic interface. They may be deleted from the random variables and computed as UDF's, – that is, as functions of (arbitrarily named) variables, using the formula editor. However, the naming conventions must be respected. Additional UDF's may also be defined. The following output variables are created automatically:

- iFj: flow from i to j,

- iMj: the marginal cost associated with arc (i, j),

- OPT: the cost of the optimum solution.

Create 2 supply nodes, $S1$ and $S2$, and one demand node $D3$ (naming is automatic) with an arc between $(S1,S2)$, $(S1,D3)$, and $(S2,D3)$. The model should look as in Figure 5.13.

Go to the variable panel and verify that the flow and cost variables have been created with default assignments. Edit the variables so that the supply to $S1 = 0$ and the cost of shipping from $S2$ to $D3$ is 10. The costs of shipping from $S1$ to $S2$, and from $S1$ to $D3$ are both 1 (default). It would be cheaper to supply $D3$ by shipping from $S2$ to $S1$, and from $S1$ to $D3$. However, the graphic model builder created only the arc from $S1$ to $S2$. From the Simulate panel, run the model with 100 samples (the minimum number) and view the report. We see that nothing is shipped from $S1$ to $S2$, and one unit flows from $S2$ to $D3$, with a cost (output variable OPT) of 10.

Now from the variable panel, delete the variable $1C2$ and add the variable $2C1$ with constant value 1. This reverses the arc so that it goes from 2 to 1. Variables $2L1$ and $2U1$ are created internally with constant values 0 and 1 respectively. Run again; the cost is now 2, since $S2$ ships to $S1$ (cost 1) and $S1$ ships to $D3$ (cost 1).

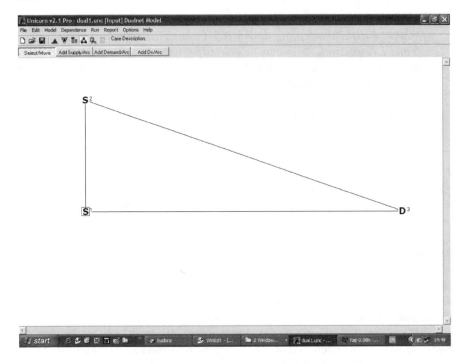

Figure 5.13 Min-cost flow network.

Project 5.2 Min-cost flow for project risk management

This is a illustration of using stochastic min-cost flow networks for managing project risks under uncertainty. The problem is to estimate the completion time of a large project, in which several sub-projects are linked. A typical example is construction: roofing cannot begin until the prefabricated materials are delivered, and the walls are built. A critical path network starts with a single source, the beginning of the project and ends with a single demand at the right. Figure 5.14 shows a simple example: This graph says, for example, that project 5 cannot be started until projects 2 and 3 are completed, project 6 cannot be started until projects 3 and 4 are completed, and the whole project is finished only when both 5 and 6 are completed. If the completion times are given, we can compute the completion time of the whole project as the length of the critical path, *that is, the path from 1 to 7 whose summed completion times are greatest. We can treat this as a min-cost flow network if we put a unit supply at node S1, and a unit demand at D7, and make the costs equal to the* negative completion times. *Then the path with minimal cost (most negative) for shipping one unit from S1 to D7 will be the path with the longest summed completion times. Enter the preceding graph. Set S2, ..., S6 = 0. Input the*

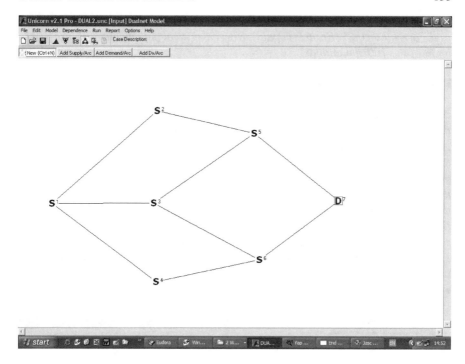

Figure 5.14 Project risk network.

following costs:

$$1c2 = -3;\ 1c3 = -5;\ 1c4 = -7;\ 2c5 = -2;\ 3c5 = -4;$$
$$3c6 = -6;\ 4c6 = -8;\ 5c7 = -9;\ 6c7 = -10.$$

Run the simulation and verify that the minimal completion time, that is, the length of the critical path, is 25.

Now assign 1C2 a beta distribution on the interval [−30, −20], with parameters 2, 2. The arc between nodes S1 and S2 was previously not on the critical path, now it is. The average completion time is 35.9 with 5- and 95-% quantiles of 32.7 and 39.2 respectively (these numbers depend on the random sample). The critical path now always includes the arc S1–S2 and the completion time is between 31 and 41.

Now assign the costs 1C4; 4C6; and 6C7 a uniform distribution on [−15,−5]. The expected project completion time hardly changes; and we see that the expectation of 1F4 is 0.11. In roughly 11 of 100 cases the critical path runs over arc S1–S4.

Make now a simple dependence tree with 4C6 and 6C7 attached to 1C4 with rank correlation 0.85 (Frank's copula). Now the critical path includes the arc S1–S4 in 33 of the 100 simulations.

Project 5.3 Social networks

Social networks interpret costs on an arc as the intensity of 'contact' between the nodes. This may be anything from the rate of phone calls between terrorists, the rate of gossiping between pairs group members, the number citations of scientists, and so on. Lets consider a gossip network. The nodes represent members; the cost of an arc is the number of gossips between the source and sink. Member nr 1 (you) want to send a message to member nr 10, and you want to minimize the number of other people who might learn the message due to gossip. From the EDIT menu, you can arrange the nodes in a circle (or other forms). Attach Dx to S10 so that the graph looks like Figure 5.15. Assign supply one to S1 and zero to all other supply nodes. Let the cost of each arc equal the index of the sink node, for example, 5C6 = 6. Running the simulation (in this case there is no uncertainty and all simulations are identical). We find that the cheapest path is 1-5-6-8-10.

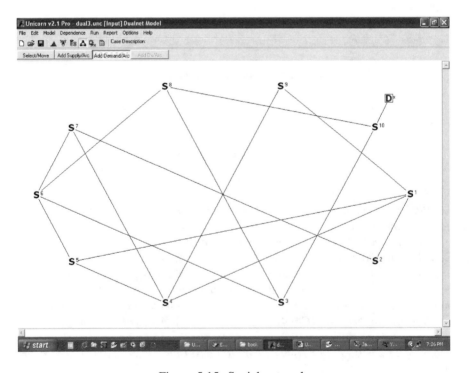

Figure 5.15 Social network.

Project 5.4 Conditional probability tables

This project shows how to use UNICORN in combination with EXCEL to create probability tables for discrete bbn's, using D-vines. We consider the bbn in Figure 5.1. Vines are used to define dependence structures for four uniform variates. When using vines in bbn's, we must therefore assume that the one dimensional distributions are known to be uniform. Of course we can transform any continuous invertible distribution function to uniform; the point is that a bbn in which influence is interpreted as conditional rank correlations via a vine will yield the joint distribution of the uniform transforms of the original variables, it will not yield the joint distribution of the variables themselves. To get these, we must apply the inverse cumulative distribution functions to the uniform variates.

UNICORN regards a discrete random variable as a transformation of a uniform random variable. If we wish to specify a probability table using vines, we must supply the transformations from uniform variates. This is equivalent to saying that we must supply the marginal probabilities of the discrete variables. Given these and a suitable vine structure, we can derive the conditional probability tables.

Using the generalized indicator functions in the formula panel, you can create discrete variables in the same way that UNICORN creates them from the variable panel. In this project, we let you create your own discrete variables in the formula panel. You may check these results against results gotten by specifying discrete variables from the variable panel.

- *Create a case named PROBTAB. Go to the variables panel and create variables $V1, V2, V3, V4$ with uniform distributions on $[0, 1]$.*

- *Go to the formula panel and create UDF's as shown in Figure 5.16. Notice that U4 has probability 0.2 of returning the value '1'. The other variables equal '1' with probability 0.5. Since we are making a* conditional *probability table given values of $U1, U2, U3$, the distributions of the latter are not important.*

- *Make a D-vine as shown in Figure 5.17.*

- *Now run this case with 1000 samples, saving the output. The output is stored in two files, one with the extension '*.sam' that is used for graphics, and*

Formula and user defined functions
Total Formulas: 4
  1 U1: i1 {0.5,V1,1}
  2 U2: i1 {0.5,V2,1}
  3 U3: i1 {0.5,V3,1}
  4 U1: i1 {0.8,V4,1}

Figure 5.16   User defined functions for bbn probability table.

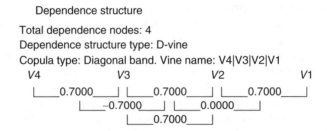

Figure 5.17 Vine for bbn probability table.

one with extension '*.sae' in comma separated EXCEL-compliant textfile for-
mat. We are going to count the number of samples in which $U4 = 1$ and
$U1, U2, U3$ take a given vector of values. This is easily done in EXCEL
and illustrates the use of EXCEL to perform special analyses of simulation
output.

- *After running, open EXCEL, choose 'Open file' in the directory where the*
  UNICORN output file lives. Type '*.sae' as the file name, the file 'Probtab.sae'
  appears, open this file. EXCEL recognizes this as a text file. EXCEL asks how
  the file is delimited, choose comma delimited. A spreadsheet is created with
  the variable names in the first row, and the sample values in succeeding rows.
  Each row is one sample.

- *You can use an 'array function' to perform the required counting. Go to an*
  empty cell and type

  =SUM(A2:A1001*B2:B1001*C2:C1001*D2:D1001)/
  SUM(A2:A1001*B2:B1001*C2:C1001).

  This is an EXCEL array function, you enter it by hitting
  'control+shift+ENTER'. The value is the ratio of number of occurrences
  of $U1 = 1, U2 = 1, U3 = 1, U4 = 1$ to number of occurrences of $U1 = 1$,
  $U2 = 1, U3 = 1$. In the next cell type

  $= SUM(A2:A1001* B2:B1001* (1-C2:C1001)*D2:D1001)/$
  SUM(A2:A1001* B2:B1001*(1-C2:C1001)).

  Using such array functions you can create a table like that shown in
  Figure 5.18.

Probability table P(U4 = 1) = 0.8								
U1	1				0			
U2	1		0		1		0	
U3	1	0	1	0	1	0	1	0
P(U4 = 1|*)	0.239316	0	0.831933	0.119658	0.045802	0	0.403226	0
P(U4 = 0|*)	0.760684	1	0.168067	0.880342	0.954198	1	0.596774	1

Figure 5.18 Probability table, P(U4 = 1) = 0.8.

Probability table P(U4 = 1) = 0.5								
U1	1				0			
U2	1		0		1		0	
U3	1	0	1	0	1	0	1	0
P(U4 = 1|*)	0.794872	0.106061	0.983193	0.581197	0.450382	0.007752	0.895161	0.21374
P(U4 = 0|*)	0.205128	0.893939	0.016807	0.418803	0.549618	0.992248	0.104839	0.78626

Figure 5.19 Probability table, P(U4 = 1) = 0.5.

Notice that the probabilities for U4=1 are zero conditional on $(1, 1, 0)$, $(0, 1, 0)$ and $(0, 0, 0)$. This is caused by the low probability of this in relation to the sample size. If you repeat the same procedure but with 10,000 samples, you will find that $P(U4 = 0 \mid U1 = 1, U2 = 0, U3 = 1) = 0.000781$.

To see the effect of the marginal probability of U4 on the conditional probability table, repeat this exercise but with the following UDF for U4: $i1\{0.5, V4, 1\}$. The resulting conditional probability table should look like that in Figure 5.19.

5.7 Supplement

Theorem 5.1 Σ *is positive (semi-)definite if and only if $v > 0(\geq 0)$. Furthermore, the rank of Σ is equal to the number of nonzero elements in v.*

Proof. We assume without loss of generality that the sequence of indices is ordered, so that the matrix $B = [b_{ij}]$, $i, j = 1, \ldots, n$ is strictly upper triangular. Let D and S be the diagonal matrices formed from the conditional variances and standard deviations respectively, that is, $D = diag(v) = S^T S$. The conditional model can be seen as a set of regression equations:

$$X_i = \mu_i + \sum_{j \in pa(i)} b_{ij}(X_j - \mu_j) + \sqrt{v}Z_j, j = 1, \ldots, n,$$

where $Z = (Z_1, \ldots, Z_n)$ are independent standard normal variables. This can be written as

$$X - \mu = B^T(X - \mu) + S^T Z.$$

From the above

$$S^T Z = (I - B^T)(X - \mu).$$

Since B is strictly upper triangular, then $(I - B^T)$ is invertible and

$$X - \mu = (I - B^T)^{-1} S^T Z = U^T S^T Z = A^T Z,$$

where $U = (I - B)^{-1}$ and $A = SU$. We get

$$\Sigma = Var(X) = Var(X - \mu) = Var(A^T Z) = A^T Z A$$
$$= A^T A = U^T S^T SU = U^T DU.$$

The preceding factorization of the covariance matrix concludes the proof. □

Theorem 5.2 *Given*

1. *a directed acyclic graph (DAG) with n nodes specifying conditional indepen-dence relationships in a bbn,*

2. *the specification of conditional rank correlations (5.4), $i = 1, \ldots, n$ and*

3. *a copula realizing all correlations $[-1, 1]$ for which correlation 0 entails independence*

the joint distribution is uniquely determined. This joint distribution satisfies the characteristic factorization (5.3) and the conditional rank correlations in (5.4) are algebraically independent.

Proof. The first term in (5.3) is determined vacuously. We assume the joint dis-tribution for $\{1, \ldots, i - 1\}$ has been determined. Term i of the factorization (5.4) involves $i - 1$ conditional variables, of which $\{i_{p(i)+1}, \ldots, i_{i-1}\}$ are conditionally independent of i given $\{i_1, \ldots, i_{p(i)}\}$. We assign

$$r(i, i_j | i_1, \ldots i_{p(i)}) = 0; \quad i_{p(i)} < i_j \leq i - 1. \tag{5.11}$$

Then the conditional rank correlations (5.4) and (5.11) are exactly those on $\mathcal{D}^i$ involving variable i. The other conditional bivariate distributions on $\mathcal{D}^i$ are already determined. It follows that the distribution on $\{1, \ldots i\}$ is uniquely determined. Since zero conditional rank correlation implies conditional independence,

$$P(1, \ldots i) = P(i | 1 \ldots i - 1) P(1, \ldots, i - 1) = P(i | i_1 \ldots i_{p(i)}) P(1, \ldots, i - 1)$$

from which it follows that the factorization (5.3) holds. □

6

Sampling Methods

6.1 Introduction

Given a joint distribution for $(X_1, \ldots, X_k)$, this has to be pushed through the model M to yield a distribution on the model output. In a few very simple cases, we can perform these calculations analytically. For example, if (X_1, X_2) is joint normal, then $M = X_1 + X_2$ is also normal. A similar relation holds for gamma-distributed variables with the same scale factor. Beyond such simple examples, we must have a recourse to simulation. We must obtain a number, say n, of random samples of the input vector and compute M on each sample, yielding $m_1, \ldots, m_n$. The distribution built up after n samples, $m_1, \ldots, m_n$, is called the *sample* or *empirical distribution* $F_n(m) = (1/n) \sum_{i=1}^{n} \mathbb{I}_{(-\infty, m]}(m_i)$, where $\mathbb{I}_A$ is the indicator function of set A. The classic Glivenko Cantelli theorem (Glivenko (1933)) states that the empirical distribution function converges uniformly to the true distribution function as $n \to \infty$. We also know that for any m, $(F_n(m) - F(m))$ is asymptotically normally distributed, with variance going to zero as $1/n$.

We can use these results to determine the distribution of M if we could draw random samples. The computer cannot really draw random samples but can simulate random numbers as *pseudo-random numbers*. If the model M is very expensive, the rate of convergence of (pseudo-) random numbers may be too slow. If we are willing to give up some features of random sampling, notably serial independence, then variance reduction techniques may be invoked. These include *quasi-random* numbers and *stratified* and *Latin hypercube* sampling. Quasi-random numbers could be used with copulae to sample dependent distributions. However, their lack of serial independence may induce spurious dependencies, the effects of which are largely unexplored. Other variance reduction techniques are also ungraceful in dealing with dependence. The first two sections of this chapter discuss pseudo-random and reduced variance sampling.

Uncertainty Analysis with High Dimensional Dependence Modelling   D. Kurowicka and R. Cooke
© 2006 John Wiley & Sons, Ltd

The lion's share of this chapter deals with sampling the structures introduced in Chapter 4. These require only independent random sampling; dependence is created by transforming these using the cumulative conditional distribution functions obtained from copulae.

6.2   (Pseudo-) random sampling

The subject of random number generators and pseudo-random numbers is large; good general texts are Ripley (1987) and Rubinstein (1981), and recent information can be obtained online from http://mathworld.wolfram.com/RandomNumber. Only a cursory treatment can be given here.

Under *random sampling*, different input vector values $(x_1, \ldots, x_k)$ are drawn 'pseudo-randomly' from the joint distribution of $(X_1, \ldots, X_k)$. We assume for the time being that $X_1, \ldots, X_k$ are independent. 'Pseudo' refers to the fact that the procedure followed by a computer is not really random, and only a finite number of distinct numbers can be generated. Common procedures depend on the choice of a seed. A large number of samples may be required to achieve results that are insensitive to the choice of seed.

The way in which distributions are represented will also affect the run time. Suppose that we wish to sample a variable X with continuous invertible univariate distribution function F. By definition, $F(x) = P(X \leq x)$; so for $r \in [0, 1]$,

$$P(F(X) \leq r) = P(X \leq F^{-1}(r)) = F(F^{-1}(r)) = r,$$

which says that $F(X)$ follows a uniform distribution on the interval $[0, 1]$. Therefore, we can sample X by calling a random number on our computer, returning a realization u of (pseudo-) uniform variable U on $[0, 1]$. We then apply the transformation $x = F^{-1}(u)$. However, if the function F^{-1} must be computed numerically, it will cost time if required repeatedly.

An important property of pseudo-random numbers is that they display serial independence. Hence, we can simulate the independent variables $X_1, \ldots, X_k$ by repeatedly calling a random number on our computer. However, it is well known that pseudo-random numbers may display clustering effects.

The most common method for generating pseudo-random numbers is the *multiplicative congruential method*: Start with a seed x_0 and compute

$$x_n = a x_{n-1} \bmod[m],$$

where a and m are the given positive integers. The *seed* is x_0. Thus, x_i takes values in $\{0, \ldots, m-1\}$ and x_i/m is approximately uniformly distributed on $[0, 1]$. The IBM System/360 Uniform Random Number Generator used in many statistics library packages takes $m = 2^{31} - 1$ and $a = 7^5$. This choice has been extensively tested and gives good results. The tests check for a variety of features including relative frequency of digits and words, serial independence, excursion length, first passage times and periodicity (Rubinstein (1981)).

Mixed congruential generators are of the type

$$x_n = (ax_{n-1} + c) \bmod[m].$$

The multiplicative generators correspond to the special case of $c = 0$. It can be shown (Greenberger (1961); Rubinstein (1981)) that for generators of this type, the Pearson correlation coefficient of X_i and X_{i+1} lies between

$$\frac{1}{a} - \frac{6c}{am}\left(1 - \frac{c}{m}\right) \pm \frac{a}{m}.$$

m is often chosen equal to the computer's word length, since this makes the computations $\bmod[m]$ very efficient.

6.3 Reduced variance sampling

We discuss a number of strategies for speeding up the convergence obtained with pseudo-random sampling.

6.3.1 Quasi-random sampling

Quasi-random number[1] sequences, otherwise known as low-discrepancy sequences, are presented as an alternative to pseudo-random numbers. They are generated to cover the k-dimensional unit cube $I^k = [0, 1)^k$ more uniformly, thereby increasing the rate of convergence at the expense of serial independence. One of the most well-known generators is that of Sobol (1967), which is based on the concept of primitive polynomials. Other generators include those of Faure (implemented in Fox (1986); Niederreiter (1988), the one-dimensional Van Der Corput sequence and its multidimensional analogue, the Halton sequence (Halton (1960))). We briefly describe the Sobol' generator here. For more information about the topic, we refer the interested readers to original papers.

Discrepancy   The discrepancy of a quasi-random sequence is a measure of the uniformity of the distribution of a finite number of points over the unit hypercube. Informally, a sequence of points is considered to be uniformly distributed in the s-dimensional unit cube $I^s = [0, 1)^s$ if, in the limit, the fraction of points lying in any measurable set of I^s is equal to the area of that set. A more formal definition follows.

Following the notation of Niederreiter (Niederreither (1992), p. 14), let P denote the set of points $x_1, x_2, \ldots, x_N \in I^s$ and B some arbitrary subset of I^s. Let $A(B; P)$ be a counting function of the number of points n ($1 \leq n \leq N$) for which $x_i \in B$. Let $\mathcal{B}^*$ be a non-empty family of subsets of I^s of the form

$$B = \prod_{i=1}^{s} [0, a_i),$$

---

[1] This section was co-authored by Belinda Chiera.

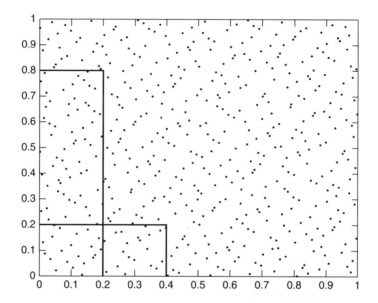

Figure 6.1   An example of taking measurable subsets to compute $D_n^*(u)$ of a set of quasi-random points.

with s-dimensional volume $a_1 a_2 \ldots a_s$; then the *star* discrepancy $D_n^*(u)$ of P is

$$D_n^*(\mathcal{B}^*; P) = \sup_{B \in \mathcal{B}^*} \left| \frac{A(B; P)}{N} - \mu_s(B) \right|, \qquad (6.1)$$

where $\mu_s(B)$ is the theoretical number of points in B and $0 \le D_n^*(\mathcal{B}^*; P) \le 1$ (Niederreither (1992)). An example of measuring discrepancy is illustrated in Figure 6.1.

The smaller the discrepancy, the more uniform the spacing of the points in the hypercube. A sharp bound on the discrepancy of quasi-random number sequences is given in Niederreither (1992),

$$D_n^*(P) = O(n^{-1}(\log n)^s),$$

although Kocis and Whiten (1997) have found that this asymptotic result may not be representative of the sample sizes N used in practice, especially in higher dimensions.

The Sobol' generator   A Sobol' sequence is based on a set of 'direction numbers', $\{v_i\}$, defined as

$$v_i = \frac{m_i}{2^i},$$

where the m_i are odd, positive integers, such that $0 < m_i < 2^i$.

The v_i must be chosen so that they satisfy a recurrence relation using the coefficients of a primitive polynomial (mod 2),[2]

$$P \equiv x^d + a_1 x^{d-1} + \cdots + a_{d-1} x + 1,$$

where each a_i is 0 or 1 and P is a primitive polynomial of degree d (mod 2).[3]

Once the primitive polynomial has been chosen, a recurrence relation for calculating v_i must be set up. We define v_i as

$$v_i = a_1 v_{i-1} \oplus a_2 v_{i-2} \oplus \cdots \oplus a_{d-1} v_{i-d+1} \oplus v_{i-d} \oplus \left\lfloor \frac{v_{i-d}}{2^d} \right\rfloor, \quad i > d,$$

where $\oplus$ is the XOR (exclusive or, binary summation) operation. An equivalent expression of this recurrence is as follows.

$$m_i = 2a_1 m_{i-1} \oplus 2^2 a_2 m_{i-2} \oplus \cdots \oplus 2^{d-1} a_{d-1} m_{i-d+1} \oplus 2^d m_{i-d} \oplus m_{i-d}.$$

To generate the nth number in the sequence, compute

$$x_n = b_1 v_1 \oplus b_2 v_2 \oplus b_3 v_3 \oplus \cdots,$$

where $\ldots b_3 b_2 b_1$ is the binary representation of n. Let us consider the following example.

Example 6.1 *Consider the primitive polynomial*

$$P \equiv x^3 + x + 1$$

of degree $d = 3$. The corresponding recurrence is as follows.

$$m_i = 4m_{i-2} \oplus 8m_{i-3} \oplus m_{i-3}.$$

Let $m_1 = 1, m_2 = 3, m_3 = 7$. Then

$$m_4 = 12 \oplus 8 \oplus 1$$
$$= 1100 \oplus 1000 \oplus 0001 (\text{in binary})$$
$$= 0101$$
$$= 5.$$

Similarly, $m_5 = 28 \oplus 24 \oplus 3 = 7$, $m_6 = 20 \oplus 56 \oplus 7 = 43$, and so forth.

We calculate nth number in the sequence as follows.

---

[2]The *order* of a polynomial $P(x)$ with $P(0) \neq 0$ is defined as the smallest integer e for which $P(x)$ divides $x^e - 1$. A polynomial of degree n with coefficients 0 and 1 is *primitive* (mod 2) if it has polynomial order $2^n - 1$.

[3]Examples of primitive polynomials: $x^2 + x + 1$ – order 3, $x^5 + x^2 + 1$ – order 31.

n	b_3	b_2	b_1	$x_n = b_1 v_1 \oplus b_2 v_2 \oplus b_3 v_3 \oplus \cdots$
1	0	0	1	$v_1 = 0.1 = \frac{1}{2}$,
2	0	1	0	$v_2 = 0.11 = \frac{3}{4}$,
3	0	1	1	$v_1 \oplus v_2 = 0.1 \oplus 0.11 = 0.01 = \frac{1}{4}$,
4	1	0	0	$v_3 = 0.111 = \frac{7}{8}$,
5	1	0	1	$v_1 \oplus v_3 = 0.111 \oplus 0.1 = 0.011 = \frac{3}{8}$,
⋮				

An example of Sobol' points is given in Figure 6.2.

The objective for a finite pseudo-random sequence is to appear as if it is a sequence of realizations of independent identically distributed uniform random variables. The objective of a quasi-random sequence is that it fills a unit hypercube as uniformly as possible. Hence, using quasi-random numbers, the convergence of the Monte Carlo method towards a solution can be much faster than with pseudo-random numbers. The difference between pseudo- and quasi-random numbers is highlighted in Figure 6.3.

6.3.2 Stratified sampling

In stratified sampling, the sampling space is divided into a number of non-overlapping subregions, called strata, of known probability. A given number of

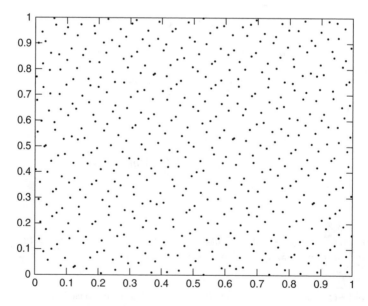

Figure 6.2 An example of the Sobol' sequence, with $N = 512$. The Sobol' primitive polynomial is $x^5 + x^2 + x^0$.

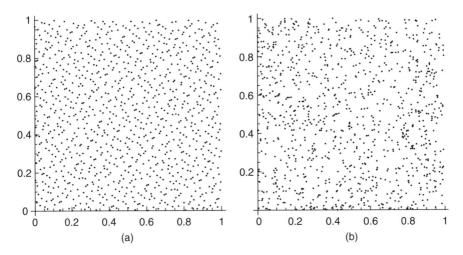

Figure 6.3 A plot of a Halton sequence (a) and a pseudo-random number sequence (b). (Sample size = 1024.)

values are randomly sampled from each strata, often only one. We illustrate the idea underlying stratified sampling as follows.

Let U be a uniform $[0, 1]$ variable. The cumulative distribution function of U is $F(r) = r, r \in [0, 1]$. Suppose we wish to determine the distribution of U by simulation, we draw, say 30 samples, from U and compute the sample cdf:

$$F_{30}(r) = (\#samples \le r)/30.$$

F_{30} converges uniformly to F, but for a small number of samples, $F \ne F_{30}$. The situation after 30 samples is pictured in Figure 6.4 (u-dark line).

Suppose we decide to sample the first 15 values uniformly from $[0, 0.5]$, and the second 15 from $[0.5, 1]$. Then, the 15th ordered sample will always be less than 0.5 and the 16th will always be greater than 0.5. The empirical distribution function obtained in this way will always converge more quickly (see stratified-light line in Figure 6.4).

Complications arise in stratified sampling of dependent distributions. Again, a very simple example best illustrates the issues. Suppose we have two independent variables, X_1, X_2, uniformly distributed on $[-1, 1]$. We wish to apply stratified sampling, such that for each variable, half the values are sampled uniformly from $[-1, 0]$ and from $[0, 1]$. This, however, does not determine a sampling strategy, rather we must decide how to sample the four cells:

$$
\begin{array}{ccc}
(-1, 0) & \times & (-1, 0), \\
(-1, 0) & \times & (0, 1), \\
(0, 1) & \times & (-1, 0), \\
(0, 1) & \times & (0, 1).
\end{array}
$$

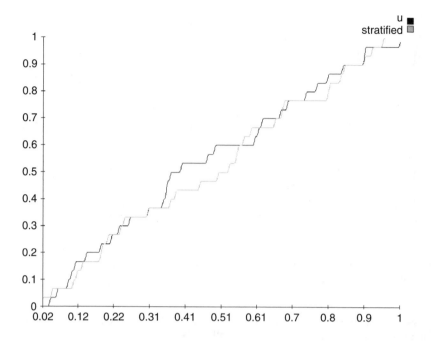

Figure 6.4 Empirical distributions functions for random sampling (u-dark line) and stratified sampling (stratified-light line).

If the variables are independent, we must ensure that our sampling strategy does not introduce a spurious correlation. If the variables are not independent, then the sampling strategy must represent the dependence.

6.3.3 Latin hypercube sampling

Latin hypercube sampling (LHS) (Iman and Helton (1988); Iman and Shortencarier (1984); McKay et al. (1979)) has found extensive application in uncertainty analysis. This sampling procedure is based on dividing the range of each variable into N intervals of equal probability. For each variable, one value is selected randomly from each interval. Then, the N values for first variable are paired at random and without replacement with the N values of second variable. These N pairs are combined in a random manner without replacement with N values of third variable, and so on.

More precisely, if $X_1, \ldots, X_n$ are mutually independent random variables with invertible distribution functions F_j, $j = 1, \ldots, n$, respectively, then the i-th Latin hypercube sample for the j-th variable can be created as

$$x_j^{(i)} = F_j^{-1} \left(\frac{\pi_{ij} - 1 + \xi_{ij}}{N} \right),$$

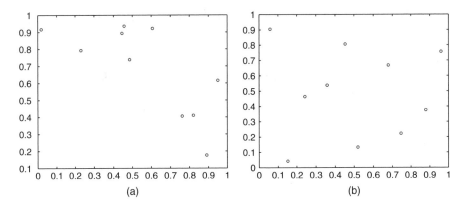

Figure 6.5   Pseudo (a) and Latin hypercube (b) sampling for two uniform on [0, 1] variables. (Sampling size = 10.)

where $P = [\pi_{ij}]$ is an $N \times n$ matrix with independent random permutations of $\{1, 2, \ldots, N\}$ in columns of P, and $\Xi = [\xi_{ij}]$ is an $N \times n$ matrix containing independent random numbers uniformly distributed on [0, 1] independent of P.

In Figure 6.5 we see 10 samples of two uniform variables on [0, 1] obtained with pseudo-random and LHS schemes. We can see that the result of LHS is more spread out and does not display the clustering effects found in pseudo-random sampling.

Samples from non-uniform variables are obtained using the inverse cumulative distribution of each variable. LHS is used when the models are computationally expensive, so that a large number of samples is not practicable. It gives unbiased estimates for means.

Correlations: the Iman and Conover method   The Latin Hypercube codes incorporate 'distribution-free' techniques for representing correlations (Iman and Conver (1982)). The idea is the same as that in Chapter 4:

- Draw N LHS samples of k variables; convert these to ranks and place them in an $N \times k$ matrix.

- Draw N samples from a k-dimensional joint normal distribution with correlation matrix R.

- Convert the normal variables to ranks.

- Permute the columns of the LHS matrix so that the ranks in each column coincide with those of the normal matrix.

- Unrank the LHS variables.

As discussed in Chapter 4, this procedure will not yield a rank correlation matrix R. The remarks in Chapter 4 apply here as well.

6.4   Sampling trees, vines and continuous bbn's

In sampling dependence structures, we will require independent uniform variables, and we will induce dependence by transforming these in a way that uses previously sampled values. The transformations will be the conditional cumulative and inverse cumulative distributions of the copulae in an appropriate tree or vine representation. The serial independence property of pseudo-random numbers is therefore essential.

6.4.1   Sampling a tree

Distributions specified using dependence trees can be sampled on the fly. In performing the conditional sampling, it is always a bivariate distribution specified by the copula that is conditionalized. In Figure 6.6 a sampling procedure for a tree on three uniform on [0, 1] variables X_1, X_2, X_3 is shown. X_1 and X_2 are joined by the diagonal band copula with the rank correlation r_{12} and X_1 and X_3 are joined by the diagonal band copula with r_{13}. First, we sample a realization of X_1, say x_1. For the diagonal copula, the conditional distributions are very simple. We see in Figure 6.6 the conditional densities $c_{2|1}$ and $c_{3|1}$. By sampling from these distributions, the samples for X_2 and X_3 can be found.

The sampling procedure involves sampling three independent uniform $(0, 1)$ variables U_1, U_2, U_3.

1. Sample U_1, U_2, U_3. Denote realizations as u_1, u_2, u_3.

2. $x_1 = u_1$.

3. Find $F_{X_2|X_1=x_1}$ and $F_{X_3|X_1=x_1}$ for shorthand-denoted $F_{r_{12};x_1}$, $F_{r_{13};x_1}$.

4. $x_2 = F_{r_{12};x_1}^{-1}(u_2)$.

5. $x_3 = F_{r_{13};x_1}^{-1}(u_3)$.

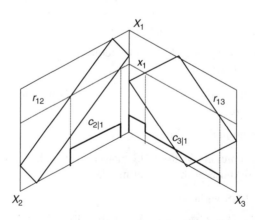

Figure 6.6   A sampling procedure for a tree on three variables.

Note that X_2 and X_3 are conditionally independent given X_1. Moreover, note that only correlations r_{12} and r_{13} are specified. The correlation value between X_2 and X_3 is determined by the tree structure and a choice of copula. For the diagonal band and the minimum information copula, r_{23} is approximately equal to (and for the elliptical copula exactly equal to) the product $r_{12} \times r_{13}$ (see Exercise 4.2).

6.4.2 Sampling a regular vine

The rank correlation specification on regular vine plus copula determines the whole joint distribution. We recall the conditional rank correlations need not be constant. The procedure of sampling such a distribution can be written for any regular vine. There are two strategies for sampling such a distribution, which we term the *cumulative* and *density* approaches. We first show the cumulative sampling procedures for the canonical and the D-vines followed by a general cumulative sampling procedure for a regular vine. The density approach will be presented in Section 6.4.3.

Canonical vine For the canonical vine, the sampling algorithm takes a simple form. We illustrate the following algorithm for a canonical vine on four variables, as shown in Figure 6.7.

The algorithm involves sampling four independent uniform $(0, 1)$ variables $U_1, \ldots, U_4$. We assume that the variables $X_1, \ldots, X_4$ are also uniform. Let $r_{i,j|k}$ denote the conditional correlation between variables (i, j) given k. Let $F_{r_{i,j|k};U_i}(X_j)$ denote the cumulative distribution function for X_j given U_i under the conditional copula with correlation $r_{i,j|k}$. The algorithm can now be stated as follows:

$$
\begin{aligned}
x_1 &= u_1; \\
x_2 &= F^{-1}_{r_{12};u_1}(u_2); \\
x_3 &= F^{-1}_{r_{13};u_1}\left(F^{-1}_{r_{23|1};u_2}(u_3)\right); \\
x_4 &= F^{-1}_{r_{14};u_1}\left(F^{-1}_{r_{24|1};u_2}\left(F^{-1}_{r_{34|12};u_3}(u_4)\right)\right).
\end{aligned}
\tag{6.2}
$$

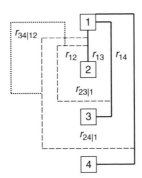

Figure 6.7 The canonical vine on four variables with (conditional) rank correlations assigned to the edges.

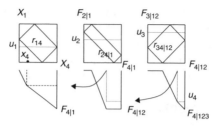

Figure 6.8 Graphical representation of sampling value of x_4 in canonical vine.

We see that the uniform variables $U_1, \ldots, U_4$ are sampled independently, and the variables $X_1, \ldots, X_4$ are obtained by applying successive inverse cumulative distribution functions.

Figure 6.8 shows the procedure of sampling the value of x_4 graphically. Notice that u_1, u_2 and u_3 are values of $X_1, F_{2|1}$ and $F_{3|12}$ respectively; hence, conditional distributions $F_{4|1}, F_{4|12}$ and $F_{4|123}$ can be easily found by conditionalizing copulae (in Figure 6.8 the diagonal band copula was used) with correlations $r_{14}, r_{24|1}$ and $r_{34|12}$ on values of u_1, u_2 and u_3, respectively. Inverting value of u_4 through $F_{4|1}, F_{4|12}$ and $F_{4|123}$ gives x_4.

In general, we can sample an n-dimensional distribution represented graphically by the canonical vine on n variables with (conditional) rank correlations

$$
\begin{array}{cccc}
r_{12}, & r_{13}, & r_{14}, & \cdots & r_{1n}, \\
 & r_{23|1}, & r_{24|1}, & \cdots & r_{2n|1}, \\
 & & r_{34|12}, & \cdots & r_{3n|12}, \\
 & & & \cdots & \\
 & & & & r_{n-1,n|12\ldots n-2},
\end{array}
$$

assigned to the edges of the vine by sampling n independent, uniform $(0, 1)$ variables, say $U_1, U_2, \ldots, U_n$, and calculating

$$x_1 = u_1;$$

$$x_2 = F^{-1}_{r_{12};u_1}(u_2);$$

$$x_3 = F^{-1}_{r_{13};u_1}\left(F^{-1}_{r_{23|1};u_2}(u_3)\right);$$

$$x_4 = F^{-1}_{r_{14};u_1}\left(F^{-1}_{r_{24|1};u_2}\left(F^{-1}_{r_{34|12};u_3}(u_4)\right)\right);$$

$$\cdots$$

$$x_n = F^{-1}_{r_{1n};u_1}\left(F^{-1}_{r_{2n|1};u_2}\left(F^{-1}_{r_{3n|12};u_3}\left(\cdots\left(F^{-1}_{r_{n-1,n|12\ldots n-2};u_{n-1}}(u_n)\right)\cdots\right)\right)\right).$$

D-vine The sampling algorithm for the D-vine is more complicated than that of the canonical vine. We illustrate sampling algorithm for a D-vine on four variables, as shown in Figure 6.9.

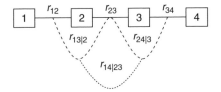

Figure 6.9 D-vine of four variables with (conditional) rank correlations.

The sampling procedure for the D-vine in Figure 6.9 is the following:

1. $x_1 = u_1$;

2. $x_2 = F_{r_{12};x_1}^{-1}(u_2)$;

3. $x_3 = F_{r_{23};x_2}^{-1}\left(F_{r_{13|2};F_{r_{12};x_2}(x_1)}^{-1}(u_3)\right)$;

4. $x_4 = F_{r_{34};x_3}^{-1}\left(F_{r_{24|3};F_{r_{23};x_3}(x_2)}^{-1}\left(F_{r_{14|23};F_{r_{13|2};F_{r_{23};x_2}(x_3)}(F_{r_{12};x_2}(x_1))}^{-1}(u_4)\right)\right)$.

Notice that the sampling procedure for D-vine uses conditional distributions and inverse conditional distributions and hence will be much slower than the procedure for the canonical vine. To shorten the notation that is used to describe the general sampling procedure for D-vine, the preceding algorithm can be stated as:

$$x_1 = u_1;$$

$$x_2 = F_{2|1:x_1}^{-1}(u_2);$$

$$x_3 = F_{3|2:x_2}^{-1}\left(F_{3|12:F_{1|2}(x_1)}^{-1}(u_3)\right);$$

$$x_4 = F_{4|3:x_3}^{-1}\left(F_{4|23:F_{2|3}(x_2)}^{-1}\left(F_{4|123:F_{1|23}(x_1)}^{-1}(u_4)\right)\right).$$

Figure 6.10 shows the procedure of sampling value of x_4 graphically. Notice that for the D-vine, values of $F_{2|3}$ and $F_{1|23}$ that are used to conditionalize copulae

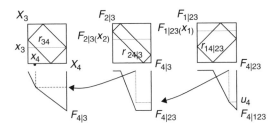

Figure 6.10 Graphical representation of sampling value of x_4 in D-vine.

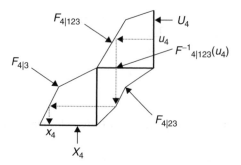

Figure 6.11 Staircase graph representation of D-vine sampling procedure.

with correlations $r_{24|3}$ and $r_{14|23}$ to obtain $F_{4|23}$ and $F_{4|123}$, respectively, have to be calculated. This is in contrast to the canonical vine, where values of u_i are used.

The 'staircase graphs' in Figure 6.11 show the sampling procedure graphically. The horizontal and vertical lines represent the $(0, 1)$ interval; the intervals are connected via conditional cumulative distribution functions. In Figure 6.11, the diagonal band copula is used.

In general, we can sample an n-dimensional distribution represented graphically by the D-vine on n variables with (conditional) rank correlations

$$
\begin{array}{cccccc}
r_{12}, & r_{13|2}, & r_{14|23}, & r_{15|234}, & \cdots & r_{1,n-1|2...n-2}, & r_{1,n|2...n-1}, \\
r_{23}, & r_{24|3}, & r_{25|34}, & \cdots & r_{2,n-1|3...n-2}, & r_{2,n|3...n-1}, \\
r_{34}, & r_{35|4}, & \cdots & r_{3,n-1|4...n-2}, & r_{3,n|4...n-1}, \\
& & \cdots & & \\
& & & r_{n-2,n-1}, & r_{n-2,n|n-1}, \\
& & & & r_{n-1,n},
\end{array}
$$

assigned to the edges of the vine as follows.

$$x_1 = u_1;$$

$$x_2 = F_{2|1:x_1}^{-1}(u_2);$$

$$x_3 = F_{3|2:x_2}^{-1}\left(F_{3|12:F_{1|2}(x_1)}^{-1}(u_3)\right);$$

$$x_4 = F_{4|3:x_3}^{-1}\left(F_{4|23:F_{2|3}(x_2)}^{-1}\left(F_{4|123:F_{1|23}(x_1)}^{-1}(u_4)\right)\right);$$

$$x_5 = F_{5|4:x_4}^{-1}\left(F_{5|34:F_{3|4}(x_3)}^{-1}\left(F_{5|234:F_{2|34}(x_2)}^{-1}\left(F_{5|1234:F_{1|234}(x_1)}^{-1}(u_5)\right)\right)\right);$$

$$\cdots$$

$$x_n = F_{n|n-1:x_{n-1}}^{-1}\left(F_{n|n-2,n-1:F_{n-2|n-1}(x_{n-2})}^{-1}\left(\cdots\right.\right.$$

$$\left.\left(F_{n|1...n-1:F_{1|2...n-1}(x_1)}^{-1}(u_n)\right)\cdots\right)\right).$$

Regular vines A regular vine on n nodes will have a single node in tree $n - 1$. It suffices to show how to sample one of the conditioned variables in this node, say n. Assuming we have sampled all the other variables, we proceed as follows.

1. By Lemma 4.6 of Chapter 4, the variable n occurs in trees $1, \ldots, n - 1$ exactly once as a conditioned variable. The variable with which it is conditioned in tree j is called its 'j-partner'. We define an ordering for n as follows: Index the j-partner of variable n as variable j. We denote the conditional bivariate constraints corresponding to the partners of n as:

$$(n, 1|\emptyset), (n, 2|D_2^n), (n, 3|D_3^n) \ldots (n, n - 1|D_{n-1}^n).$$

Again by Lemma 4.6 of Chapter 4, variables $1, \ldots, n - 1$ first appear as conditioned variables (to the left of '|') before appearing as conditioning variables (to the right of '|'). Also,

$$0 = \#D_1^n < \#D_2^n < \cdots < \#D_{n-1}^n = n - 2.$$

2. Assuming we have sampled all variables except n, sample one variable uniformly distributed on the interval $(0, 1)$, denoted as u_n. We use the general notation $F_{a|b,C}$ to denote $F_{a,b|C:F_{b|C}}$, which is the conditional copula for $\{a, b|C\}$ conditional on a value of the cumulative conditional distribution $F_{b|C}$. Here, $\{a, b|C\}$ is the conditional bivariate constraint corresponding to a node in the vine.

3. Sample x_n as follows.

$$x_n = F_{n|1,D_1^n}^{-1} \left(F_{n|2,D_2^n}^{-1} \left(\cdots \left(F_{n|n-1,D_{n-1}^n}^{-1} (u_n) \right) \ldots \right) \right) \qquad (6.3)$$

The innermost term of (6.3) is:

$$F_{n|n-1,D_{n-1}^n}^{-1} = F_{n,n-1|D_{n-1}^n:F_{n-1|D_{n-1}^n}}^{-1}$$

$$= F_{n,n-1|D_{n-1}^n:F_{n-1,(n-2)'|D_{(n-2)'}^{n-1}:F_{(n-2)'|D_{(n-2)'}^{n-1}}}}^{-1}.$$

Example 6.2 *We illustrate the sampling procedure for the regular vine on five variables in Figure 6.12, which is neither a canonical nor a D-vine. The top node is $\{34|125\}$. Assuming we have sampled variables 2,5,1,3 already, the sampling procedure for x_4 is:*

$$x_4 = F_{4|2}^{-1} \left(F_{4|52}^{-1} \left(F_{4|152}^{-1} \left(F_{4|3152}^{-1}(u_4) \right) \right) \right).$$

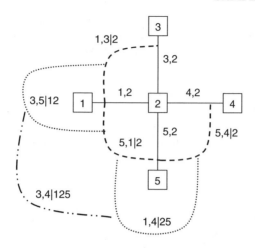

Figure 6.12 Regular vine with five variables.

6.4.3 Density approach to sampling regular vine

When the vine-copula distribution is given as a density, the density approach to sampling may be used. For tree T_i, let E_i be the edge set, and for $e \in E_i$ with conditioning set D_e and conditioned set $\{j, k\}$, let $c_{jk|D_e}$ be the copula density associated with e, then the density for a distribution specified by the assignment of copulae to the edges of $\mathcal{V}$ is given by the expression in Theorem 4.2, which we reproduce here for convenience.

$$\prod_{i=1}^{n-1} \prod_{e \in E_i} c_{jk|D_e}(F_{j|D_e}(x_j), F_{k|D_e}(x_k)) f_1(x_1) \ldots f_n(x_n) \tag{6.4}$$

$$= \prod_{i=1}^{n-1} \prod_{e \in E_i} c_{jk|D_e}(F_{j|D_e}(x_j), F_{k|D_e}(x_k)), \tag{6.5}$$

where, by uniformity, the density $f_i(x_i) = 1$.

This expression may be used to sample the vine distribution; that is, a large number of samples $(x_1, \ldots, x_n)$ is drawn uniformly and then these are resampled with probability proportional to (6.5). This is less efficient than the general sampling algorithm given previously; however, it may be more convenient for conditionalization.

6.4.4 Sampling a continuous bbn

In Chapter 5, we have shown simple examples of continuous Bayesian belief nets (bbn's) and the relationship between continuous bbn's and vines. To sample a continuous bbn, we use the sampling procedure for the D-vine presented earlier.

We start with the bbn in Figure 5.1. This simple bbn can be represented as one D-vine, as in Figure 5.3.

Since variables 1, 2, 3 are mutually independent, the sampling procedure can be simplified to

1. $x_1 = u_1$;

2. $x_2 = u_2$;

3. $x_3 = u_3$;

4. $x_4 = F_{r_{34};x_3}^{-1} \left(F_{r_{24|3};x_2}^{-1} \left(F_{r_{14|23};x_1}^{-1} (u_4) \right) \right).$

Most bbn's cannot be represented as one D-vine. In this case, the sampling algorithm consists of sampling the i-th variable in the ordering according to $\mathcal{D}^i$ and possibly calculating some conditional distributions. Independencies present in bbn lead to simplifications of this sampling procedure. We show how the sampling procedure can be created for the crew alertness model in Figure 5.5 as follows.

The sampling order: 1, 2, 3, 4, 5, 6, 7, 8.

Factorization: $P(1)P(2|\underline{1})P(3|21)P(4|\underline{321})P(5|4\underline{321})$

$$P(6|54\underline{321})P(7|6\underline{54321})P(8|637\underline{5421}).$$

Variables 1, 2 are independent, hence

$$x_1 = u_1 \text{ and } x_2 = u_2.$$

To sample the variable 3, we use the sampling procedure for $\mathcal{D}^3(3, 2, 1)$

$$x_3 = F_{r_{23};x_2}^{-1} \left(F_{r_{13|2};F_{r_{12};x_2}(x_1)}^{-1} (u_3) \right).$$

Since variables 1 and 2 are independent, $F_{r_{12};x_2}(x_1) = x_1$ and the preceding expression can be simplified to

$$x_3 = F_{r_{23};x_2}^{-1} \left(F_{r_{13|2};x_1}^{-1} (u_3) \right).$$

Further, since the variable 4 is independent of 1, 2, 3, sampling $\mathcal{D}^4(4, 3, 2, 1)$ simplifies to $x_4 = u_4$. The variable 5 depends only on 4, so

$$x_5 = F_{r_{54};x_4}^{-1} (u_5).$$

By sampling variable 6 according to $\mathcal{D}^6(6, 5, 4, 3, 2, 1)$ and by noticing that 6 depends only on 4 and 5, we get

$$x_6 = F_{r_{64};x_4}^{-1} \left(F_{r_{65|4};F_{5|4}(x_5)}^{-1} (u_6) \right).$$

The variable 7 is sampled independently, $x_7 = u_7$, and 8 is sampled in the following way

$$x_8 = F_{r_{86};x_6}^{-1}\left(F_{r_{83|6};x_3}^{-1}\left(F_{r_{87|36};x_7}^{-1}(u_8)\right)\right).$$

This sampling procedure does not require any additional calculations. All necessary distributions are given.

The results of sampling the crew alertness model in Figure 5.5, conditional on certain values of variables 'Hour of sleep', 'Fly duty period', and so on, were obtained using the density approach. The density of the copula for the crew alertness model is:

$$g(u_1, \ldots, u_8) = c_{23}(u_2, u_3)c_{13|2}(u_1, F_{3|2}(u_3))c_{45}(u_4, u_5)c_{46}(u_4, u_6)$$
$$c_{65|4}(F_{6|4}(u_6), F_{5|4}(u_5))c_{86}(u_6, u_8)c_{83|6}(F_{8|6}(u_8), u_3)$$
$$c_{87|36}(F_{8|36}(u_8), u_7).$$

Frank's copula was used. Updating is done by resampling the network each time a new policy is evaluated. This approach would be time-consuming for large networks . Moreover, as shown in Example 6.3, sampling may require some additional calculations. The following example illustrates the complications that can arise in sampling continuous bbn's.

Example 6.3 *Consider the following bbn on five variables.*

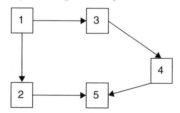

Sampling order: 1, 2, 3, 4, 5.

Factorization: $P(1)P(2|1)P(3|\underline{2}1)P(4|3\underline{2}1)P(5|4\underline{3}\underline{2}1)$.

The following rank correlations must be assessed:

$$r_{21}, r_{31}, r_{43}, r_{54}, r_{52|4}.$$

In this case, $\mathcal{D}^4 = D(4, 3, 1, 2)$, but the order of variables in $\mathcal{D}^5$ must be $D(5, 4, 2, 3, 1)$. Hence, this bbn cannot be represented as one vine.

Using conditional independence properties of the bbn, the sampling procedure can be simplified as follows.

$x_1 = u_1;$

$x_2 = F_{r_{21};x_1}^{-1}(u_2);$

$$x_3 = F^{-1}_{r_{31};x_1}(u_3);$$

$$x_4 = F^{-1}_{r_{43};x_3}(u_4);$$

$$x_5 = F^{-1}_{r_{54};x_4}\left(F^{-1}_{r_{52|4};F_{2|4}(x_2)}(u_5)\right).$$

Notice that the conditional distribution $F_{2|4}(x_2)$ is not given and must be found by calculating:

$$f(x_2, x_4) = \iint_{[0,1]^2} c_{12}(x_1, x_2)c_{13}(x_2, x_3)c_{34}(x_3, x_4)\, dx_1\, dx_3,$$

$$F_{2|4}(x_2) = \int_0^{x_2} f(v, x_4)\, dv.$$

This is caused by the different order of variables in $\mathcal{D}^4$ and $\mathcal{D}^5$.

For bbn's of moderate size, we may use the technique of Section 4.4.5 to circumvent these problems. We realize the rank correlation vine via a normal vine and transform the normal vine to a normal vine with re-ordered variables by computing the re-ordered partial correlations. Hence, using normal vine in Example 6.3, we may calculate correlations in $\mathcal{D}^5 = D(5, 4, 2, 3, 1)$ by reordering the vine $\mathcal{D}^4$ and calculating the required re-ordered partial correlations. Notice that no integration is now required.

Updating with NETICA   Constructing and quantifying a large model takes a long time. One can wait for a few months to have a good model quantified with traceable and defensible methods. One would even be prepared to run such a model for a few days. However, if the client wants to consider new polices, check 'what would happen if...' he/she would not be prepared to wait for days or even hours for the results. In these cases, the advantages of fast-updating algorithms for discrete *bbn's* are decisive. We would like to combine the reduced assessment burden and modelling flexibility of the continuous bbn with the fast-updating algorithms of discrete bbn's. This can be done using vine sampling with existing discrete bbn software:

- Quantify nodes of a *bbn* as continuous univariate (not necessarily uniform) random variables and arcs as parent–child rank correlations.

- Sample this structure, creating a large sample file.

- Use this sample file to build conditional probability tables, not too crude, for a discretized version of the continuous *bbn*. (This can be done automatically in the *bbn* program NETICA using option 'Relation/Incorporate Case File'.)

- Perform fast-updating for the discretized *bbn*.

Figure 6.13 shows the crew alertness model in NETICA. Each variable was discretized to 10 states. A specially prepared case file with 800,000 samples, obtained

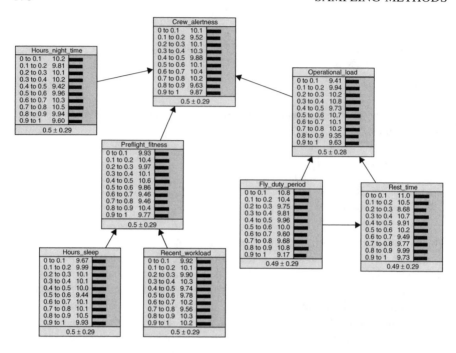

Figure 6.13 The crew alertness model in NETICA (Uniform (0, 1) distribution discretized to 10 states).

using the sampling procedure presented in the preceding text, was used to generate the probability tables for the discretized *bbn*.

The quantification of the discretized *bbn* requires more than 12,000 probabilities. The conditional probability table of Crew alertness (8) given Operational load (6), Preflight fitness (3) and Hours of night-time flight (7) alone requires 10,000 probabilities. The quantification with continuous nodes requires only eight algebraically independent conditional rank correlations and eight continuous invertible distribution functions.

Figure 6.14 compares the updating results in three cases:

1. Hours of sleep (2) is equal to 0.25 and Flight duty period (4) is equal to 0.8 (vines conditionalized on point values);

2. Hours of sleep from [0.2, 0.3] and Flight duty period from [0.8, 0.9] (vines conditionalized on intervals);

3. conditionalizing in NETICA, where the distributions were conditionalized on intervals as in the case 2 in the preceding text.

The entire procedure involves two approximations: the discretization and the conditionalization on discretized values, which is actually conditionalization on

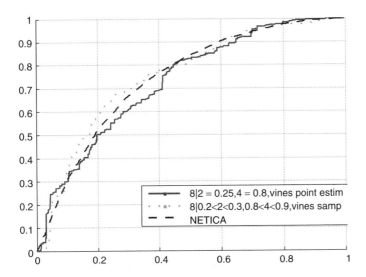

Figure 6.14 Comparison of updating results in vines and NETICA (conditional probability tables in NETICA obtained using 800,000 samples).

the interval. Thus, conditionalizing on the discretized value 0.25 corresponds to conditionalizing on the interval [0.2, 0.3].

In Figure 6.14 the probability tables were built with 800,000 samples, and the conditional distribution from NETICA agrees admirably with that obtained from vines with sampling in case 2. Figure 6.15 is similar to Figure 6.14 except that the

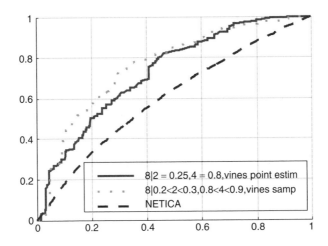

Figure 6.15 Comparison of updating results in vines and NETICA (conditional probability tables in NETICA obtained using 10,000 samples).

conditional probability tables in NETICA were built using only 10,000 samples. The agreement is considerably degraded, and the reason is not hard to find. There are 1000 different input vectors for node 8, each requiring 10 probabilities for the distribution of 8 given the input. With 10,000 samples, we expect each of the 1000 different inputs to occur 10 times, and we expect a distribution on 10 outcomes to be very poorly estimated with 10 samples. If our *bbn* contains nodes with large numbers of inputs, then we must make very large sample files. The good news is that this only needs to be done once.

6.5 Conclusions

The subjects of stipulating, sampling and communicating high-dimensional dependence information are intimately related. One way to stipulate a distribution is to stipulate all its moments and to stipulate conditions on the distribution's support, which entail that the moments uniquely determine the distribution. This might be useful for some purposes, but it would not tell us how to sample this distribution. Giving a sampling algorithm is another way to stipulate a distribution. Obviously, when dealing with complex objects, it is useful to view it from different perspectives.

Trees, vines and bbn's, in combination with copulae, provide practical and flexible ways of stipulating high-dimensional distributions.

The emphasis here has been on sampling distributions specified in this way. It is worth mentioning that these techniques can also be used for prediction. If we can infer a graphical model from a data set, and if we subsequently observe some values of variables sampled from the same distribution, we may use the conditional distributions, given the observed values, to predict unobserved values.

6.6 Unicorn projects

Project 6.1 Stratified sampling

This example illustrates stratified sampling with conditionalization and user-defined quantiles. We must estimate the probability of flooding in a region protected by five dike sections. Each section has a design height of 6 m, and flooding occurs if the effective water height exceeds 6 m at one or more sections. Effective water height is the sum of contributions from the North Sea, wave action and the Rhine discharge. Sophisticated models compute these effects and translate them into effective water heights; we will consider them simply as random variables affecting all sections at once during the yearly maximum water level in the North Sea above the baseline level. In addition, each section may sink, and the amount of sinkage at different sections are independent.

Create a case with variables

- *h (dike height); constant 6;*

- *ns (yearly maximum north sea) lognormal; median 2, error factor 1.5;*

- *w (wave) lognormal; median 0.2, error factor 2;*

- *r (Rhine); lognormal; median 0.1, error factor 2;*

- *sink1,...,sink5 (sinkages at sections 1,...,5); lognormal; median 0.1, error factor 2.*

Create user-defined functions (UDFs)

- *dike1: h-ns-w-r-sink1;*

- *dike2: h-ns-w-r-sink2;*

- *dike3: h-ns-w-r-sink3;*

- *dike4: h-ns-w-r-sink4;*

- *dike5: h-ns-w-r-sink5;*

- *flood: min{dike1, dike2, dike3, dike4, dike5};*

- *nrflrs: $prev(0) + i1\{\ll, flood, 0\}$;*

- *condition: $i1\{3, ns, \gg\}$.*

This last UDF contains UNICORN's generalized indicator function. The outer numbers (3 and $\gg$) are (inclusive) bounds. The function returns the integer value '1' if the middle variable is between the bounds ($\gg$ denotes infinity). The bounds may themselves be random variables. If we had used $ik\{L, v1, \ldots, vn, H\}$, this would return '1' if at least k of $v1, \ldots, vn$ are between L and H.

In UNICORN, conditional sampling is accomplished by having a UDF named condition. A sample is retained only if 'condition' takes the value 1 and is rejected otherwise.

Beware*: 'condition' must be integer valued. Real numbers on a computer sometimes behave unpredictably many places after the decimal point.*

Flooding occurs if the variable flood is less than or equal to zero. We will need 1,000,000 samples to reliably estimate the probability of flooding. This will take some time. Drawing and processing 1,000,000 samples takes about 20 minutes on an average computer. We can reduce runtime by observing that flooding is extremely unlikely if the water height in the North Sea is less than 3 m. We therefore condition on ns $\geq$ 3 with the UDF, where '$\gg$' denotes infinity:

$$condition : i1\{3, ns, \gg\}.$$

Only samples in which ns > 3 are retained for further computing and processing. With conditionalization, the simulation takes 3 minutes.

The variable nrflrs counts the number of failures, starting with 0; it adds 1 if $flood \leq 0$.

Before running the simulation with 10,000 runs (1,000,000 samples), check the box User Defined Quantiles. After hitting Run, a dialogue box appears, in which the user can select one or more output variables and stipulate the extreme quantiles to be computed. For the variable flood, choose Q1 = 0.0001, Q2 = 0.0002, Q3 = 0.0003, Q4 = 0.0004 and Q5 = 0.0005. These quantiles can then be included in the report.

From the report, we see that 50,534 samples are selected from the 1,000,000. In these samples, the maximal value of nrflrs is 8. The percentage of failures in the retained sample is thus 0.000158. We see from the user-defined quantiles that flood ≤ 0 is between 0.0001 and 0.0002. With respect to the unconditional sample, the probability of flooding is 8E−6.

Project 6.2 Powergrid: vines in trees

This is a highly stylized version of a planning problem inspired by Papefthymiou et al. (2004–2005). It is used to demonstrate the technique of building complex high-dimensional distributions by treating a vine as a variable in a tree. When a vine is attached to a tree, the first node is rank correlated with the other nodes in the tree. This technique is useful in building large dependence structures with localized complexity, as illustrated in this project.

The design of electric power-generating systems must increasingly cope with the power generated by 'non-dispatchable technologies' (i.e. renewables). These complicate the design problem in two ways: (1) their contribution to the grid is stochastic, depending on what power engineers like to call the 'prime mover' (i.e. wind, insolation, rainfall) and (2) their uncertainties are coupled because of (a) coupling to the prime mover and (b) similarities of design and operation. We are interested in being able to generate more power than the load requires (we consider only load not covered by base – load units and 'dispatchable generators').

Our grid consists of three Metropolitan Areas – 1, 2 and 3. Each area has four wind parks of either the old or the new design. The windspeed distributions, as measured at one meteo tower in each area, are:

- *W1: Weibull shape 0.85, scale 25;*

- *W2: Weibull shape 1.2, scale 25;*

- *W3: Weibull shape 1.2 scale 20.*

The power output of a windpark is basically proportional to windspeed if the latter does not exceed a maximal capacity value. At windspeeds above this maximal capacity value, the windparks must shut down for safety reasons, and no power is generated. We arrange that power output is equal to windspeed for values below maximal capacity. At the old units, the maximal capacity is 50 km/hour; for the new units this is 60 km/hour.

The actual wind at each park is strongly correlated with the wind at the area's meteo tower Wi, and the area windspeeds are correlated to the prime mover. The total generable power at windpark j in area i (not taking maximum capacity into account) has the same distribution as Wi and is rank correlated with Wi. Units with the same design will have an additional rank correlation given Wi.

The total generable power from the old park j in area i is denoted by Nij; new parks are notated Nnij. At area 1, we have the variables N11, N12, Nn13 and Nn14, each with the distribution of W1. Similar variables characterize area 2 and 3.

We model the situation at area 1 with a C-vine, with W1 as root. All parks are rank correlated with W1 at 0.9. Given W1, N12 and N13 are rank correlated at 0.5. Similarly, given W1, N11, N12 and the new parks Nn13 and Nn14 are rank correlated at 0.5. It is not difficult to show (see Exercise 6.1) that this is also the rank correlation of Nn13 and Nn14 given W1.

Load is identically distributed (Normal, mean 30 std 7) in all areas and is moderately correlated with Wi (0.7). Conditional on Wi, the load and the generation variables are independent.

Hence, the C-vine for area 1, named 'area1', will look like the one in Figure 6.16. Similar C-vines apply for areas 2 and 3.

The vines are connected via their root Wi to the prime mover. They are also connected to the load in area i. Indeed, when wind is high, temperatures are lower, people stay indoors and consume more energy. We capture all this in a tree with prime mover as root, as shown in Figure 6.17.

Notice that the vines are treated just as variables in the PM dependence tree.

We are interested in the reserve, that is, the difference between the total generated power and the total load. We distinguish the old units with maximal capacity of 50 km/hour and new units with maximal capacity of 60 km/hour. Indicator functions truncate the generation variables at maximal capacity. Hence, we enter the following three UDFs:

```
W1
 ├ (0.90) N11    N11|W1
 ├ (0.90) N12    ├ (0.50) N12    N12|N11,W1
 ├ (0.90) Nn13   ├ (0.00) Nn13   ├ (0.00) Nn13   Nn13|N12,N11,W1
 └ (0.90) Nn14   └ (0.00) Nn14   └ (0.00) Nn14   └ (0.50) Nn14
```

Figure 6.16   C-vine for area 1.

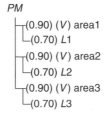

Figure 6.17   Tree with prime mover.

oldunits:
i1{0,N11,50}*N11+i1{0,N12,50}*N12+i1{0,N21,50}*N21
    +i1{0,N22,50}*N22+i1{0,N31,50}*N31;
newunits:
i1{0,Nn13,60}*Nn13+i1{0,Nn14,60}*Nn14+i1{0,Nn23,60}*Nn23
    +i1{0,Nn23,60}*Nn24+i1{0,Nn33,60}*Nn33+i1{0,Nn34,60}*Nn34;
reserve:
oldunits+newunits-L1-L2-L3.

If we run this case with 5000 samples, we find that the reserve is negative with probability about 20% (look at percentiles of reserve). If we had ignored the global correlation to the prime mover, this would be about 7%.

Cobweb plots tell the full story. We select only the UDFs, the area variables and the prime mover and arrange the variables as shown in the unconditional cobweb plot in Plate 4.

Observe that the loads are normal and thus concentrated in the middle of their range, whereas the Wi are Weibull. Notice that W1 has a much higher maximal value than W2 or W3.

Now, when we conditionalize on the samples for which reserve is negative, we see in Plate 5 that there are two scenarios. Negative reserve can arise with very high wind or with very low wind. Note that in these cases load and area wind tend to be anti-correlated.

6.7   Exercise

Ex 6.1 *Consider a C-vine, C(1,2,3,4), with Frank's or diagonal band copula with $r_{12} = r_{13} = r_{14} = r_{34|12} = 0.9$ and all other conditional correlations equal to 0. Use the density formulation of the vine distribution to show that the density satisfies $f(v3, v4|v1, v2) = f(v3, v4)$.*

Try this in UNICORN and look at the scatter plots. The plots for (v1, v2) and (v3, v4) are the same. This is NOT true for the elliptical copula; the scatter plot for (v3, v4) is not elliptical. Why not?

Ex 6.2 *If you have access to a mathematical package (MATLAB, MAPLE), write the distribution function for x_4 conditional on x_1, x_2, x_3 for the C-vine in Figure 6.7 and for the D-vine in Figure 6.9 using Frank's copula with parameter $\theta = \frac{1}{2}$ for all nodes. Plot the conditional distribution function for $x_1 = x_2 = x_3 = 0.5$ and for $x_1 = x_2 = x_3 = 0.95$.*

Ex 6.3 *Show that every regular vine on three variables is a D-vine and also a C-vine. Show that every regular vine on four variables is either a C- or a D-vine.*

7

Visualization

7.1 Introduction

'A picture is worth a thousand words'[1] is true in many areas of statistical analysis and in modelling. Graphs contribute to the formulation and construction of conceptual models and facilitate the examination of underlying assumptions. Data visualization is an area of considerable scientific challenges, particularly, when faced with high-dimensional problems characteristic of uncertainty analysis.

A literature search reveals very little in the way of theoretical development for graphical methods in sensitivity and uncertainty analysis apart from certain reference books, for example, Cleveland (1993). Perhaps it is the nature of these methods that one simply 'sees' what is going on. Cleveland studies visualizing univariate, bivariate and general multivariate data. Our focus, however, is not visualizing data in general, but rather visualization to support uncertainty and sensitivity analysis. The main sources for graphical methods are software packages. Standard graphical tools such as scatter plots and histograms are available in almost all packages, but the more challenging multidimensional visualization tools are less widely available.

In this chapter, we first choose a simple problem for illustrating generic graphical techniques. The virtue of a simple problem is that we can easily understand what is going on, and therefore we can appreciate what the various techniques are and are not revealing. Then in subsequent sections, we will discuss more generic techniques that can be used for more complex problems. The generic techniques discussed here are tornado graphs, radar plots, matrix and overlay scatter plots and cobweb plots. Where appropriate, the software producing the plots and/or analysis will be indicated.

In the last sections of this chapter, having grasped what graphical techniques can do, it is instructive to apply them to real problems where we do not immediately

---

[1] This chapter was adapted from Cooke and van Noortwijk (2000).

Uncertainty Analysis with High Dimensional Dependence Modelling D. Kurowicka and R. Cooke
© 2006 John Wiley & Sons, Ltd

'see' what is going on. First, a problem concerning dike ring reliability is used to illustrate the use of cobweb plots in identifying local probabilistically important parameters. The second problem from internal dosimetry illustrates the use of radar plots to scan a very large set of parameters for important contributors to the overall uncertainty. The detailed discussion of these problems can be found in the original reports; in this chapter, we concentrate rather on the usefulness of the different graphical techniques in providing insights into the model behaviour.

7.2   A simple problem

The following problem serves to illustrate the generic techniques (see Project 7.2).

Suppose, we are interested in how long a car will start after the headlights have stopped working. We build a simple reliability model of the car consisting of three components:

- the battery (bat),
- the headlight lampbulb (bulb),
- the starter motor (strtr).

The headlight fails when either the battery or the bulb fail. The car's ignition fails when either the battery or the starter motor fail. Thus considering bat, bulb and strtr as life variables:

headlite = min(bat, bulb),

ignitn = min(bat, strtr).

In other words, the headlight lives until either the battery or the bulb expires, and similarly, the ignition lives until either the battery or the starter expires.

The variable of interest is then

ign-head = ignitn − headlite.

Note that this quantity may be either positive or negative, and that it equals zero whenever the battery fails before the bulb and before the starter motor.

We shall assume that bat, bulb and strtr are independent exponential variables with unit expected lifetime. The question is, which variable is most important to the quantity of interest, ign-head? We shall now present some more specialized graphical tools, using this simple example to illustrate their construction and interpretation.

7.3   Tornado graphs

Tornado graphs are simply bar graphs where (in this case) the rank correlation between the factors and the model response is arranged vertically in order of

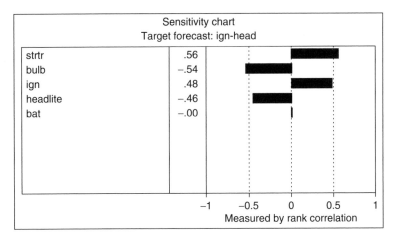

Figure 7.1   Tornado graph.

descending absolute value. The spreadsheet add-on Crystal Ball performs uncertainty analysis and gives graphic output for sensitivity analysis in the form of tornado graphs (without using this designation). After selecting a 'target forecast variable', in this case ign-head, Crystal Ball shows the rank correlations of other input variables and other forecast variables as in Figure 7.1.

The values, in this case, the rank correlation coefficients, are arranged in decreasing order of absolute value. Hence the variable strtr with rank correlation 0.56 is first, and bulb with rank correlation −0.54 is second, and so on. When influence on the target variable, ign-head, is interpreted as rank correlation, it is easy to pick out the most important variables from such graphs. Note that bat is shown as having rank correlation 0 with the target variable ign-head. This would suggest that bat was completely unimportant for ign-head. Obviously, any of the other global sensitivity measures mentioned in Chapter 8 could be used instead of rank correlation (see also Kleijnen and Helton (1999)).

7.4   Radar graphs

Continuing with this simple example, another graphical tool, radar graphs is introduced. A radar graph provides another way of showing the information in Figure 7.1. Figure 7.2 shows a radar graph made in EXCEL (1995) by entering the rank correlations from Figure 7.1.

Each variable corresponds to a ray in the graph. The axis on each ray spans the absolute value of the correlation values (−0.6, 0.6), and the value of the rank correlation for each factor is then plotted on the corresponding ray and connected. The variable with the highest rank correlation is plotted furthest from the midpoint, and the variable with the lowest rank correlation is plotted closest to the midpoint.

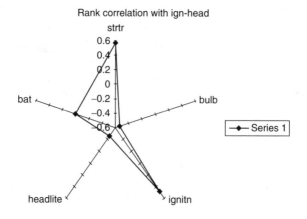

Figure 7.2 Radar plot.

Thus, bulb and headlite are plotted closest to the midpoint and strtr and ignitn plotted furthest. The discussion of the internal dosimetry problem in Section 7.8 shows that the real value of radar plots lies in their ability to handle a large number of variables.

7.5 Scatter plots, matrix and overlay scatter plots

Scatter plots simply plot bi-variate data in $\mathbb{R}^2$ with the variables as axes. Figure 7.3 shows a scatter plot for Bat and Ign-head, from which we can easily see why the

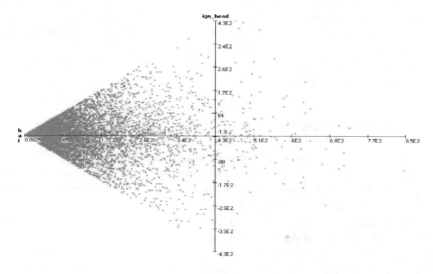

Figure 7.3 Scatter plot for Bat and Ign-Head.

correlation between these two is zero; the plot is symmetric around Ign-head $= 0$. (The transformation to ranks would not disturb this symmetry, hence the rank correlation is also zero.) On the other hand, it is immediately apparent that these variables are not independent: the conditional distribution for Ign-head given Bat depends on the value of Bat.

The simple scatter plot has also been the subject of some development for use in multivariate cases. Such extensions include the matrix scatter plot and the idea of overlaying multiple scatter plots on the same scale. Many statistical packages support these (and other) variations of scatter plots. For example, SPSS (1997) provides a matrix scatter plot facility. Simulation data produced by UNICORN for 1000 samples has been read into SPSS to produce the matrix scatter plot shown in Figure 7.4. We see pairwise scatter plots of the variables in our problem. The first row, for example, shows the scatter plots of ign-head and, respectively, bat, bulb, strtr, ignitn and headlite. The matrix scatter plot is symmetrical and we need only focus on the top right plots, since they are replicated in the lower left.

Figure 7.4 summarizes the relationships between each pair of variables in the problem (15 such individual plots). Let (a,b) denote the scatter plot in row a and column b, thus (1, 2) denotes the second plot in the first row with ign-head on the vertical axis and bat on the horizontal axis; as in Figure 7.3. Note that (2, 1) shows the same scatter plot, but with bat on the vertical and ign-head on the horizontal axes.

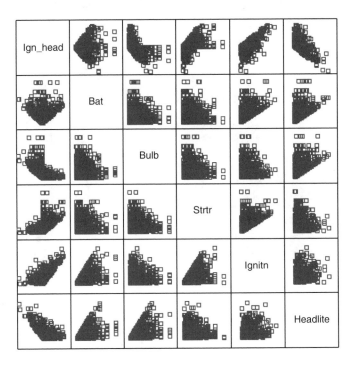

Figure 7.4 Multiple scatter plots.

Figure 7.4 shows that the value of bat can say a great deal about ign-head. Thus, if bat assumes its lowest possible value, then the values of ign-head are severely constrained. This reflects the fact that if bat is smaller than bulb and strtr, then ignitn = headlite, and ign-head = 0. From (1, 3), we see that large values of bulb tend to associate with small values of ign-head. If the bulb is large, then the headlight may live longer than the ignition making ign-head negative. Similarly, (1, 4) shows that large values of strtr are associated with large values of ign-head. These latter facts are reflected in the rank correlations of Figure 7.1.

Inspite of the preceding remarks, the relation between rank correlations depicted in Figure 7.1 and the scatter plots of Figure 7.4 is not direct. Thus, bat and bulb are statistically independent, but if we look at (2, 3), we might infer that high values of bat tend to associate with low values of bulb. This however is an artifact of the simulation. There are very few, very high values of bat, and as these are independent of bulb, the corresponding values of bulb are not extreme. If we had a scatter plot of the rank of bat with the rank of bulb, then the points would be uniformly distributed on the unit square. This simple example again emphasizes the importance of graphically examining relationships and not simply relying on summary statistics such as the correlation coefficient.

Continuing with the developments for simple scatter plots, we consider the idea of overlaying separate plots (on the same scale) and using different symbols to convey additional information, see Figure 7.5 for an example:

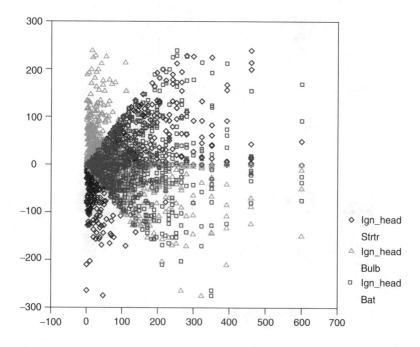

Figure 7.5 Overlay scatter plots.

Ign-head is depicted on the vertical axis, and the values for bat, bulb and strtr are shown as squares, triangles and diamonds respectively. Figure 7.5 is a superposition of plots (1, 2), (1, 3) and (1, 4) of Figure 7.4. However, Figure 7.5 is more than just a superposition. Inspecting Figure 7.5 closely, we see that there are always a square, triangle and diamond corresponding to each realized value on the vertical axis. Thus, at the very top there is a triangle at ign-head = 238 and bulb slightly greater than zero. There is a square and diamond also corresponding to ign-head = 238. These three points correspond to the same sample. Indeed, ign-head attains its maximum value when strtr is very large and bulb is very small. If a value of ign-head is realized twice, then there will be two triples of squares-triangle-diamonds on a horizontal line corresponding to this value, and it is impossible to resolve the two separate data points. For ign-head = 0, there are about 300 realizations.

The joint distribution underlying Figure 7.4 is six dimensional. Figure 7.4 does not show this distribution, but rather shows 30 two-dimensional (marginal) projections from this distribution. Figure 7.5 shows more than a collection of two-dimensional projections, as we can sometimes resolve the individual data points for bat bulb, strtr and ign-head, but it does not enable us to resolve all data points. The full distribution can however be shown in cobweb plots.

7.6   Cobweb plots

The uncertainty analysis program UNICORN contains a graphical feature that enables interactive visualization of a moderately high-dimensional distribution. Our sample problem contains six random variables. Suppose, we represent the possible values of these variables as parallel vertical lines[2]. One sample from this distribution is a six-dimensional vector. We mark the six values on the six vertical lines and connect the marks by a jagged line. If we repeat this 200 times, we get Figure 7.6 below:

The number of samples (200) is chosen for black-white reproduction. On screen, the lines may be colour coded according to the leftmost variable: for example, the bottom 25% of the axis is yellow, the next 25% is green, then blue, then red. This allows the eye to resolve a greater number of samples and greatly aids visual inspection. From the cobweb plot, we can recognize the exponential distributions of bat, bulb and strtr. Ignitn and headlite, being the minimum of independent exponentials, are also exponential. Ign-head has a more complicated distribution. The graphs at the top are the 'cross densities'; they show the density of line crossings midway between the vertical axes. The role of these in depicting dependence becomes clear when we transform the six variables to ranks or percentiles, as in Figure 7.7.

A number of striking features emerge when we transform to the percentile cobweb plot. First of all, there is a curious hole in the distribution of ign-head.

---

[2]Wegman (1990) introduced parallel plots, of which cobweb plots are an independent implementation incorporating extended user interaction.

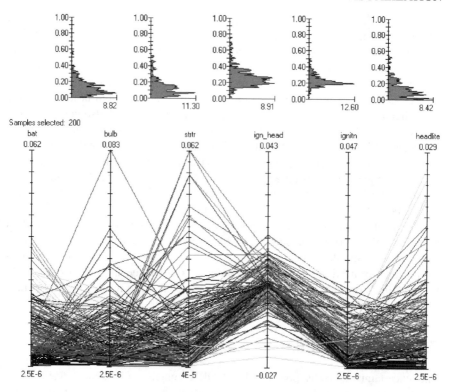

Figure 7.6 Cobweb plot, natural scale.

This is explained as follows. On one-third of the samples, bat is the minimum of bat, bulb, strtr. On these samples ignitn = headlite and ign-head = 0. Hence the distribution of ign-head has an atom at zero with weight 0.33. On one-third of the samples strtr is the minimum and on these samples ign-headlite is negative, and on the remaining third, bulb is the minimum and ign-head is positive. Hence, the atom at zero means that the percentiles 0.33 up to 0.66 are all equal to zero. The first positive number is the 67th percentile. If there is no atom, then the points at which the jagged lines intersect the vertical lines corresponding to the variables will be uniformly distributed.

Note the cross densities in Figure 7.7. One can show the following for two adjacent continuously distributed variables X and Y in a (unconditional) percentile cobweb plot (see exercise 7.2):

- If the rank correlation between X and $Y = 1$ then the cross density is uniform.

- If X and Y are independent (rank correlation 0), then the cross density is triangular.

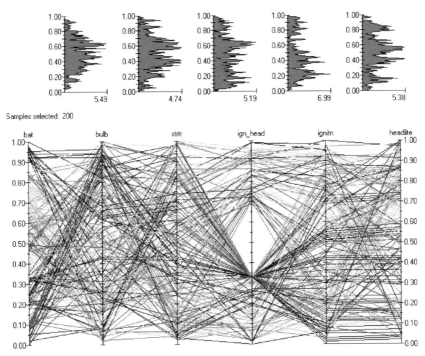

Figure 7.7 Cobweb plot, percentile scale.

- If the rank correlation between X and $Y = -1$, then the cross density is a spike in the middle.

Intermediate values of the rank correlation yield intermediate pictures. The cross density of ignitn and headlite is intermediate between uniform and triangular; and the rank correlation between these variables is 0.42.

Cobweb plots support interactive conditionalization; that is, the user can define regions on the various axes and select only those samples that intersect the chosen region. Figure 7.8 shows the result of conditionalizing on ign-head = 0.

Notice that if ign-head = 0, then bat is almost always the minimum of bat,bulb,strtr, and ignitn is almost always equal to headlite. This is reflected in the conditional rank correlation between ignitn and headlite almost equals to 1. We see that the conditional correlation as in Figure 7.8 can be very different from the unconditional correlation of Figure 7.7. From Figure 7.8, we also see that bat is almost always less than bulb and strtr.

Cobweb plots allow us to examine local sensitivity. Thus, we can say, suppose ign-head is very large, what values should the other variables take? The answer is obtained simply by conditionalizing on high values of ign-head. Figure 7.9 shows conditionalization on high values of ign-head, Figure 7.10 conditionalizes on low values (the number of unconditional samples has been increased to 1000).

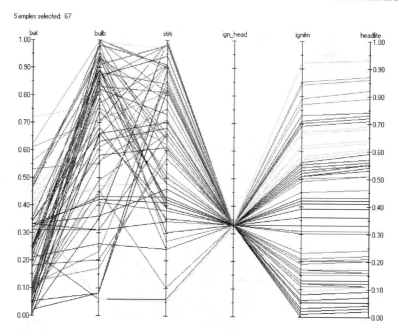

Figure 7.8 Conditional cobweb plot, ign-head = 0.

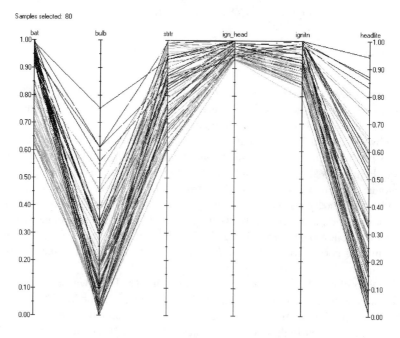

Figure 7.9 Conditional cobweb plot, ign-head high.

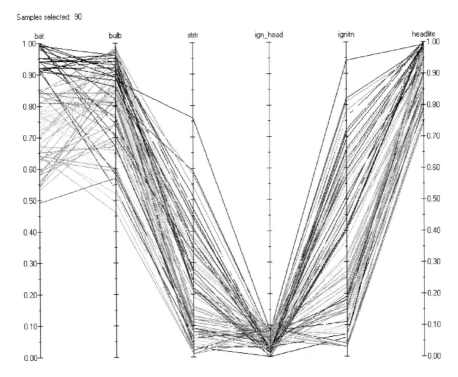

Figure 7.10 Conditional cobweb plot, ign-head low.

If ign-head is high, then bat, strtr and ignitn must be high, and bulb must be low. If ign-head is low, then bat, bulb and headlite must be high. Note that bat is high in both cases. Hence, we should conclude that bat is very important both for high values and for low values of ign-head. This is a different conclusion than we would have drawn if we considered only the rank correlations of Figures 7.1, 7.2.

These facts can also be readily understood from the formulae themselves. Of course, the methods come into their own in complex problems where we cannot see these relationships immediately from the formula. The graphical methods then draw our attention to patterns that we must then seek to understand. The following three sections illustrate graphical methods used in anger, that is, for real problems where our intuitive understanding of the many complex relationships is assisted by the graphical tools introduced earlier.

7.7 Cobweb plots local sensitivity: dike ring reliability

In this section, we discuss a recent application in which graphical methods were used to identify important parameters in a complex uncertainty analysis. This application concerns the uncertainty in the dike ring reliability and was discussed in Cooke and Noortwijk (1998). The dike ring in question is built up of more than 30 dike sections. The reliability of each dike section i is modelled as:

$$Reliability_i = Strength_i - Load_i.$$

The reliability of the dike ring is:

$$relia = Reliability\, ring = min\{Reliability_i\}.$$

The dike ring fails when relia $\leq$ 0. Plate 6 shows the unconditional[3] percentile cobweb plot for relia and 10 explanatory variables. From left to right the variables are: roughness ('rough'), storm length ('storm'), model factors for load, strength, significant wave period, significant wave height and local water level ('mload', 'mstrn', 'mwvpr', 'mwvht' and 'mlwat', respectively), wind ('wind'), North Sea ('nsea') and Rhine discharge ('rhine'). For a further discussion of these variables and their role in determining reliability, we refer to Cooke and Noortwijk (1998).

The unconditional cobweb plot is not terribly revealing. Indeed, all these variables have product moment correlation with 'relia' between 0.05 and −0.05. Note from the cross densities that nsea and rhine are negatively correlated, whereas nsea and wind are positively correlated. The conditional cobweb plot Figures 7.11, 7.12 and 7.13 show the results of conditionalizing, respectively, on the upper 5% of relia, the 6th–8th percentiles, and the bottom 3%. Failure occurs at the 2 percentiles, hence Figure 7.13 shows conditionalization on 'dangerous' values of the dike ring reliability.

We make the following observations:

- Very large values of relia are associated with very high values of mstrn and low values of storm; other variables are hardly affected by this conditionalization, and their conditional distributions are practically equal to their unconditional distributions (i.e. uniformly distributed over percentiles).

- For values of relia between the 8th and 6th percentile, mstrn must be rather low, Storm must be high, and other variables are hardly affected by the conditionalization.

- For dangerous values of relia, the lowest 3%, nsea and wind must be high, other variables are unaffected by the conditionalization.

- For dangerous values of relia, the correlations wind-nsea and nsea-rhine are weaker than in the unconditional sample.

---

[3]The dikes in this study are between 5 and 6 m above 'standard' sea level; all these simulations are all based on a local water level above 3 m.

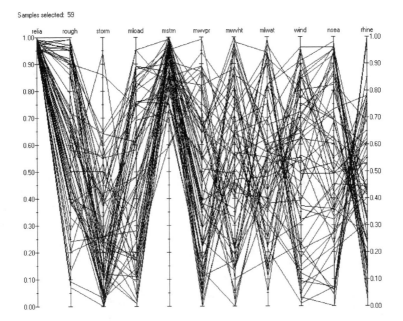

Figure 7.11 Cobweb for dike ring, relia high.

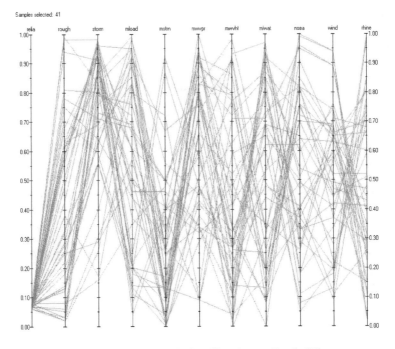

Figure 7.12 Cobweb for dike ring, relia 6–8%.

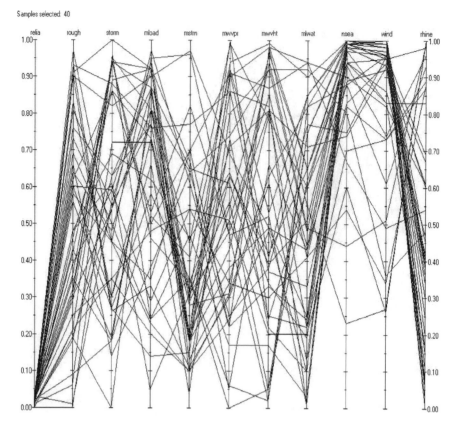

Figure 7.13 Cobweb for dike ring, relia 0–3%.

We interpret 'unaffected by the conditionalization' in this context as unimportant for the values of 'relia' on which we conditionalize, for example, knowing that reliability is very high, we should expect that storm length is low and the model factor for strength is very high. With regard to other variables, knowing that relia is very high does not affect our knowledge about what values these variables might take.

Something else in the preceding figures illustrates one of the more curious features of probabilistic reasoning, called *probabilistic dissonance*. The variable *relia* is monotonically decreasing in *storm*; in any given sample, if we would increase the storm length, the reliability must go down. Now compare Figures 7.12, 7.13. *Stochastically, storm* is greater in Figure 7.12 than in 7.13, even though reliability is lower in the latter figure than in the former. *relia* is functionally decreasing and stochastically increasing in the variable storm. How is that possible? A project at the end of this chapter discusses and shows, how this can arise.

This example shows that importance in the preceding sense is local. The variables that are important for high values of relia are not necessarily the same variables as those, which are important for very low values of relia.

Intuitively, when we conditionalize an output variable on a given range of values, those input variables which are important for the given range are those whose conditional distributions differ most markedly from their unconditional distributions. One convenient measure of this is the derivative of the conditional expectation of the input variable (Cooke and Noortwijk (1998) see also Chapter 8). More precisely, the local probabilistic sensitivity measure (LPSM) of variable X for model G when G takes value g is the rate of change of the expectation of X conditional on $G = g$. In the special case that G is a linear combination of independent normals, $LPSM(X, G)$ is just the product moment correlation between X and G (i.e. it does not depend on g). If $LPSM(X, G) = 0$, then the conditional expectation of X given $G = g$ is not changing in the neighbourhood of g, which is taken to mean that X is probabilistically not important for $G = g$. Conversely large absolute values of $LPSM(X, G)$ suggest that X is important for $G = g$. These notions are applied in the following example.

7.8   Radar plots for importance; internal dosimetry

An extended joint study of the European Union and the US Nuclear Regulatory Commission quantified the uncertainty for accident consequence models for nuclear power plants, based on structured expert judgment (Goossens et al. (1997)). The number of uncertain variables is in the order of 500. Not all of these can be considered in the Monte Carlo uncertainty analysis. In this study there are a large number of output variables. In the example discussed here, there are six output variables corresponding to collective dose to six important organs. Moreover, we are not interested in all values of these collective doses, rather, we are interested in those variables that are important for high values of collective dose, for some organ. Hence, the $LPSM(X, cd_i)$ (see Chapter 8) is applied to measure the sensitivity of input variable X for high values of collective dose to organ i. There are 161 uncertain variables, which might in principle be important for high collective dose.

With this number of input variables and output variables, and given current screen sizes, cobweb plots are not useful. An EXCEL bar chart in Figure 7.14 shows $LPSM(X, cd_i)$, for the first 30 variables (the full set of variables requires a 7-page bar chart).

This provides a good way of picking out the important variables. Bars extending beyond, say, -0.2 or beyond 0.2 indicate that the corresponding variable is important for high values of a collective dose to some organ. Of course, the entire bar chart is too large for presentation here. Instead the radar charts may be used to put all the data on one page, as in Plate 7. Although compressed for the current page size, this figure enables us to compare all variables in one view. The 6 different output variables corresponding to collective dose to six organs are colour

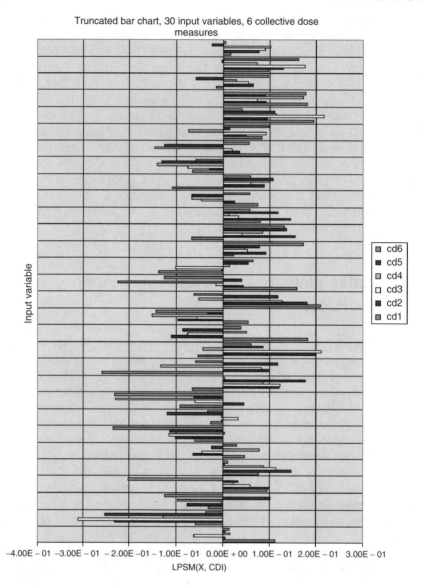

Figure 7.14 Bar chart for collective dose.

coded. Each ray corresponds to an input variable, and each point corresponds to the LPSM for that variable, with respect to one of the six output variables. The same information in Bar chart form requires 7 pages.

In principle, from such a plot, we are able to identify the most important variables.

7.9 Conclusions

This chapter has discussed a number of graphical methods; tornado graphs, radar graphs, scatter plots, matrix scatter plots, overlay scatter plots and cobweb plots. These tools have different strengths and weaknesses, and lend themselves to different problems. The results deduced from the graphical analysis should always be backed up by the more formal tools which are described in other chapters. Together, they provide powerful tools for exploring model behaviour.

For presenting a large number of functional relationships, the radar plots of Plate 7 are the most powerful technique. For studying arbitrary stochastic relations between two variables, scatter plots are the most familiar technique, and therefore require no explanations. However, extensions of scatter plots to multivariate data, such as overlay scatter plots and matrix scatter plots, do require explanation and do not always give the full picture. For multivariate problems with, say, less than 15 variables, cobweb plots give a full picture. Perhaps the most powerful feature of cobweb plots is user-interaction. The UNICORN implementation supports scale transformations, variable rearrangement and selection and conditionalization. Conditionalization can be used to discover relationships, which are not reflected in global sensitivity measures. Cobweb plots are more complex and less familiar than other techniques, and their use therefore requires explanation. In our experience, however, they are much more intuitive than multiple or overlay scatter plots, and the interactive possibilities are very helpful in this regard. For problems of very high dimensionality there are, as yet, no generic graphical tools. It is a question of 'flying by the seat of our pants', and finding subsets of variables of lower dimension, which can be studied with the preceding techniques.

Finally, we note that graphical tools facilitate communication with decision makers, users and stakeholders. All of the tools discussed here are used not only for analysis but also for communicating results.

7.10 Unicorn projects

Project 7.1 Probabilistic dissonance

One of the many subtleties of probabilistic reasoning involves the phenomenon of probabilistic dissonance: A function G of X and Y can be strictly increasing in each variable, yet for some values of G, it can be stochastically decreasing. X and Y are independent, Y is uniform $[0,1]$ and X is uniform on $[0, 1] \cup [2, 3]$. $G(X, Y) = X + Y$. G is continuous and monotonically increasing in X and Y; the partial derivatives in each argument are unity, suggesting that a small increase in Y or X is associated with an equal increase in G. For $\delta > 0, \delta \rightarrow 0$ the conditional distribution of $Y|G = 2 - \delta$ becomes concentrated at 1; while the conditional distribution of $Y|G = 2 + \delta$ becomes concentrated at 0. Probabilistically, Y is decreasing in G at $G = 2$ in the sense that for all r

$$P(Y > r|G = 2 - \delta) \le (P(Y > r|G = 2 + \delta)).$$

(In prose, the probability that Y is large goes down as G increases.)

X is a mixture of a uniform variate on [0,1] and a uniform variate on [2,3]. If $v1, v2, v3$ are uniform [0,1], then X can be written as

$$v1 * i1\{0, v3, 0.5\} + (v2 + 2) * i1\{0.5, v3, 1\}.$$

Determine $E(Y|G(X,Y) \in (1.95, 2))$ and $E(X|G(X,Y) \in (2, 2.1))$ and explain the result.

Project 7.2 Strtlite

The ignition system in your car depends on both the battery and the starting motor. The headlights depend on both the battery and the lamp bulb. The lifetimes of the battery, the bulb and the starter motor are exponentially distributed with failure rate 0.01. Consider the following questions:

- *What is the dependence between the ignition and the headlights?*

- *What is the dependence between ignition, headlights and the battery?*

- *What is the dependence between the ignition and the length of time, which the engine keeps starting after the headlights have failed?*

Create three variables named

    BAT, BULB, STRTR

and assign them each an exponential distribution with failure rate 0.01.

Create a UDFs:
'headlite': $min\{BAT, BULB\}$,
'ignitn': $min\{BAT, STRTR\}$,
'ign_head': ignitn - headlite.

Save this as STRTLITE. The difference between rank and product moment correlation is small in this case and is not the issue. The point is whether the numbers in the correlation table reveal the significant features of the dependencies in this example. Notice that for 1000 samples:

- *the correlation between ign_head and BAT is 0.05, suggesting that Bat is not important for ign_head;*

- *ign_head and ignitn have correlation 0.61, suggesting a modest tendency to covary;*

- *ignitn and headlite both have correlation about 0.5 with BAT, again suggesting mild covariation; (see exercise)*

The correlations do not suggest any interesting structure. Now go to UNIGRAPH and make a cobweb plot with 200 samples for the following sequence of variables,

(IN THIS ORDER):

    bat; bulb; strtr; ign_head, ignitn; headlite.

    There is a hole in ign_head! Why? ign_head is the difference between ignitn and headlite. Each of these minimizes over bat. If bat is smaller than bulb and strtr, then both ignitn and headlite will in fact equal bat. ign_head is the difference ignitn − headlite, and when bat is the smallest, ign_headlite will be zero. This is most likely to happen when bat is very small. Note that this simple problem contains only continuous functions and continuous distributions, yet results in a discontinuous distribution. The hole in ign_head is caused by a discontinuity in the distribution of ign_head. ign_head has about 1/3 of its mass concentrated at zero. Zero corresponds to the 33rd percentile of its distribution. Since 33% of all samples are equal to the 33rd percentile, the next percentile is the 67th. If you switch to the natural scale, the discontinuity disappears. Note also the triangular cross density between bat and bulb; this is characteristic of independent variables (see Figure 7.7).

    Whenever we have an interesting point in a cobweb plot, the thing to do, is to look at just those lines passing through that point. Conditionalize this cobweb plot by using the mouse. Click just above the 33rd percentile of ign_head and just below this percentile. The region selected should be highlighted. Now double click. Only the lines passing through the highlighted region are shown (Figure 7.8).

    We see that on those samples passing through the atom at ign_head = 33%, the variables ignitn and headlite are almost perfectly correlated. This is because in almost all of these samples, both are equal to bat. Notice also that most of these lines go though the bottom part of the percentiles of bat. This accords with the explanation given in the preceding text. Unconditionalize on ign_head by right clicking and choosing 'remove filter'. Now conditionalize on the top and bottom 10% of ign_headlite (Figure 7.9, 7.10). We see the complex relations between ign_headlite and headlite. When ign_headlite is conditioned on low values, headlite is forced to be high. However, when ign_headlite conditioned on high values, headlite's distribution is not strongly affected.

7.11 Exercises

Ex 7.1 *Show that if $\rho_r(X, Y) = 1$, then when these variables are adjacent on a percentile cobweb plot, the cross density is uniform. Show that if $\rho_r(X, Y) = -1$, then when these variables are adjacent on a percentile cobweb plot, the cross density has unit mass at 0.5.*

Ex 7.2 *Show that if X and Y are independent and continuously distributed, then when these two variables are adjacent in a percentile cobweb plot, the cross density is triangular* (hint, show that this cross density is the density of $X + Y$).

Ex 7.3 *Prove that* $\rho(headlite, Bat) = 0.5$. *Hint, the problem is unchanged if you assume that the failure rate of Bat and Bulb = 1. Then headlite has the marginal distribution of an exponential variable with failure rate 2. Put X = headlite, Y = Bat, Z = Bulb. It suffices to show that*

$$\int_z \left(\int_{y \geq z} z \min\{y, z\} e^{-y} e^{-z} \, dy + \int_{y < z} z \min\{y, z\} e^{-y} e^{-z} \, dy \right) dz = 0.75.$$

8

Probabilistic Sensitivity Measures

8.1 Introduction

Sensitivity analysis is concerned with identifying 'important parameters'. This may be used either in a *pre-* or *post-analysis* mode. Before conducting an analysis, we may want to filter out unimportant parameters to reduce modelling effort. *Screening techniques* are designed for this purpose. After an analysis has been carried out, we want to identify important parameters to support subsequent decisions. Our emphasis in this chapter is on probabilistic sensitivity analysis; we explore techniques which utilize distributions over input parameters. These techniques may be either global or local. By their nature, screening techniques will not typically utilize distributional information, as this is not usually available prior to an analysis. We briefly introduce two popular screening techniques in the first section. Subsequent sections treat global and local probabilistic sensitivity analysis. For non-probabilistic sensitivity analysis see (Saltelli et al. (2000, 2004); Scott and Saltelli (1997)).

We let $G = G(X_1, \ldots, X_n)$ denote the quantity of interest, which is a function of random variables $X_1, \ldots, X_n$. If we knew the true values of the uncertain arguments of G, say $x_1, \ldots, x_n$, then we could express the sensitivity of G to input x_i, or to combinations of inputs, simply by computing partial derivatives. When we do not have this knowledge, we must search for other ways of representing sensitivity.

8.2 Screening techniques

8.2.1 Morris' method

This method was presented in Morris (1991). It is a randomized experimental plan which is global in the sense that it covers the entire space over which factors may vary. It allows us to determine which factors have negligible effects, linear or additive effects or non-linear or interaction effects. This design is one of the so-called one at a time (OAT) designs.

We first must scale all factors X_i, $i = 1, \ldots, n$ to take values in the interval [0,1]. Hence the region of interest for our experiment is the n-dimensional unit hypercube. The n-dimensional unit hypercube can be discretized to an n-dimensional p-level grid, where each element X_i can take values x_i from $\{0, 1/(p-1), 2/(p-1), \ldots, 1\}$. Let Δ be a predetermined multiple of $1/(p-1)$, then for all $\mathbf{x} = [x_1, \ldots, x_n]$ such that $x_i < 1 - \Delta$ we can define the *elementary effect* of the i-th factor as

$$d_i(\mathbf{x}) = \frac{G(x_1, \ldots, x_{i-1}, x_i + \Delta, x_{i+1}, \ldots, x_n) - G(\mathbf{x})}{\Delta}, \qquad (8.1)$$

where G denotes output. Notice that d_i is just an approximation to the partial derivative of G with respect to x_i at the point $\mathbf{x}$. Morris estimates the main effect of a factor by computing r elementary effects at different points $\{\mathbf{x}_1, \mathbf{x}_2, \ldots, \mathbf{x}_r\}$. We obtain r samples of i-th elementary effect, and we summarize these in terms of their mean μ_i and standard deviation σ_i. A large absolute mean indicates a factor with a strong influence on the output. A large standard deviation indicates either a factor interacting with other factors or a factor whose effect is non-linear. An illustration of the sampling strategy for 3-dimensional parameter space is shown in Figure 8.1.

We can see in (8.1) that each elementary effect requires the evaluation of G twice, hence the total computational effort to estimate all elementary effects with r samples is $2rn$. Morris proposed a design that requires only $(n+1)r$ evaluations of G. This more economical design uses the fact that some runs are used to calculate more then one elementary effect. Let us assume that p is even and $\Delta = p/[2(p-1)]$. We start by constructing an $(n+1) \times n$ matrix B of elements '0' and '1' with

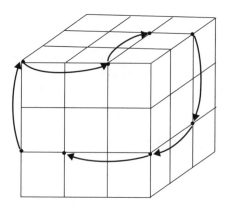

Figure 8.1 Two samples of elementary effects of each parameter in case of 3-dimensional parameter space with 4-level grid.

the property that for every column there are two rows of B that differ only in that column. The matrix B could be chosen for example, as follows:

$$B = \begin{bmatrix} 0 & 0 & 0 & \ldots & 0 \\ 1 & 0 & 0 & \ldots & 0 \\ 1 & 1 & 0 & \ldots & 0 \\ 1 & 1 & 1 & \ldots & 0 \\ \ldots & \ldots & \ldots & \ldots & \ldots \\ 1 & 1 & 1 & \ldots & 1 \end{bmatrix}$$

Now, taking a randomly chosen base value $\mathbf{x}^*$ of $\mathbf{x}$, we obtain a design matrix

$$B' = J_{n+1,1}\mathbf{x}^* + \Delta B,$$

where $J_{n+1,1}$ is a $(n+1) \times 1$ matrix of ones. Model evaluations corresponding to B' would provide n elementary effects in one run based on only $n+1$ model evaluations but they would not be randomly chosen. A randomized version of the design matrix is given by

$$B^* = (J_{n+1,1}\mathbf{x}^* + (\Delta/2)[(2B - J_{n+1,n})D^* + J_{n+1,n}])P^*$$

where

D^* is a n-dimensional diagonal matrix with either $+1$ or -1 with equal probability,

P^* is a $n \times n$ random permutation matrix in which each column contains one element equal to 1 and all the others equal to 0 and no two columns have 1's in the same position, where each of such matrices has equal probability of selection.

It is shown in Morris (1991) that the orientation matrix B^* provides samples of elementary effects for each factor that are randomly selected.

Example 8.1 *Let $n = 3$, $p = 4$ and $\Delta = 2/3$. In Figure 8.1, two random samples of the elementary effect of each parameter in this case are shown. Our three factors can take values from $\{0, 1/3, 2/3, 1\}$. We start with the matrix B*

$$B = \begin{bmatrix} 0 & 0 & 0 \\ 1 & 0 & 0 \\ 1 & 1 & 0 \\ 1 & 1 & 1 \end{bmatrix}$$

and the randomly chosen

$$x^* = (1/3, 1/3, 0), D^* = \begin{bmatrix} 1 & 0 & 0 \\ 0 & 1 & 0 \\ 0 & 0 & -1 \end{bmatrix}, P^* = \begin{bmatrix} 1 & 0 & 0 \\ 0 & 1 & 0 \\ 0 & 0 & 1 \end{bmatrix}.$$

Then the orientation matrix is given by

$$B^* = \begin{bmatrix} 1/3 & 1/3 & 2/3 \\ 1 & 1/3 & 2/3 \\ 1 & 1 & 2/3 \\ 1 & 1 & 0 \end{bmatrix}.$$

The matrix B^* gives one sample for each elementary effect. Choosing randomly other $r - 1$ orientation matrices, we can get r independent samples for each elementary effect.

In Figures 8.2 and 8.3 two graphical representations of the results of the Morris' method are given. We considered two models with four variables X_1, X_2, X_3, X_4 taking values in the interval $[0,1]$
- First model: $G_1 = 2X_1 + \sin(X_2) + 3X_3X_4$,
- Second model: $G_2 = X_1^{10} - X_2^2 + X_3X_4$.
In both cases $p = 4$, $\Delta = 2/3$ and $r = 9$. For each factor we plot, the estimated elementary effect's mean against its standard deviation. Two lines are also graphed that relate to the estimated mean $\bar{d}_i$ and standard deviation S_i in the following way

$$\bar{d}_i = \pm 2SEM_i = \pm 2\frac{S_i}{\sqrt{r}},$$

where SEM_i is a standard error of the mean. If the coordinates $(\bar{d}_i, S_i)$ of i-th effect lie outside of the wedge formed by these two lines, this may suggest significant evidence that the expectation of the elementary effect is not zero.

For the first model, Figure 8.2, $(\bar{d}_1, S_1) = (2, 0)$ clearly indicates that G_1 is linear with respect to X_1. The output G_1 is not linear with respect to X_2 but the much shorter distance of the point $(\bar{d}_2, S_2)$ from the origin indicates less importance of X_2. $(\bar{d}_3, S_4)$ and $(\bar{d}_4, S_4)$ lie within the wedge. Their standard deviations are large, which indicates that X_3 and X_4 are involved in interactions or their effects are highly non-linear.

For the second model, Figure 8.3 shows that the importance of input X_1 is not negligible, X_2 is quite important but not linear and X_3, X_4 have non-linear effects or involve interactions.

The Morris' method allows us to rank inputs with respect to their importance. This can be done by introducing a measure of importance, for example, as distance of $(\bar{d}_i, S_i)$ from the origin. It is one of many screening techniques, for the review and references consult Saltelli et al. (2000).

8.2.2   Design of experiments

Design of experiments was first introduced by Fisher (1935). This technique allows us to choose input variables (*factors*) that mostly effect output by providing a pattern of factor combinations (*design points*) that give the most information about the input–output relationship. We explain this technique on an example with three

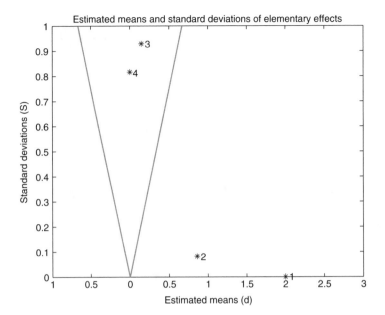

Figure 8.2 The graphical representation of the results of Morris method for the model $G_1 = 2X_1 + \sin(X_2) + 3X_3X_4$, ($n = 4$, $p = 4$, $\Delta = 2/3$ and $r = 9$).

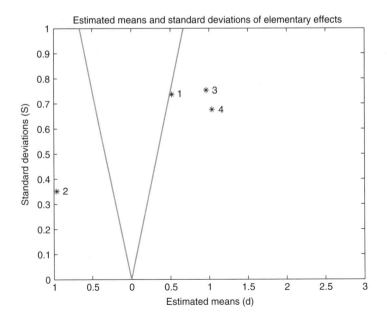

Figure 8.3 The graphical representation of the results of Morris method for the model $G_2 = X_1^{10} - X_2^2 + X_3X_4$ ($n = 4$, $p = 4$, $\Delta = 2/3$ and $r = 9$).

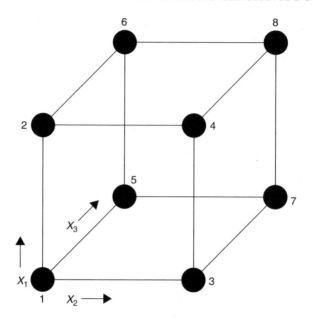

Figure 8.4 A 2^3 full factorial design.

factors X_1, X_2, X_3 that can take two possible values (*levels*) $(-1, 1)$ (low and high). This design is called two-level full factorial design or 2^3 design.

Two-level full factorial design - 2^3 design.   The 2^3 design requires 8 runs. Graphically, this can be represented by the cube in Figure 8.4. The arrows show the directions of increase of the factors. The numbers assigned to the corners correspond to the run number in the 'standard order', shown also in Table 8.1. This design can be presented in the following tabular form called *design table* or *design matrix*.

Run	X_1	X_2	X_3
1	−1	−1	−1
2	1	−1	−1
3	−1	1	−1
4	1	1	−1
5	−1	−1	1
6	1	−1	1
7	−1	1	1
8	1	1	1

Table 8.1 A 2^3 full factorial design table.

I	X_1	X_2	X_3	X_1X_2	X_1X_3	X_2X_3	$X_1X_2X_3$	G
1	−1	−1	−1	1	1	1	−1	g_1
1	1	−1	−1	−1	−1	1	1	g_2
1	−1	1	−1	−1	1	−1	1	g_3
1	1	1	−1	1	−1	−1	−1	g_4
1	−1	−1	1	1	−1	−1	1	g_5
1	1	−1	1	−1	1	−1	−1	g_6
1	−1	1	1	−1	−1	1	−1	g_7
1	1	1	1	1	1	1	1	g_8

Table 8.2   Analysis matrix for a 2^3 experiment.

The 8×3 matrix in Table 8.1 can be extended by columns that represent the interactions between variables, which are created by multiplying respective columns. This extended matrix is called the *model matrix* or *analysis matrix*. In the Table 8.2, we see the analysis matrix extended to include a column of ones (I) and a column of outcomes (G).

Notice that the analysis matrix has columns that are all pairwise orthogonal and sum to 0. Let us denote the analysis matrix by $A = [A_{ij}]$. To estimate main and interaction effects of the factors, we must multiply the corresponding column of the analysis matrix (say $[A_{\cdot j}]$) by column G and average this result by the number of plus signs in the column. Hence

$$E_j = \frac{\sum_{i=1}^{n} A_{ij} \times g_j}{n/2}. \tag{8.2}$$

The main effect of e.g. X_2 can be calculated as

$$E_2 = \frac{-g_1 - g_2 + g_3 + g_4 - g_5 - g_6 + g_7 + g_8}{4},$$

which is equivalent to calculating the difference between the average response for the high (1) level and average response for the low (−1) level of X_2. Analogously to 8.1, this can also be seen as an approximation to an average derivative. A geometric representation is shown in Figure 8.5.

In the full factorial design, we must run all possible factor combinations. This allows us to estimate all main and interaction effects, that is, all beta coefficients $\{\beta_0, \beta_1, \ldots, \beta_{123}\}$ that appear in the full model

$$\begin{aligned} G = \ & \beta_0 + \beta_1 X_1 + \beta_2 X_2 + \beta_3 X_3 \\ & \beta_{12} X_1 X_2 + \beta_{13} X_1 X_3 + \beta_{23} X_2 X_3 \\ & \beta_{123} X_1 X_2 X_3. \end{aligned} \tag{8.3}$$

In practice, to improve this design one can add some centre points and randomized experimental runs to protect the experiment against extraneous factors possibly effecting the results.

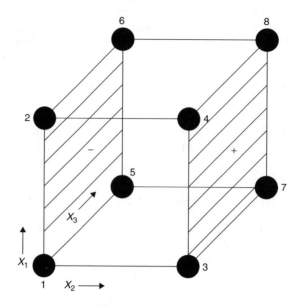

Figure 8.5 The graphical representation of the main effect for X_2.

Full factorial designs even without centre points can require many runs. For six factors, a two-level full factorial design requires $2^6 = 64$ runs. In some cases, we can use only a fraction of the runs specified by the full factorial design. The right choice of design points is then very important. In general, we pick a fraction such as $1/2, 1/4$, and so on, of the runs.

Fractional factorial design.   To construct 2^{3-1} half fraction two-level factorial design, we start with 2^2 full factorial design and then assign the third factor to the interaction column.

The design shown in Table 8.3 chooses for dark points is Figure 8.6. If we assign $X_3 = -X_1 * X_2$, then we obtain the design in Table 8.4. Both these designs allow us to estimate main effects of the factors. Of course, our estimates can differ in these designs and will not be as good as for full factorial design.

Run	X_1	X_2	$X_1X_2 = X_3$
1	-1	-1	1
2	1	-1	-1
3	-1	1	-1
4	1	1	1

Table 8.3 A 2^{3-1} factorial design table.

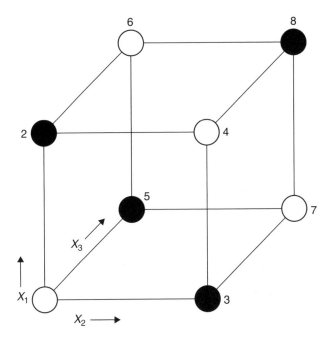

Figure 8.6 A 2^3 factorial design with dark and light points representing different choices of 2^{3-1} design.

Run	X_1	X_2	$-X_1X_2 = X_3$
1	−1	−1	−1
2	1	−1	1
3	−1	1	1
4	1	1	−1

Table 8.4 A 2^{3-1} factorial design table.

By assigning $X_1X_2 = X_3$, we reduced numbers of required runs but we are not able now to calculate interaction effect for X_1X_2. The assumption $X_1X_2 = X_3$ is acceptable if β_{12} is small compared to β_3 in (8.3). Our computation of the main effect for X_3 in a 2^{3-1} design in fact computes a sum of main effect for X_3 and interaction effect for X_1X_2. This is called *confounding* or *aliasing*.

In our example of fractional factorial design, we assumed that the column for X_3, that is 3, is equal to the column X_1X_2, that is $1 \cdot 2$, in full factorial design. We can write this as $3 = 1 \cdot 2$. By simple algebraical operations, remembering that multiplication means multiplying columns, we can find the complete set of aliases: $\{1 = 2 \cdot 3, 2 = 1 \cdot 3, 3 = 1 \cdot 2, I = 1 \cdot 2 \cdot 3\}$ where I is a column consisting only of ones.

8.3   Global sensitivity measures

Global sensitivity measures are based on the entire distribution of the input variables. Several measures of uncertainty contribution have been proposed in the literature:

Product moment correlation coefficients

Log correlation coefficients

Rank correlation coefficients

Linear regression coefficients

Loglinear regression coefficients

Rank regression coefficients

Logrank correlation coefficients

Partial correlation coefficients

Partial regression coefficients

Partial rank correlation coefficients

Partial rank regression coefficients

Correlation ratios/Variance decompositions

Sobol indices/Total effect indices.

Evidently, there is a need to adopt a unifying viewpoint. The idea underlying all these is prediction; how well can we predict the output (G) on the basis of input. If we restrict to linear predictors, then the linear regression coefficient measures the rate of change in the prediction, relative to a unit increase in the (single) input variable. Considering a set of input variables gives the partial regression coefficients. Since this depends on the unit of measuring the input variables, we may standardize the 'unit increase' of the input variables by scaling them to have unit standard deviation. This leads to the product moment correlation and partial correlation. Alternatively, we may first apply transformations before attempting linear prediction, this leads to log- and rank regression or correlation.

Let us consider a function f of input X, such that $f(X)$ yields the best linear prediction of output G. The correlation ratio is the square correlation of G and $f(X)$, and thus unifies most of the ideas represented in the above list. It also is related to variance decomposition. We therefore restrict the discussion of global sensitivity to the correlation ratio, and their close relatives, Sobol indices. For a rich exposition of global measures see Hamby (1994); Helton (1993); Iman and Davenport (1982); Kleijnen and Helton (1999); Saltelli et al. (2000).

8.3.1 Correlation ratio

The problem of finding the most important parameters can be viewed as decomposing the variance of the output according to the input variables. We would like to be able to apportion the uncertainty of the output G that 'comes' from uncertainties of model inputs $X_i, i = 1, \ldots, n$. If we could fix X_i at a certain value and calculate how much the variance of G decreased, this would give us an indication about the importance of X_i. Hence the following quantity could be considered

$$Var(G|X_i = x_i^*).$$

However, we do not know which value x_i^* should be chosen. Moreover, for non-linear models $Var(G|X_i = x_i^*)$ can be bigger than $Var(G)$. To overcome these problems, one may average over all values of X_i. Hence we could consider the following measure of importance of the variable X_i:

$$E(Var(G|X_i)), \tag{8.4}$$

where the expectation is taken with respect to distribution of X_i. The smaller (8.4) the more important X_i. It is well known that (see exercise 8.3):

$$E\left(Var(G|X_i)\right) + Var\left(E(G|X_i)\right) = Var\left(G\right). \tag{8.5}$$

Instead of (8.4), we could as well use the quantity $Var\left(E(G|X_i)\right)$. The higher $Var\left(E(G|X_i)\right)$ the more important X_i. From the above the following definition suggests itself:

Definition 8.1 (Correlation ratio) *For random variables $G, X_1, \ldots X_n$, the correlation ratio of G with X_i is*

$$CR(G, X_i) = \frac{Var\left(E(G|X_i)\right)}{Var\left(G\right)}.$$

Unlike correlation, the correlation ratio is *not* symmetric; that is $CR(G, X) \neq CR(X, G)$. We motivated the correlation ratio from the perspective of variance decomposition. We can also motivate it from the viewpoint of prediction. For simplicity, we suppress subscripts and consider three random variables, X, Y, G with $\sigma_G^2 < \infty$. We may ask for which function $f(X)$ with $\sigma_{f(X)}^2 < \infty$ is $\rho^2(G, f(X))$ maximal? The answer is given in Proposition 8.1.

Proposition 8.1 *If $\sigma_G^2 < \infty$ then*

(i) $Cov(G, E(G|X)) = \sigma_{E(G|X)}^2$,

(ii) $max_{f; \sigma_{f(X)}^2 < \infty} \rho^2(G, f(X)) = \rho^2(G, E(G|X)) = \frac{\sigma_{E(G|X)}^2}{\sigma_G^2}$.

The function that maximizes $\rho^2(G, f(X))$ is conditional expectation of G given X and $\rho^2(G, E(G|X)) = CR(G, X)$. If $E(G|X)$ is linear then $\rho^2(G, X)$ is equal to the correlation ratio of G with X. Hence the ratio of the square correlation to the correlation ratio can be taken as the measure of linearity of $E(G|X)$, called the *linearity index*.

The correlation ratio is one of the most important non-directional measures of uncertainty contribution. Note that the correlation ratio is always positive, and hence gives no information regarding the direction of the influence. In the next two propositions, we explore some properties of the correlation ratio.

Proposition 8.2 *Let* $G(X, Y) = f(X) + h(Y)$, *where* f *and* h *are invertible functions with* $\sigma_f^2 < \infty$, $\sigma_h^2 < \infty$ *and* X, Y *are not both simultaneously constant* $(\sigma_G^2 > 0)$. *If* X *and* Y *are independent then:*

$$\rho^2(G, E(G|X)) + \rho^2(G, E(G|Y)) = 1.$$

Proposition 8.3 *Let* $G = G(X, Y)$, *with* $Cov(E(G|X), E(G|Y)) = 0$ *then*

$$\rho^2(G, E(G|X)) + \rho^2(G, E(G|Y)) \leq 1.$$

Computing correlation ratio's   Computing correlation ratios may be tricky, as it involves a conditional expectation, which is not generally available in closed form. However, if we can sample Y' from the conditional distribution $(Y|X)$ independently of Y, and if the evaluation of G is not too expensive, then the following simple algorithm may be applied (Ishigami and Homma (1990)):

1. Sample (x, y) from (X, Y),

2. Compute $G(x, y)$,

3. Sample y' from $(Y|X = x)$ independent of $Y = y$,

4. Compute $G' = G(x, y')$

5. Store $Z = G * G'$

6. Repeat

The average value of Z will approximate $E(E^2(G|X))$, from which the correlation ratio may be computed as

$$\frac{E(E^2(G|X)) - E^2(G)}{\sigma_G^2}.$$

If Y and X are independent, then this algorithm poses no problems. If Y and X are not independent, then it may be difficult to sample from $(Y|X)$.

'Pedestrian' method

If Y and X are not independent, then problems arise in computing the variance of a conditional expectation. An obvious, although bad, idea is the 'pedestrian' method:

- Save N samples of (G, X, Y);

- Order the X values $x_{(1)}, \ldots, x_{(N)}$ from smallest to largest;

- where $N = k \times m$, divide the samples into m cells $C_1, \ldots, C_m$, where C_1 contains the samples with the k smallest X values, C_2 contains samples with the k smallest X values which are bigger than those in C_1, and so on.

- compute $E(G|X \in C_i), i = 1, \ldots, m$;

- compute the variance of these conditional expectations.

This is an intuitive transliteration of the mathematical definition, and yet it is a bad idea. The good news is that its badness is illuminating. The problem lies in the choice of m. If $m = N$, then C_i contains exactly one sample, say (g', x', y') and $E(G|X \in C_i) = g' = G(x', y')$. Taking the variance of these numbers will simply return the unconditional variance of G. On the other hand, if we take $m = 1$, then all sample values (g', x', y') will satisfy $x' \in C_1$ and $E(G|X \in C_1) = E(G)$, so the variance of the conditional expectation will be zero. Appropriately choosing m we traverse the values between $Var(G)$ and 0.

Kernel estimation methods

Kernel estimation methods are more sophisticated versions of the pedestrian method, for an overview of this and related methods see Wand and Jones (1995). Assume that samples $(x_1, g_1), \ldots, (x_N, g_N)$ of X and G are given. We will estimate the conditional expectation at point x as the average of observations in some neighbourhood of x. The samples close to x should get a higher weight than those further from x.

$$E(G|X = x) = \frac{\sum_{i=1}^{N} K_h(x_i - x)g_i}{\sum_{i=1}^{n} K_h(x_i - x)}, \tag{8.6}$$

where $K_h(\cdot) = K(\cdot/h)/h$ for some weight function $K(\cdot)$ and parameter $h > 0$. The parameter h is called the *bandwidth*. The *kernel function* $K(\cdot)$ is usually a symmetric probability density. Commonly used kernel functions are

a. the Gaussian kernel

$$K(x) = \frac{1}{\sqrt{2\pi}} \exp\left(-\frac{x^2}{2}\right)$$

b. the Epanechnikov kernel

$$K(x) = \begin{cases} \frac{3}{4}(1 - x^2), & \text{if } |x| < 1, \\ 0, & \text{otherwise}; \end{cases}$$

c. the tricube kernel

$$K(x) = \begin{cases} \frac{70}{81}(1 - |x|^3)^3, & \text{if } |x| < 1, \\ 0, & \text{otherwise}. \end{cases}$$

The bandwidth h controls the local character of kernel regression. If h is small, only samples close to x get non-negligible weights, if h is large, samples further from x are also included. The estimator (8.6) depends on a choice of bandwidth and on the choice of kernel function. The discussion of the pedestrian method should alert us that this freedom of choice is *not* an advantage.

Maximize squared correlation

This method uses the interpretation of the correlation ratio presented in Proposition 8.1. To calculate $CR(G, X)$ we find a function of X whose square correlation with G is maximal. This maximum is attained for $f(X) = E(G|X)$. We will not try to estimate $E(G|X = x_i)$ for each value of x_i in the sample file as in the previous method. Rather, we assume that the conditional expectation can be approximated by a (possibly high order) polynomial $P(x)$ and use optimization to find the coefficients of $P(x)$ that maximize $\rho^2(G, P(X))$. Suppose we choose third-degree polynomials: $P(x) = a + bx + cx^2 + dx^3$. We first generate a sample file of values for G, X, Y, and then solve

$$\text{Maximize: } \rho^2(G, a + bx + cx^2 + dx^3).$$

This problem is equivalent to minimizing sum-squared error (Exercise (8.5)):

$$\text{Minimize: } \sum (G - a + bx + cx^2 + dx^3)^2.$$

If the approximation $E(G|X = x) \sim P(x)$ is poor, the result may not be satisfactory. Unlike the previous approaches, however, we know how to improve the approximation: choose higher order polynomials. If we increase the degree of $P(x)$ without increasing the sample size, then we will eventually start overfitting. Indeed, for fixed sample size N, we can always find a polynomial P_N of degree N such that for each sample $i; i = 1 \dots N$, $P_N(x_i) = y_i$. Obviously $\rho(y, P_N(x)) = 1$.

The virtue of maximizing squared correlation is that it affords a prophylactic against overfitting. Suppose that we have found P_k which correlates very strongly with G. If we have overfit, then if we draw a *second* independent set of N samples, it is most likely that the P_k we fit from the first sample would *not* exhibit strong correlation with G on the second sample set. This indeed is what 'overfitting' means. On the other hand, if P_k really *was* the regression function $P_k(X) = E(G|X)$, then we should expect the performance on the two samples to be statistically equivalent.

There are many ways to implement this idea; we sketch two which work well in practice (Duintjer-Tebbens (2006)).

Rank test

- Split the sample into two equal pieces $N = N_1 \cup N_2$,

- Fit polynomial $P(X)$ on N_1,

- Divide N_1 and N_2 each into 10 equal pieces $N_{1,1} \dots N_{2,10}$,

- Compute $\rho_{i,j}^2(G, P(X))$ on $N_{i,j}$, $i = 1, 2; \ j = 1 \dots 10$,

- Use the Wilcoxon (or some similar rank test) to test the hypothesis that the numbers $\rho_{1,j}^2$ are not significantly higher ranked than $\rho_{2,j}^2$,

- If the hypothesis is rejected, then $P(X)$ is overfit.

Inflection criterion

- Split the sample into two equal pieces $N = N_1 \cup N_2$,

- Set $k = 1$,

- Fit polynomials $P_k(X)$ and $P_{k+1}(X)$ of order k and $k+1$ on N_1,

- If on N_2, $\rho^2(G, P_k(X)) > \rho^2(G, P_{k+1}(X))$ then stop; otherwise set $k := k + 1$ and repeat previous step,

UNICORN implements the first procedure with user control over the degree k.

Generalization of CR   The multivariate generalization of the correlation ratio is straightforward:

Definition 8.2 (Correlation ratio) *Let* $G, X_1, X_2, \ldots, X_n$ *be random variables with* $Var(G) < \infty$ *and* $i_1, i_2, \ldots, i_s \in \{1, 2, \ldots, n\}$*; then the correlation ratio* G *with* $X_{i_1}, \ldots, X_{i_s}$ *is*

$$CR(G, \{X_{i_1}, \ldots, X_{i_s}\}) = \frac{Var(E(G|\{X_{i_1}, \ldots, X_{i_s}\}))}{Var(G)}.$$

The computation of higher dimensional correlation ratio's in the manner described above poses no special problems. The interpretation in terms of maximal squared correlation in Proposition 8.1 generalizes in a straightforward manner, as do the results on polynomial fitting.

8.3.2   Sobol indices

The Sobol method is a variation of the correlation ratio method. Let $Y = G(\mathbf{X})$ be a function of vector $\mathbf{X} = (X_1, \ldots, X_n)$, where $X_i, i = 1, \ldots, n$ are uniformly distributed on $[0, 1]$ and independent.

The function G can be decomposed in the following way (Sobol (1993)):

$$G(X_1, \ldots, X_n) = g_0 + \sum_{i=1}^{n} g_i(X_i) + \sum_{1 \leq i < j \leq n} g_{ij}(X_i, X_j) + \cdots + \tag{8.7}$$
$$g_{1,2,\ldots,n}(X_1, \ldots, X_n).$$

g_0 must be a constant and the integrals of every summand over any of its own variables must be equal to zero. Hence

$$\int_0^1 g_{i_1,\ldots,i_s}(X_{i_1}, \ldots, X_{i_s}) \, dX_{i_l} = 0 \quad 1 \leq l \leq s.$$

It can be shown that

$$g_0 = \int_0^1 \cdots \int_0^1 g(X_1, \ldots, X_n) \, dX_1 \ldots dX_n.$$

Moreover, all summands in (8.7) are orthogonal, that is, if $(i_1, \ldots, i_s) \neq (j_1, \ldots, j_s)$, then

$$\int_{K^n} g_{i_1,\ldots,i_s} g_{j_1,\ldots,j_s} \, d\mathbf{X} = 0.$$

The decomposition (8.7) is unique whenever $G(\mathbf{X})$ is integrable over K^n and all 2^n summands can be evaluated in the following way

$$g_0 = E(G),$$

$$g_i(X_i) = E(G|X_i) - g_0, \quad 1 \leq i \leq n$$

$$g_{ij}(X_i, X_j) = E(G|X_i X_j) - E(G|X_i) - E(G|X_j) + g_0, \quad 1 \leq i < j \leq n$$

$$g_{ijk}(X_i, X_j, X_k) = E(G|X_i X_j X_k) - E(G|X_i X_j) - E(G|X_i X_k) - E(G|X_k X_j)$$

$$+ E(G|X_i) + E(G|X_j) + E(G|X_k) - g_0$$

$$\ldots$$

Since all summands in (8.7) are orthogonal then variance of G is a sum of variances of all elements of this decomposition. Hence, if we denote $D = Var(G)$ and $D_{i_1,\ldots,i_s} = Var(g_{i_1,\ldots,i_s})$, then

$$D = \sum_{i=1}^n D_i + \sum_{1 \leq i < j \leq n} D_{ij} + \cdots + D_{1,\ldots,n}.$$

The measure of the uncertainty contribution of variables $X_{i_1}, \ldots, X_{i_s}$ to the model can be now defined as:

$$S_{i_1,\ldots,i_s} = \frac{D_{i_1,\ldots,i_s}}{D}. \tag{8.8}$$

It can be easily checked that

$$S_i = \frac{D_i}{D} = CR(G, X_i), \quad 1 \leq i \leq n \tag{8.9}$$

$$S_{ij} = \frac{D_{ij}}{D} = CR(G, \{X_i, X_j\}), \quad 1 \leq i < j \leq n \tag{8.10}$$

$$\ldots \tag{8.11}$$

Computation of Sobol indices   Since the variables are assumed independent, the idea of (Ishigami and Homma (1990)) can be applied to compute Sobol indices.

The sensitivity indices S_i can be calculated via Monte Carlo methods. For a given sample size N, the following estimates can be obtained (the hat denotes the estimate)

$$\hat{g}_0 = \frac{1}{N} \sum_{k=1}^{N} G(\mathbf{x}_k)$$

where $\mathbf{x}_k$ is a sampled point in $[0, 1]^n$; similarly,

$$\hat{D}_i + \hat{g}_0^2 = \frac{1}{N} \sum_{k=1}^{N} G(\mathbf{u}_k, x_{ik}) G(\mathbf{v}_k, x_{ik}), \qquad (8.12)$$

where $\mathbf{u}_k, \mathbf{v}_k \in [0, 1]^{n-1}$. In other words, $\hat{D}_i$ is obtained by summing products of two values of the function G; one with all the variables sampled and the other with all the variables resampled except the variable X_i. The second-order terms D_{ij} can be estimated as (Homma and Saltelli (1996))

$$\hat{D}_{ij} + \hat{D}_i + \hat{D}_j + \hat{g}_0^2 = \frac{1}{N} \sum_{k=1}^{N} G(\mathbf{s}_k, x_{ik}, x_{jk}) \times G(\mathbf{t}_k, x_{ik}, x_{jk}), \qquad (8.13)$$

where $\mathbf{s}_k, \mathbf{t}_k \in [0, 1]^{n-2}$. Similar expressions can be derived for higher order terms (see Homma and Saltelli (1996)).

When calculating the estimators of $D_{i_1,\dots,i_s}$, it is important that the resampled variables are always generated using the same random numbers. This can be accomplished by sampling a random matrix of size $N \times (2n)$. (This matrix can be generated using pseudo, quasi, stratified or Latin Hypercube sampling see Chapter 6.) The first n columns of this matrix can be used for sampled values and last n columns for resampled values in (8.12) and (8.13).

To compute each of $S_{i_1,\dots,i_s}$, one separate sample of size N is needed. Since, there are $2^n - 1$ elements in (8.9) and one sample is needed to estimate g_0, then $N \times 2^n$ model evaluations must be computed. For large number of variables this can be too expensive.

Total effect indices One variation of the above method uses the total effect indices (Homma and Saltelli (1996)). In this case the function G is decomposed in the following way:

$$G(\mathbf{X}) = g_0 + g_i(X_i) + g_{\backslash i}(\mathbf{X}_{\backslash i}) + g_{i,\backslash i}(X_i, \mathbf{X}_{\backslash i})$$

where $g_i(X_i)$ denotes terms only involving X_i, $g_{\backslash i}(\mathbf{X}_{\backslash i})$ denotes the terms not containing X_i and $g_{i,\backslash i}(X_i, \mathbf{X}_{\backslash i})$ denotes the terms representing the interaction between X_i and the other variables $\mathbf{X}_{\backslash i}$. A sensitivity index S_{T_i} is introduced which gives the 'total' effect of the variable X_i. It adds the fraction of variance accounted for by variable X_i alone and the fraction accounted for by any combination of X_i with the remaining variables. Using the above notation, the total effect index of X_i can be calculated as

$$S_{T_i} = S_i + S_{i,\backslash i} = 1 - S_{\backslash i}.$$

This can be computed very efficiently with Monte Carlo calculations by $N \times (n + 1)$ model evaluations as follows:

$$\hat{S}_{T_i} = \frac{1}{D} \frac{1}{N} \sum_{k=1}^{N} G(\mathbf{u}_k, x_{ik}) G(\mathbf{u}_k, x'_{ik}) - \hat{g}_0^2,$$

where D denotes the total variance, $\mathbf{u}_k \in [0, 1]^{n-1}$ and prime denotes resampling.

8.4   Local sensitivity measures

When our goal is to determine sensitivity in a given region of the output, then global sensitivity measures are not indicated. The paradigm example of this occurs in structural reliability. The reliability of a structure is defined as *strength – load*, where *strength* and *load* are random variables. Normally, the reliability is positive. When the structure is weakened or threatened, the reliability decreases, and when the reliability becomes zero the structure fails. Failure is a rare event.

We are typically interested in the factors that drive reliability when the structure is threatened. This requires a local sensitivity analysis in the region of zero reliability. If the model is linear, and the input variables are independent, then there is no difference between local and global behaviour. In general, however, the results of local and global analyses may differ greatly. The dike ring reliability example in the previous chapter gave striking visual evidence of strong local behaviour in combination with weak global behaviour. Two methods will be discussed here, namely:

First Order Reliability Method (FORM)

Local Probabilistic Sensitivity Measure (LPSM)

FORM methods involve linearizing a reliability model at a 'design point' and hence can mask the differences between local and global sensitivity. Local methods can reveal the differences, and pose challenges for computations.

8.4.1   First order reliability method

We suppose that we are interested in a given point x^*, called the design point, of the input space. Let $G = G(\mathbf{X})$ be a function of vector $\mathbf{X} = (X_1, \ldots, X_n)$. In typical applications, x^* is a point of maximum probability in the set where G takes a specified value, say 0. Assuming that G can be represented as a Taylor series, we can linearize it in the neighbourhood of some point $x^* = (x_1^*, \ldots, x_n^*)$

$$G(\mathbf{X}) = G(x^*) + \sum_{i=1}^{n} \partial_i G(x^*)(X_i - x_i^*) + \text{HOT}, \qquad (8.14)$$

where ∂_i denotes $\frac{\partial}{\partial X_i}$ and *HOT* means 'higher order terms'.

Let μ_i and σ_i denote mean and standard deviation of X_i, respectively. Neglecting the higher order terms in (8.14) we obtain

$$G(\mathbf{X}) \sim G(x^*) + \sum_{i=1}^{n}(X_i - x_i^*)\partial_i G(x^*),$$

$$E(G) \sim G(x^*) + \sum_{i=1}^{n}(\mu_i - x_i^*)\partial_i G(x^*),$$

$$Var(G) \sim \sum_{i,j=1}^{n} Cov(X_i, X_j)\partial_i G(x^*)\partial_j G(x^*).$$

If X_i are all uncorrelated then

$$Cov(G, X_i) = \sigma_i^2 \partial_i G = \rho(G, X_i)\sigma_G \sigma_i.$$

Hence, in the linear uncorrelated model, the rate of change of G with respect to X_i may be expressed as

$$\partial_i G = Cov(G, X_i)/\sigma_i^2. \tag{8.15}$$

We note that the left-hand side depends on the point x^* whereas the right-hand side does not. This of course reflects the assumption of non-correlation and the neglect of HOT's. A familiar sensitivity measure involves a 'sum square normalization':

$$\alpha_i = \rho(G, X_i) = \frac{\partial_i G(x^*)\sigma_i}{\sigma_G}.$$

The factor α_i gives the influence of the standard deviation of variable X_i on the standard deviation of G. It depends on the slope of the tangent line of G in the design point. For the linear model and when X_is are uncorrelated,

$$R^2 = \sum_{i=1}^{n}\alpha_i^2 = 1.$$

This can be considered as a measure of the variance of G explained by the linear model. If R^2 is less then one, this may be caused either by dependencies among X_is or by the contribution of higher order terms neglected in (8.14).

8.4.2 Local probabilistic sensitivity measure

LPSM were introduced in Cooke and Noortwijk (1998) to describe the importance of an input variable X to a given contour of an output variable G:

$$LPSM(X) = \frac{\sigma_G}{\sigma_X}\frac{\partial E(X|G = g)}{\partial G}|_{g=g_o} = \frac{\sigma_G}{\sigma_X}\frac{\partial E(X|g_o)}{\partial g_o}. \tag{8.16}$$

The local sensitivity measure (8.16) is intended to measure the rate of change with respect to G of 'some function' of $X|G$ at a given point. For the uncorrelated linear model, 'global' and 'local' are equivalent, hence the global and local measures should coincide. This motivates choosing 'some function' as a normalized conditional expectation in (8.16). In fact, local probabilistic and global sensitivity measures may be seen as dual, in the following sense. Using the Taylor expansion of $E(X|G)$:

$$Cov(G, X) = Cov(G, E(X|G))$$

$$\sim Cov\left(G, E(X|g_o) + (G - g_o)\frac{\partial E(X|g_o)}{\partial g_o}\right)$$

$$= \sigma_G^2 \frac{\partial E(X|g_o)}{\partial g_o}.$$

Thus, if the regression of X on G is linear, then higher order terms vanish and

$$\frac{\partial E(X|g_o)}{\partial g_o} = \frac{Cov(G, X)}{\sigma_G^2} \tag{8.17}$$

which may be compared with (8.15).

The obvious way to approximate $LPSM(X)$ in Monte Carlo simulations is to compute

$$\frac{E(X|G \in (g_o, g_o + \epsilon)) - E(X|G \in (g_o - \epsilon, g_o))}{E(G|G \in (g_o, g_o + \epsilon)) - E(G|G \in (g_o - \epsilon, g_o))}. \tag{8.18}$$

In some cases this is very unstable. Consider the following example[1]:

Example 8.2 *X and Y are independent standard normal*

$$G(X, Y) = \min(3 - X, 3 - Y)$$

then

$$E(X|G = g) = E(X|G = g, X < Y)P(X < Y)$$

$$+ E(X|G = g, Y \leq X)P(Y \leq X)$$

$$= \frac{E(X|Y = 3 - g, X < Y) + E(X|X = 3 - g, X \geq Y)}{2}$$

$$= \frac{E(X|X < 3 - g) + E(X|X = 3 - g, Y < 3 - g)}{2}$$

$$= \frac{(E(X|X < 3 - g) + 3 - g)}{2}.$$

---

[1] We are grateful to Ton Vrouwenvelder for this example.

where

$$E(X|X < 3 - g) = \frac{\int_{-\infty}^{3-g} x\phi(x)\,dx}{\int_{-\infty}^{3-g} \phi(x)\,dx}$$

and ϕ is the standard normal density, with cumulative distribution function Φ. The partial derivative of the right hand side at $g = 0$ is

$$\frac{-3\phi(3)\Phi(3) + \phi(3)\int_{-\infty}^{3} x\phi(x)\,dx}{2\Phi(3)^2} - 0.5 = -0.507.$$

On a Monte Carlo simulation with 5,000,000 samples and $\epsilon = 0.1$ the above method 8.18 yields the estimates

$$\frac{\partial}{\partial g} E(X|g = 0)_{simulation} = -0.517,$$

$$\frac{\partial}{\partial g} E(Y|g = 0)_{simulation} = -0.807.$$

Of course, by symmetry these two derivatives must be equal. The number of samples used is unrealistically large, and still performance is poor. This is explained by a number of factors. First if high accuracy is desired, ϵ must be chosen small in (8.18). On the other hand, the difference in conditional expectations must be large enough to be statistically significant. In the above example, this difference was barely significant at the 5% level for Y and was not significant for X. In this case, the difference in conditional expectations in (8.18) is small, because, roughly speaking, X feels the effect of conditionalizing on $G = 0$ on only one half of the samples. Finally, conditionalizing on extreme values of G, as in this case, can introduce strong correlations between the input variables. In this case the conditional correlations are negative (as can be checked with UNICORN). This means that sampling fluctuations in the estimates of the conditional expectations in (8.18) will be correlated. Indeed, it required an unrealistically large number simply to obtain estimates whose signs were both negative. Clearly, alternative methods of calculating the LPSM are needed.

8.4.3 Computing $\frac{\partial E(X|g_o)}{\partial g_o}$

We discuss two methods for computing the derivative of a conditional expectation. Both are better than the (8.18), but neither is wholly satisfactory. To evaluate performance, we must have examples for which the conditional expectation can be put in closed form.

Assume that X, Y are independent and uniformly distributed on $[0, 1]$, and let $G(X, Y)$ be sufficiently differentiable in both arguments. To compute the expectation of X given $G = g_o$, we define a density along the contour $G = g_o$ which is proportional to arc length. If the contour is simple, we may parametrize arc length in terms of x and write $g_o = G(x, y(x))$. The arc length element, ds and conditional expectation are given by

$$ds = \sqrt{dx^2 + dy^2} = dx\sqrt{1 + (dy/dx)^2}.$$

$$E(X|g_o) = \int xf(X|g_o)\,dx$$

$$= \frac{\int x\sqrt{1 + (dy/dx)^2}\,dx}{\int \sqrt{1 + (dy/dx)^2}\,dx}.$$

The reader may verify the following examples:

Example 8.3

$$G(X, Y) = 2X + Y; \quad f(x|g_o) = 2/g_o; \qquad\qquad\quad 0 < x < g_o/2,$$

$$G(X, Y) = XY; \quad f(x|g_o) = \frac{\sqrt{1 + g_o^2/x^4}}{\int_{g_o}^1 \sqrt{1 + g_o^2/x^4}}; \qquad \begin{array}{l} 0 \le g_o \le 1, \\ g_o < x < 1, \end{array}$$

$$G(X, Y) = X^2 Y; \quad f(x|g_o) = \frac{\sqrt{1 + 4g_o^2/x^6}}{\int_{\sqrt{g_o}}^1 \sqrt{1 + 4g_o^2/x^6}}; \qquad \begin{array}{l} 0 \le g_o \le 1, \\ \sqrt{g_o} < x < 1, \end{array}$$

$$G(X, Y) = X^2 + Y^2; \quad f(x|g_o) = \frac{\sqrt{1 + x^2/(g_o^2 - y^2)}}{\int_0^{\sqrt{g_o}} \sqrt{1 + x^2/(g_o^2 - y^2)}}; \qquad \begin{array}{l} 0 \le g_o \le 1, \\ 0 < x < \sqrt{g_o}. \end{array}$$

Method 1: Linearization via reweighted Monte Carlo simulation   The following is an example of new approaches to calculating $\frac{\partial E(X|g_o)}{\partial g_o}$ (Cooke et al. (2003)). The idea is to make the duality relation (8.17) approximately true by reweighting the sample emerging from a Monte Carlo simulation. Since $E(X|G)$ can be expanded around g_o as

$$E(X|G) = E(X|g_o) + (G - g_o)\frac{\partial E(X|g_o)}{\partial g_o} + \frac{1}{2}(G - g_o)^2\frac{\partial^2 E(X|g_o)}{\partial g_o^2} + HOT$$

then

$$Cov(X, G) = Cov(E(X|G), G)$$

$$= \sim \frac{\partial E(X|g_o)}{\partial g_o}\,Var\,G + \frac{1}{2}\frac{\partial^2 E(X|g_o)}{\partial g_o^2}\{E(G - g_o)^3$$

$$+ (g_o - E(G))(E(G - g_o)^2)\}.$$

We assign a 'local distribution' to G such that the terms between curly brackets are equal to zero, then

$$\frac{\partial \overline{E}(X|g_o)}{\partial g_o} = \frac{\overline{Cov}(X, G)}{\overline{Var}(G)}.$$

To achieve this, the local distribution should be chosen so that

$$\overline{E}G = g_o$$

and

$$\overline{E}(G - g_o)^3 = 0$$

where $\overline{G}$ means G with a local distribution. We want this distribution to be as close as possible to the distribution of G. In our case, we take the distribution which minimizes the relative information with respect to the original distribution of G.

With regard to the example $G = \min\{3 - X, 3 - Y\}$, the results are better than those obtained with (8.18), but not overwhelming. With 5,000,000 samples, we find

$$\frac{\partial E(X|G = g)}{\partial g}\Big|_{g=0} = -0.5029,$$

$$\frac{\partial E(Y|G = g)}{\partial g}\Big|_{g=0} = -0.5038.$$

Needless to say, this number of samples is not realistic in practice. With only 10,000 samples, the results were not acceptable.

Method 2: Conditional expectation calculated via reweighted Monte Carlo samples. A different approach uses Monte Carlo reweighting to calculate the conditional expectation directly[2]. Monte Carlo simulations have been performed to obtain N samples $g_i = G(x_i, y_i)$. The conditional expectation of X given $G = g_0$ can be estimated:

$$E(X|g_0) = \frac{1}{\sum_{i=1}^{N} w(d_i)} \sum_{i=1}^{N} w(d_i) x_i$$

with some weighted function $w \in [0, 1]$ that depends on $d_i = |g_i - g_0|$. $\frac{\partial}{\partial g_o} E(X|g_o)$ can now be calculated as

$$\frac{\partial}{\partial g_o} E(X|g_o) = \frac{1}{\sum_{i=1}^{N} w(d_i)} \left(\sum_{i=1}^{N} \frac{\partial w(d_i)}{\partial g_o} (x_i - E(X|g_o)) \right).$$

With regard to the example $G = \min\{3 - X, 3 - Y\}$, the results are better than those obtained with (8.18) and better then with the above reweighting scheme. Using the weight function of the form $K(x) = \frac{1}{1+x^4}$ and calculating weights as $w(d_i) = K(d_i/h)$ ($h = 0.1$ is the bandwidth) with 500,000 samples we got satisfactory results.

8.5 Conclusions

In the last years, there has been considerable convergence of opinion regarding the choice of sensitivity measures. Proposition 8.1 provides a strong motivation for

---

[2]We are grateful to Ulrich Callies for this method

the choice of the correlation ratio to measure global, no-directional influence. Its main disadvantages are computational in nature. UNICORN's sensitivity analysis satellite program tested many computational approaches. Best results for a generic program were obtained with optimization-based methods. We select a degree of polynomial to approximate the regression function, and find the coefficients by optimization, and check against overfitting.

Calculating the correlation ratio, as with other regression-or correlation-based methods, typically requires Monte Carlo simulation. When evaluation of the functions is very expensive, Monte Carlo methods become infeasible. DOE and Morris' method may be seen as shortcuts when Monte Carlo simulation is not on.

The Local Probabilistic Sensitivity measures have a strong intuitive appeal in certain applications, but we still lack good overall methods of computation.

8.6   Unicorn projects

The projects in this section are used to illustrate features of the sensitivity analysis satellite program. This program reads a UNICORN output file and computes many of the measures discussed in this chapter. The program has a great many features and an elaborate help file which contains definitions and calculational details. The projects below simply step through the analysis of some of the cases presented in previous chapters.

Project 8.1   VineInvest

Use the Invest_Dvine created in Chapter 4. Create a sample file with 20 runs. After sampling, the button 'Sen.Analysis' is available, hit it. The sensitivity analysis satellite program is opened, and asks which variables you wish to import. For large cases, it is convenient to select the subset of interest. In this case accept all.

The main panel now opens, showing two check lists. The 'predicted variables' are those whose behaviour you want to explain in terms of other variables, called the 'base variables'. We are interested in the variable '5yrreturn', and we want to see how this depends on the interest rate in each of the five years. Choose V1, V2, ..., V5 as base variables and hit RUN. You should see the screen shot in Figure 8.7.

The right-hand panel is now filled with results. Each line shows sensitivity indices for a given base variable, for the indicated predicted variable. If you go to TOOLS/Options, you can influence the choice of sensitivity indices which are displayed.

The help file contains the definitions of all these measures, and the user is encouraged to explore all the program's features. We simply take the first few steps here. Note that V3 has the highest correlation to '5yrreturn', V1 has the smallest. All linearity indices are close to 1, though V5 is the smallest[3].

---

[3]Theoretically the linearity index is less than or equal to one. In practice, we can get values greater than one owing to approximations in computing the correlation ratio.

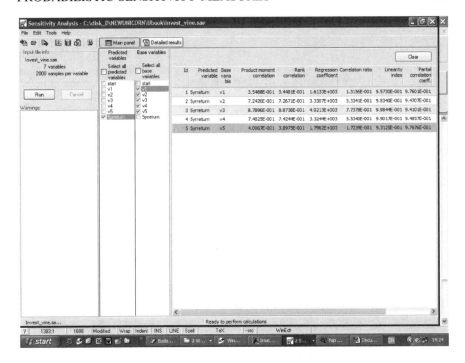

Figure 8.7 Sensitivity analysis Main Panel.

Hit the button 'detailed results'; the selector under 'item id' allows you to see the effect of each base variable on the predicted variable. The first graph plots the predicted variable against the base variable, and shows the linear regression line. The second graph plots the regression of the predictor against the base variable. This is an approximation gotten by polynomial fitting.

Select the other variables and view their regressions. This is a case where the relation between the predictor and base variables is very well behaved, and the global measures give a good picture of the relationship.

Project 8.2 Strtlite sensitivity

Open the file 'Strtlite' constructed in the previous chapter, simulate with 20 runs and choose sensitivity analysis. Select all variables, and use 'ign_headlite' as predicted variable, all others being base variables. The main sensitivity indices are shown in (Figure 8.8). These results are copied to the clipboard and pasted into any suitable document. This case offers more interesting regression functions; Figure 8.9 shows the regression of ign_headlite against strtr. Note the non-linearity in the regression function. In such cases, the correlation ratio is a better index of influence than product moment correlation or regression coefficients.

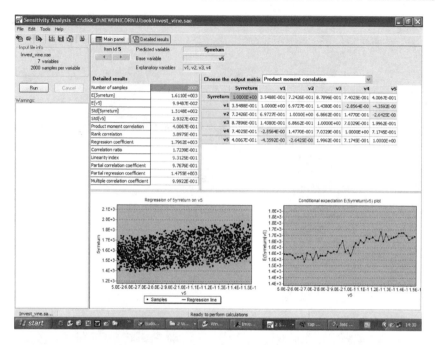

Figure 8.8 Sensitivity analysis Main Panel.

Project 8.3 Shock and awe

This one is for fun. Take the case 'Probdiss' and add the following UDFs:

$$x : v1 * i1\{0, v3, .5\} + (2 + v2) * i1\{.5000001, v3, 1\};$$

$$g : x + y;$$

$$h : x^2 + y^2;$$

$$k : sin(h);$$

$$m : x * y;$$

$$n : m/k;$$

$$q : m * k;$$

The regressions of some interesting variables are shown in Plate 8 and in Figure 8.10. The cobweb of this case is also quite fetching.

8.7 Exercises

Ex 8.1 *Construct a 2^3 full factorial design for the following model*

$$G(v1, v2, v3) = v2^{10} - v1 * v2 + \max v1, v2, v3, \qquad (8.19)$$

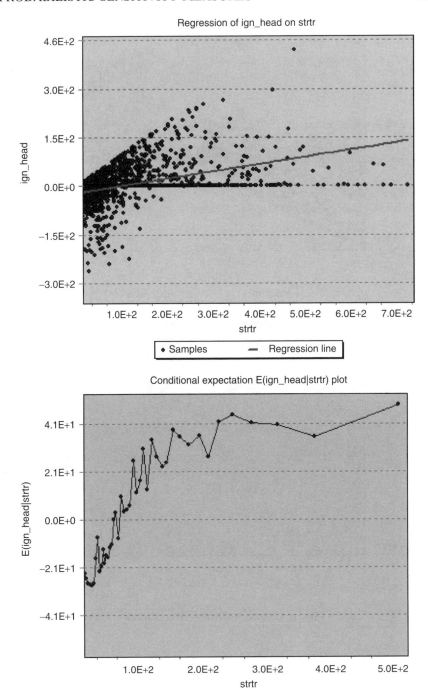

Figure 8.9 Regression ign_headlite against strtr.

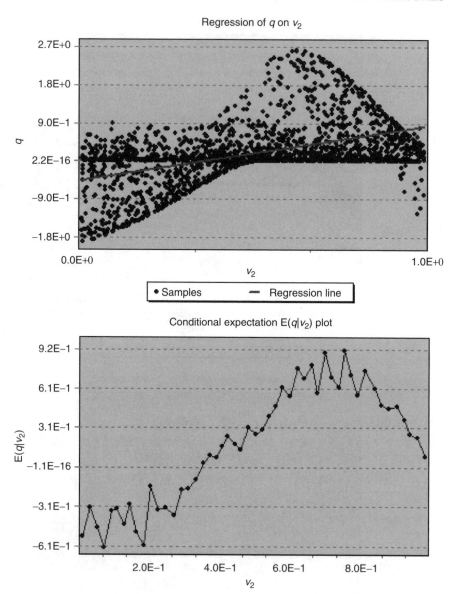

Figure 8.10 Regressions from Probdiss.

where v1, v2, v3 are independent variables uniform on [0, 1]. Calculate main and interaction effects and interpret the results.

Ex 8.2 *Find the product moment correlation and the rank correlations of G with vi as well as $CR(G, vi)$, $i = 1, 2, 3$ for the model 8.19, where v1, v2 and v3 are the*

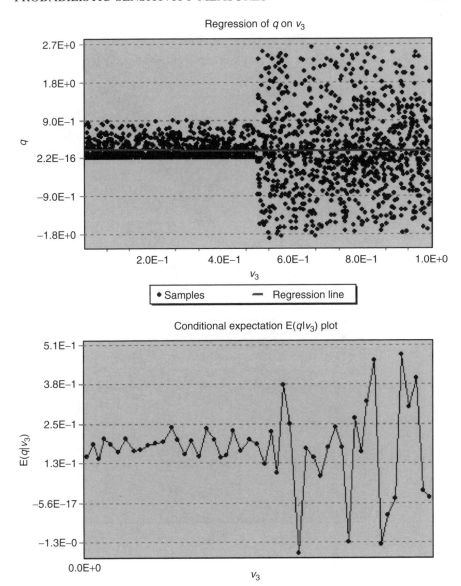

Figure 8.10 *(continued)*

independent standard normal variables. Explore conditional cobweb plots for high and low values of G. Interpret the results.

Ex 8.3 *Prove that*

$$Var(G) = Var(E(G|X)) + E(Var(G|X)).$$

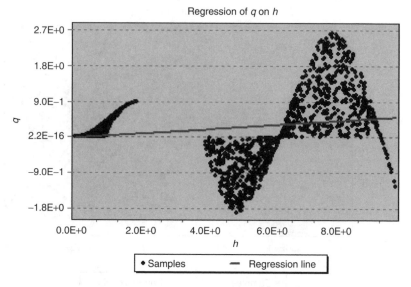

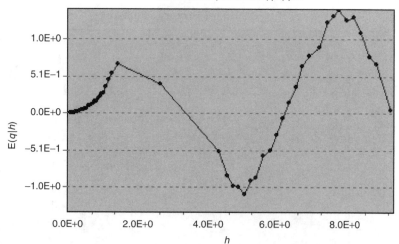

Figure 8.10 (*continued*)

Hint: $Var(G|X) = E(G^2|X) - E^2(G|X)$; *take expectations of both sides.*

Ex 8.4 *Verify that*

$$\iiint (E(G|x_1, x_2) - E(G|x_1) - E(G|x_2) + g_0)(E(G|x_1, x_3) - E(G|x_1)$$
$$- E(G|x_3) + g_0) \, dx_1 \, dx_2 \, dx_3 = 0.$$

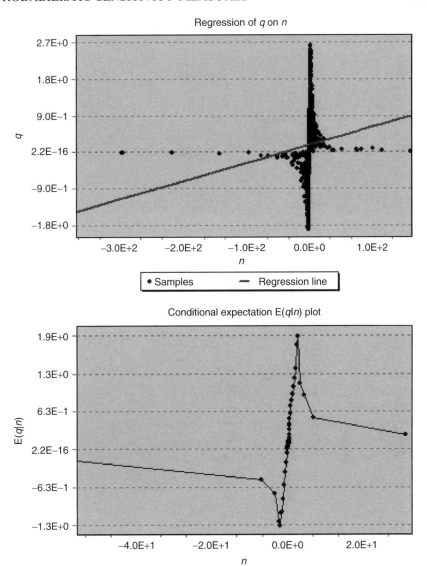

Figure 8.10 (*continued*)

Ex 8.5 *Suppose, we have a sample* (G_i, x_i); $i = 1 \ldots N$. *Show that the following two problems are equivalent:*

$$\text{Maximize: } \rho^2(G, a + bx + cx^2 + dx^3).$$

$$\text{Minimize: } \sum_i (G_i - a + bx_i + cx_i^2 + dx_i^3)^2.$$

Hint: maximizing $\rho^2(G, a + bx + cx^2 + dx^3)$ *is equivalent to maximizing* $\ln(\rho^2(G, a + bx + cx^2 + dx^3))$. *Show that both problems entail solving*

$$E\left(G \times \frac{\partial f}{\partial \alpha}\right) = E\left(f \times \frac{\partial f}{\partial \alpha}\right)$$

where $f(x) = a + bx + cx^2 + dx^3$ *and* α *is any of the parameters in* $f(x)$.

Ex 8.6 *Compute the regression function* $E(headlite|Bat)$ *in the project SRTLITE of Chapter 7. Compare your result with the graph generated by UNICORN's sensitivity analysis package.*

Ex 8.7 *Prove Exercise 8.3.*

8.8   Supplement

8.8.1   Proofs

Proposition 8.1 *If* $\sigma_G^2 < \infty$ *then*

(i)$Cov(G, E(G|X)) = \sigma^2_{E(G|X)}$,

(ii) $\max_{f;\sigma^2_{f(X)} < \infty} \rho^2(G, f(X)) = \rho^2(G, E(G|X)) = \frac{\sigma^2_{E(G|X)}}{\sigma_G^2}$.

Proof.
(i) $Cov(G, E(G|X)) = E(E(GE(G|X)|X)) - EGE(E(G|X)) = E(E^2(G|X)) - E^2(E(G|X))$.
(ii) Let $\delta(X)$ be any function with finite variance.
Put $A = \sigma^2_{E(G|X)}$, $B = Cov(E(G|X), \delta(X))$, $C = \sigma_G^2$, and $D = \sigma_\delta^2$. Then

$$\rho^2(G, E(G|X) + \delta(X)) = \frac{(A + B)^2}{C(A + D + 2B)}; \qquad (8.20)$$

$$\frac{\sigma^2_{E(G|X)}}{\sigma_G^2} = \frac{A}{C}. \qquad (8.21)$$

$$\frac{(A + B)^2}{C(A + D + 2B)} \leq \frac{A}{C} \iff B^2 \leq AD. \qquad (8.22)$$

The latter inequality follows from the Cauchy Schwarz inequality. This is similar to a result in (Whittle (1992)). $\square$

Proposition 8.2 *Let* $G(X, Y) = f(X) + h(Y)$, *where* f *and* g *are invertible functions with* $\sigma_f^2 < \infty$, $\sigma_h^2 < \infty$ *and* X, Y *are not both simultaneously constant* ($\sigma_G^2 > 0$). *If* X *and* Y *are independent then:*

$$\rho^2(G, E(G|X)) + \rho^2(G, E(G|Y)) = 1.$$

Proof.

We have $E(G|X) = E(G|f(X))$, and $h(Y) \perp E(G|f(X))$, $f(X) \perp E(G|h(Y))$; therefore,

$$\sigma_G^2 = Cov(G, G) = Cov(G, f(X) + h(Y))$$
$$= Cov(G, f(X)) + Cov(G, h(Y))$$
$$= Cov(E(G|f(X)), f(X)) + Cov(E(G|h(Y)), h(Y))$$
$$= Cov(E(G|f(X)) + E(G|h(Y)), f(X) + h(Y))$$
$$= Cov(E(G|f(X)) + E(G|h(Y))), G)$$
$$= Cov(E(G|X) + E(G|Y), G) = \sigma_{E(G|X)}^2 + \sigma_{E(G|Y)}^2$$

The result now follows with Fact (8.1). $\square$

9

Probabilistic Inversion

9.1 Introduction

Chapter 1 discussed a recent application in which probabilistic inversion played a central role. In this chapter, we define probabilistic inversion problems, discuss existing algorithms for solving such problems and study new algorithms based on iterative re-weighting of a sample.

We know what it means to invert a function at some value in its range. When we invert the function at a random variable, this is called *probabilistic inversion*. More precisely, a probabilistic inversion problem may be characterized as follows: Given a random vector $\mathbf{Y}$, taking values in $\mathbb{R}^M$ and a measurable function $G : \mathbb{R}^N \to \mathbb{R}^M$, find a random vector $\mathbf{X}$ such that $G(\mathbf{X}) \sim \mathbf{Y}$, where $\sim$ means that $G(\mathbf{X})$ and $\mathbf{Y}$ share the same distribution. $\mathbf{X}$ is sometimes termed the *input* to model G and $\mathbf{Y}$ the output. If $G(\mathbf{X}) \in \{\mathbf{Y}|\mathbf{Y} \in \mathcal{C}\}$, where $\mathcal{C}$ is a subset of random vectors on $\mathbb{R}^M$, then $\mathbf{X}$ is called *a probabilistic inverse* of G at $\mathcal{C}$. $\mathbf{X}$ is sometimes termed the *input* to model G and $\mathbf{Y}$ the *output*. If the problem is feasible, it may have many solutions, and we require a preferred solution; if it is infeasible, we seek a random vector $\mathbf{X}$ for which $G(\mathbf{X})$ is 'as close as possible' to $\mathbf{Y}$. Such problems arise in quantifying uncertainty in physical models with expert judgment (Kraan and Cooke (2000a)). We wish to quantify the uncertainty on parameters of some model using expert judgment, but the parameters do not possess a clear physical meaning and are not associated with the physical measurements with which experts are familiar. Often the models are new and do not command universal assent. We must then find the observable quantities $\mathbf{Y}$ functionally related with $\mathbf{X}$ that can be assessed by experts. Inferring uncertainties on $\mathbf{X}$ from uncertainties on $\mathbf{Y}$, as specified by experts, is clearly an inverse problem.

In practical applications, the vector $\mathbf{Y}$ is characterized in terms of some percentiles or quantiles of the marginal distributions $Y_1, \ldots, Y_M$. In this case, we seek a random vector $\mathbf{X}$ such that $G(\mathbf{X})$ satisfies quantile constraints imposed

Uncertainty Analysis with High Dimensional Dependence Modelling   D. Kurowicka and R. Cooke
© 2006 John Wiley & Sons, Ltd

on $Y_1, \ldots, Y_M$. There may be other constraints, which may reflect mathematical desiderata when we require independence between variables, as in Section 9.6. Physical considerations may also impose constraints on $\mathbf{X}^1$. A few algorithms for solving such problems are available in literature, namely, conditional sampling, PARFUM (PARameter Fitting for Uncertain Models) (Cooke (1994); Hora and Young algorithm Harper et al. (1994)) and PREJUDICE (Kraan and Bedford (2005); Kraan and Cooke (2000a,b)). We summarize existing approaches to probabilistic inversion in Section 9.2.

The function G to be inverted will not generally be given in closed form; indeed, it may be given as a large computer code. Inverting this function may therefore be quite expensive. The most attractive algorithms for probabilistic inversion do *not* actually invert the function G but use a technique called *sample re-weighting*. Special knowledge about the problem at hand and complicated heuristic steering on the part of the user are not required. Moreover, the operations on the sample can be performed one variable at a time, so that the entire sample need not be held in memory. This means that there is virtually no size limitation.

One type of algorithm is known from the literature as *iterative proportional fitting* (IPF) (IPF) (Kruithof (1937)). Starting from a given sample distribution, we may consider the starting point of the IPF algorithm as the uniform distribution over the sample points. IPF iteratively re-weights these sample points. IPF need not converge, but if it does, it converges to a solution that is minimally informative with respect to the starting distribution (Csiszar (1975)).

A variation on this is an iterative version of the PARFUM algorithm. We show that this algorithm has fixed points minimizing an information functional even if the problem is infeasible. If the problem is feasible, the fixed points are solutions. We discuss IPF and PARFUM algorithms in Section 9.3 and then show how they can be applied to solve probabilistic inversion problems.

9.2 Existing algorithms for probabilistic inversion

The algorithms discussed here look for an inverse of a partially specified random vector $\mathbf{Y} = (Y_1, \ldots, Y_M)$. Typically, we specify quantiles of the marginal random variables $Y_1, \ldots, Y_M$. In other words, for each variable Y_i, we specify an ordered set of numbers and the probability that Y_i falls between these numbers. The first algorithm applies only when $M = 1$.

9.2.1 Conditional sampling

Let $\mathbf{Y}$ consist of only one variable Y. A simple conditional sampling technique can be used on the basis of the following result (Kraan and Cooke (1997)).

---

[1]In some cases, probabilistic inversion problems may have trivial solutions, for example, if under G, $\mathbf{X}$ makes the coordinates of $\mathbf{Y}$ completely rank correlated. Such solutions may be rejected on physical grounds; hence, physical constraints may stipulate the support of $\mathbf{X}$. Other physical constraints are discussed in the example in Section 9.5.

Proposition 9.1 *Let X and Y be independent random variables with range $\{1, \ldots, k\}$. $P(X = i) > 0$, $P(Y = i) > 0$, $i = 1, \ldots, k$. Let $X_{X=Y}$ denote a random variable with distribution*

$$P(X_{X=Y} = i) = P(X = i | X = Y).$$

Then, $X_{X=Y} \sim Y$ if and only if X is uniformly distributed.

Proof. Put $p_i = P(X = i)$, $q_i = P(Y = i)$, $i = 1, \ldots, k$. Then,

$$P(X_{X=Y} = i) = \frac{p_i q_i}{\sum p_j q_j}.$$

For all $i = 1, \ldots k$, $p_i q_i / \sum p_j q_j = q_i$ if and only if $p_i = \sum p_j q_j$; that is, p_i does not depend on i. $\square$

Consider a random vector $\mathbf{X} = (X_1, X_2, \ldots, X_n)$ and function $G : \mathbb{R}^n \to \mathbb{R}$ such that $G(\mathbf{X})$ and Y are concentrated on an interval I with invertible cumulative distribution functions F_G and F_Y, respectively. Let G^* be a discretized version of $G(\mathbf{X})$ such that the cumulative distribution function F_{G^*} is concentrated on $I_d = \{1/k, 2/k, \ldots, 1\}$. In other words, define G^* as:

$$G^*(\mathbf{X}) = \sum_{j=1}^{k} F_G^{-1}\left(\frac{j}{k}\right) \mathbb{I}_{\left(\frac{j-1}{k}, \frac{j}{k}\right]}(F_G(\mathbf{X})), \tag{9.1}$$

where $\mathbb{I}_A$ denotes the indicator function of a set A.

$F_{G^*}(G(\mathbf{X}))$ is uniformly distributed on I_d. $F_{G^*}(Y)$ is concentrated on I_d but is not necessarily uniform. The following procedure is used to find the conditional distribution of $G(\mathbf{X})$ that approximates the distribution of Y:

1. Sample $\mathbf{X}$ and Y independently of $\mathbf{X}$.

2. If $F_{G^*}(G(\mathbf{X})) = F_{G^*}(Y)$, then retain the sample, otherwise discard the sample.

3. Repeat steps 1 and 2.

Since $F_{G^*}(G(\mathbf{X}))$ is uniformly distributed in the unconditional sample, the preceding proposition shows that in the conditional sample, $F_{G^*}(G(\mathbf{X})) \sim F_{G^*}(Y)$. The conditional distribution, P_{cond} of $G(\mathbf{X})$, approximates the distribution of Y in the sense that

$$P_{cond}\left(G(\mathbf{X}) \in \left\{F_G^{-1}\left(\frac{i}{k}\right), F_G^{-1}\left(\frac{i+1}{k}\right)\right\}\right) = P\left(Y \in \left\{F_G^{-1}\left(\frac{i}{k}\right), \right.\right.$$

$$\left.\left. F_G^{-1}\left(\frac{i+1}{k}\right)\right\}\right).$$

The advantage of this technique is its simplicity. Its disadvantage is that it works only for one output variable.

9.2.2 PARFUM

Let $\mathbf{Y} = (Y_1, Y_2, \ldots, Y_M)$ be a random vector with marginal densities $(f_1, \ldots, f_M)$ and let $G_m : \mathbb{R}^n \to \mathbb{R}, m = 1, 2, \ldots, M$ be the measurable functions. The PARFUM algorithm can be described in the following steps Cooke (1994):

1. Choose a finite set $\mathcal{X} \subset \mathbb{R}^n; \#(X) = K$, where $\#$ means the number of points.

2. Define the conditional mass function Q_m of Y_m on the image $G_m(\mathcal{X})$ of $\mathcal{X}$ under G_m, where $\mathbf{x} \in \mathcal{X}$:

$$Q_m(G_m(\mathbf{x})) = \frac{f_m(G_m(\mathbf{x}))}{\sum_{\mathbf{z} \in \mathcal{X}} f_m(G_m(\mathbf{z}))}.$$

3. Define the minimally informative distribution on $\mathcal{X}$, the push-forward distribution P_m on $G_m(\mathcal{X})$ of which agrees with Q_m, that is, for $\mathbf{x} \in \mathcal{X}$

$$P_m(\mathbf{x}) = \frac{Q_m(G_m(\mathbf{x}))}{\#\{\mathbf{z} \in \mathcal{X} | G_m(\mathbf{z}) = G_m(\mathbf{x})\}}.$$

4. Find a distribution P on $\mathcal{X}$ that minimizes the relative information $\sum_{m=1}^{M} I(P_m | P)$, where

$$I(P_m | P) = \sum_{\mathbf{x} \in \mathcal{X}} P_m(\mathbf{x}) \ln\left(\frac{P_m(\mathbf{x})}{P(\mathbf{x})}\right).$$

Let

$$\mathcal{S}^K = \left\{ s \in \mathbb{R}^K | s_k \geq 0, \sum_{k=1}^{K} s_k = 1 \right\}. \tag{9.2}$$

It is not difficult to show (Cooke (1994)) the following proposition:

Proposition 9.2 Let $P_m \in \mathcal{S}^K$, $m = 1, \ldots, M$. Then

$$\min_{P \in \mathcal{S}^K} \sum_{m=1}^{M} I(P_m | P) = \sum_{m=1}^{M} I(P_m | P^*)$$

if and only if $P^* = (1/M) \sum_{m=1}^{M} P_m$.

The advantage of this method is that it is always feasible and easily implemented. One disadvantage is that the conditional distributions Q_m might be different to those of Y_m, but this may be steered by a judicious choice of $\mathcal{X}$. More serious is the fact that the push forward of P need not have marginal distributions that agree with those of the Y_m. This can also be influenced by steering but is more difficult. When this algorithm is iterated, all fixed points are feasible if the set of feasible points is non-empty (Theorem 9.3).

9.2.3 Hora-Young and PREJUDICE algorithms

Instead of basing the fitting on the conditional measure Q_m, which may be different from F_m, the Hora-Young method constrains the choice of P to the set $\mathbb{P}$, whose margins for $G_m(\mathcal{X})$ satisfy the quantile constraints

$$P\{G_m(\mathcal{X}) \in [y_{mk-1}, y_{mk}]\} = F_m(y_{mk}) - F_m(y_{mk-1}), k = 2, \ldots, K,$$

where $y_{m1}, \ldots, y_{mK}$ are in the range of Y_m, $m = 1, \ldots, n$, $P(G_m(\mathcal{X}))$ is the push-forward measure on the range of Y_m induced by the probability measure P on $\mathcal{X}$ and F_m is the cumulative distribution function of Y_m. The disadvantage of this method is that it may be infeasible; that is, for a given choice of $\mathcal{X}$, there may be no measure P satisfying the constraints.

The algorithm PREJUDICE is an elaboration of this algorithm. Using duality theory for constrained optimization, it performs model inversions, thereby augmenting the set $\mathcal{X}$ in a way that optimally reduces infeasibility. Performing model inversions can be very expensive. Although PREJUDICE represents the most sophisticated method to date for probabilistic inversion, the use of model inversions makes it unsuitable for a generic uncertainty analysis system. For some models, the inversion step may simply be too difficult. For more information about this method we refer the reader to Kraan (2002); Kraan and Bedford (2005).

9.3 Iterative algorithms

In this section, we introduce two iterative algorithms applied to solve the probabilistic problem in the next section. First, we introduce the necessary notation, definitions and simple facts and present IPF and PARFUM algorithms for the discrete distributions. We formulate the problem and show results only for the two-dimensional case ($M = 2$). We indicate the results that can be generalized. The generalizations of these algorithms are presented in the Supplement.

Let $K \in \mathbb{N}$; $p_{.j} = \sum_{i=1}^{K} p_{ij}$, $j = 1, \ldots, K$ and $p_{i.} = \sum_{j=1}^{K} p_{ij}$, $i = 1, \ldots, K$. Let

$$S^{K \times K} = \left\{ p \in \mathbb{R}^{K \times K} | p_{ij} \geq 0, \sum_{i,j=1}^{K} p_{ij} = 1 \right\},$$

$$S^{*K \times K} = \{ p \in S^{K \times K} | p_{i,.} > 0, p_{.,j} > 0, i, j = 1, \ldots, K \}$$

be the sets of probability vectors in $\mathbb{R}^{K \times K}$ and in $\mathbb{R}^{K \times K}$ with non-degenerate margins, respectively. Our problem can be now formulated as follows.

For given $p \in S^{*K \times K}$ and $a, b \in S^K$, find a distribution $q \in S^{*K \times K}$ such that

$$I(q|p) = \sum_{i,j=1}^{K} q_{ij} \log \frac{q_{ij}}{p_{ij}} \quad \text{is minimum,}$$

subject to the following constraints

1. $q_{.,j} = b_j$, $j = 1, \ldots, K$;

2. $q_{i,.} = a_i$, $i = 1, \ldots, K$.

Let

$$Q_1 = \{q \in S^{*K \times K} | q_{i,.} = a_i, \ i = 1, \ldots, K\},$$

$$Q_2 = \{q \in S^{*K \times K} | q_{.,j} = b_j, \ j = 1, \ldots, K\}.$$

We shall assume throughout that $a_i > 0$, $b_j > 0$; $i, j = 1, \ldots, K$. As in Csiszar (1975), we define the I-projection of p onto the set of distributions with one fixed margin as a closest distribution, in sense of relative information, with this margin fixed.

Definition 9.1 *Let Q_m, $m = 1, 2$ be as defined in the preceding text. An I-**projection** p^{Q_m} **of p on** Q_m is*

$$p^{Q_m} = argmin_{q \in Q_m} I(q|p).$$

Since Q_m is convex and $p \in S^{*K \times K}$, it follows from Theorem 2.2 of Csiszar (1975) that p^{Q_m} is unique and is of the form

$$p_{ij}^{Q_1} = p_{ij} \frac{a_i}{p_{i,.}}, \quad \left(p_{ij}^{Q_2} = p_{ij} \frac{b_j}{p_{.,j}} \right)$$

for $i, j = 1, \ldots K$, respectively.

9.3.1 Iterative proportional fitting

The IPF algorithm (Kruithof (1937)) projects a starting measure onto the set with fixed first margins (Q_1), then projects this projection onto the set with fixed second margins (Q_2), projects this again onto the set Q_1, and so on. Hence, if we have arrived at vector p by projecting onto Q_1, the next iteration is

$$p' = p^{Q_2}.$$

This algorithm was first used to estimate cell probabilities in a contingency table, subject to certain marginal constraints. It is easy to see that if p satisfies the constraints, then p is a fixed point of this algorithm.

The convergence of the IPF algorithm has been studied by many authors; see Bishop (1967); Brown (1959); Fienberg (1970); Haberman (1974, 1984) and Csiszar (1975). Csiszar shows that if the IPF algorithm converges, then it converges to the I-projection of the starting probability vector on the set of probability vectors satisfying the constraints. He further showed that starting with a probability vector p, IPF converges if there is a vector r satisfying the constraints and having zeros in the cells where p is zero

$$p_{ij} = 0 \Rightarrow r_{ij} = 0, i, j = 1, \ldots, K. \tag{9.3}$$

The reverse implication need not hold.

For the two-dimensional case, the IPF algorithm is equivalent to an alternating minimization procedure studied in Csiszar and Tusnady (1984). It is proven there that if sequences $\{P_n\}$ and $\{Q_n\}$ from $\mathcal{Q}_1$ and $\mathcal{Q}_2$, respectively, are obtained by alternating minimization of $I(P_n|Q_n)$ with respect to P_n resp. Q_n, then $I(P_n|Q_n)$ converges to the infimum of $I(P|Q)$ on $\overline{\mathcal{Q}_1} \times \mathcal{Q}_2$, where $\overline{\mathcal{Q}_1}$ is a set of all $P \in \mathcal{Q}_1$ such that $I(P|Q_n) < \infty$ for some n. Moreover, the convergence of the sequences $\{P_n\}$ and $\{Q_n\}$ is proven. Csiszar and Tusnady (1984) proved this result for a general case where the sets $\mathcal{Q}_1$ and $\mathcal{Q}_2$ are convex sets of finite measures, and the function that is minimized in alternating minimization procedure is an extended real-valued function.

9.3.2 Iterative PARFUM

The iterative PARFUM algorithm is an algorithm based on the result in Proposition 9.4. In contrast to IPF, the iterative PARFUM algorithm projects a starting distribution on $\mathcal{Q}_1$ and on $\mathcal{Q}_2$ and takes the average of these two distributions. If we have arrived at the probability vector p, we define the next iteration p' as follows.

$$p' = \frac{p^{\mathcal{Q}_1} + p^{\mathcal{Q}_2}}{2}. \tag{9.4}$$

Each measure $p^{\mathcal{Q}_i}$ adapts the measure p to have the ith margin fixed. We see that p' is the probability vector that is 'closest', in the sense of relative information, to both the measures $p^{\mathcal{Q}_i}$ (Proposition 9.4). If p satisfies the constraints, then $p' = p$. In other words, the iterative PARFUM algorithm has fixed points at all feasible probability vectors.

The following theorem shows that the relative information functional of the iterates of this algorithm always converge even when the constraints cannot be satisfied.

Theorem 9.1 *Let $\mathcal{Q}_1, \mathcal{Q}_2$ be closed convex subsets of $S^{K \times K}$. For $p(j) \in S^{*K \times K}$, let $q(j)^m$ be the I-projection of $p(j)$ on $\mathcal{Q}_m$, $m = 1, 2$. Let $p(j+1) = \frac{q(j)^1 + q(j)^2}{2}$. Then $I(p(j+1)|p(j)) \to 0$ as $j \to \infty$.*

The set $\mathcal{Q}_m$ may be regarded as the set of vectors satisfying the mth constraint. The term (9.7) (in Supplement) can be written as

$$J(q(j)^1, q(j)^2) = \sum_{m=1}^{2} I(q(j)^i | p(j+1))$$

$$= \sum_{m=1}^{2} I\left(q(j)^m \left| \frac{\sum_{m=1}^{2} q(j)^m}{2}\right.\right).$$

The iterative PARFUM algorithm may be seen as minimizing the function J, and the minimal value of J may be taken as a measure of 'how infeasible' the problem is.

If $Q_1 \cap Q_2 = \emptyset$, then the problem is infeasible. If $Q_1 \cap Q_2 = Q \neq \emptyset$, then the algorithm converges to an element of Q and $J(q(j)^1, q(j)^2) \to 0$ (see Theorem 9.3 below).

A sufficient condition for feasibility of the problem is given in the following theorems:

Theorem 9.2 *Let p be a fixed point of the PARFUM algorithm with $p_{ij} > 0$ for all $i, j = 1, \ldots, K$ then p is feasible, that is, $p_{i,\cdot} = a_i$, $i = 1, \ldots, K$ and $p_{\cdot, j} = b_j$, $j = 1, \ldots, K$.*

Theorem 9.3 *Let $Q_1 \cap Q_2 = Q \neq \emptyset$. If p is a fixed point of the PARFUM algorithm, then $p \in Q$.*

For easier presentation, we have assumed that both margins are from S^K. This can be trivially extended. Moreover, the results presented in this section can be generalized to higher dimensions ($M > 2$). This generalization will be used in the next section to solve probabilistic inversion problem.

9.4   Sample re-weighting

In this section, we show how sample re-weighting combined with iterative algorithms presented in Section 9.3 can solve probabilistic inversion problems. This yields generic methods for probabilistic inversion that do not require model inversion. The idea of re-weighting a sample to perform probabilistic inversion can be sketched roughly as follows. Starting with a random vector $\mathbf{X}$, we generate a large sample from $\mathbf{X} : (X_1, \ldots, X_n)$, $Y_1 = G_1(\mathbf{X}), \ldots, Y_M = G_M(\mathbf{X})$. Let the i-th sample be denoted as $s_i \in \mathbb{R}^{N+M}$. Obviously, each sample $s_i = (x_1^{(i)}, \ldots, x_n^{(i)}, y_1^{(i)}, \ldots, y_M^{(i)})$ has the same probability of occurrence. If N samples have been drawn, then the sampling distribution assigning $p(s_i) = 1/N$ approximates the original distribution from which the samples were drawn.

The idea is now to change the probabilities $p(s_i)$ so as to ensure that the distributions of $Y_1, \ldots, Y_M$ satisfy the specified quantile constraints. No model inversions are performed; that is, we do not invert the function G. To achieve reasonable results, the sample size N may have to be very large. This places strong restrictions on the type of methods that can be implemented in a generic uncertainty analysis program.

9.4.1   Notation

Since the variables $X_1, \ldots, X_n$ play no role in choosing the weights $p(s_i)$, we may leave them out of the problem description, hence $s_i = (y_1^{(i)}, \ldots, y_M^{(i)})$. We denote $p(s_i)$ as p_i, and introduce:

Data matrix: M variables, all with N samples, are grouped in matrix $\mathcal{Y}$. Hence, $\mathcal{Y} = [Y_1, Y_2, \ldots, Y_M]$, where $Y_m = [Y_{m1}, Y_{m2}, \ldots, Y_{mN}]^T$, $m = 1, 2, \ldots, M$.

Inter-quantile vector: We consider a vector $q = [q_1, q_2, \ldots, q_K]$ of lengths of inter-quantile, or inter-percentile, intervals. If we specify the 5%, 50% and 95% quantiles, then $K = 4$ and $q = [0.05, 0.45, 0.45, 0.05]$.

Constraints: A matrix $R = [r_{jm}]$, $j = 1, \ldots, K - 1$; $m = 1, \ldots, M$, contains the percentiles that we want to impose: r_{jm} is the number for which $P\{Y_m \leq r_{jm}\} = q_1 + \cdots + q_j$. Thus, we want the probability vector $[p_1, \ldots, p_N]$ to satisfy the *constraint set* $\mathcal{C}$:

$$\text{for all } k = 1, 2, \ldots, K \text{ and all } m = 1, 2, \ldots, M$$

$$\sum_{i=1}^N p_i \mathbb{I}_{J_{k,m}}(Y_{i,m}) = q_k$$

where $\mathbb{I}_A$ denotes the indicator function of a set A and the interval $J_{k,m}$ is defined as:

$$J_{1m} = (-\infty, r_{1,m}],$$

$$J_{km} = (r_{k-1,m}, r_{km}], \quad k = 2, \ldots, K - 1,$$

$$J_{K,m} = (r_{Km}, \infty),$$

for all $m = 1, 2, \ldots, M$.

We note that these constraints do not really say, for example, r_{m1} is the q_1-th quantile of Y_m. Indeed, the q_1-th quantile is defined as the *least* number, a, satisfying the constraint $P(Y_m < a) = q_1$. If Y_m is concentrated on a few points, then there may be many values satisfying the preceding constraints.

Partitions of samples To each variable Y_m, we associate a partition of the samples

$$\mathcal{A}^m = \{A_k^m\}_{k=1}^K; \quad m = 1, \ldots, M, \tag{9.5}$$

where

$A_k^m = $ set of samples for which Y_m falls in inter-quantile interval k.

The output of a re-weighting scheme is a vector $p = [p_1, p_2, \ldots, p_N]$ of weights. After re-weighting the samples with these weights, the constraints $\mathcal{C}$ are satisfied 'as nearly as possible'.

9.4.2   Optimization approaches

It is natural to approach the problem of probabilistic inversion as a constrained non-linear optimization problem. The constraints $\mathcal{C}$ are linear, and a convex objective function will have a unique solution if the constraints are consistent. Minimal information is a natural choice of objective function. This yields the following optimization problem:

Find $p = [p_1, p_2, \ldots, p_N]$ such that p has minimum information with respect to the uniform distribution, that is,

$$\text{minimize} \quad \sum_{i=1}^{N} p_i \ln(p_i)$$

under the constraints C.

Good performance for this problem has been obtained with an interior point solver *if* the problem is feasible.

If the problem is not feasible – and this is frequently the case – then several lines of attack are suggested, of which we mention two. First, an objective function must be defined, whose optimization will minimize infeasibility. For example, one could minimize the quantity Δ:

$$\Delta = \sum_{m=1}^{M} \sum_{k=1}^{K} \left[\sum_{i=1}^{N} p_i \mathbb{I}_{(r_{k-1,m}, r_{k,m}]}(Y_{i,m}) - q_k \right]^2.$$

Of course, this quantity has nothing to do with minimum information and has nothing in particular to recommend it. Many other choices would be equally defensible. Solvers tested thus far have been unable to handle very large problems.

Another approach would be to relax the constraints C by replacing equality with 'equality up to ϵ'. In this case, the choice of ϵ will be driving the solution, and this choice will be largely heuristic.

To date, experience with infeasible problems has not been wholly satisfactory.

9.4.3   IPF and PARFUM for sample re-weighting probabilistic inversion

Iterative algorithms involve successively updating a probability vector $[p_1, \ldots, p_N]$ so as to approach a solution satisfying constraints C, or satisfying these constraints as closely as possible. Note that samples s_i and s_j, which fall in the same inter-quantile intervals, for $Y_1, \ldots, Y_M$, will be treated in exactly the same way. Hence, the weights for these samples must be the same. Starting with the uniform vector $p_i = 1/N$ for s_i, we may redistribute all samples over K^M cells and obtain K^M-dimensional discrete distribution $[p_{i_1, i_2 \ldots, i_M}]$ in the following way.

$$p_{i_1, i_2 \ldots, i_M} = \frac{1}{N} \sum_{n=1}^{N} \mathbb{I}_{A_{i_1}^1 \cap \cdots \cap A_{i_M}^M}(s_n),$$

where $i_m \in \{1, 2, \ldots, K\}, m = 1, 2, \ldots, M$. $p_{i_1, i_2 \ldots, i_M}$ must be changed in a minimum information manner so that all one-dimensional marginal distributions are $[q_1, q_2, \ldots, q_K]$.

Taking $p_{i_1,i_2...,i_M}$ as a starting distribution, the IPF or PARFUM algorithms discussed in Section 9.3 may be applied to change this distribution so as to satisfy the quantile constraints if these are feasible. Otherwise, a fixed point of PARFUM minimizes infeasibility in the sense of (9.5). The probability mass assigned to the $(i_1, i_2 ..., i_M)$ cell of the stationary distribution must be distributed uniformly over weights corresponding to samples in $A_{i_1}^1 \cap A_{i_2}^2 \cap ... \cap A_{i_M}^M$.

9.5   Applications

We first illustrate PARFUM and IPF with a simple example involving dispersion coefficients from Harper et al. (1994), which is also extensively used in Kraan (2002) to explain the steps of probabilistic inversion technique involving model inversion. We then present results from a recent study involving a chicken processing line.

9.5.1   Dispersion coefficients

The lateral spread σ of an airborne plume dispersing in the downwind direction x is modelled as the power law function of distance x from the source of release:

$$\sigma(x) = Ax^B, \tag{9.6}$$

where the coefficients A and B depend on the stability of atmosphere at the time of release. Of course, there may be more variables such as wind, surface roughness, and so on. It is however recognized that model (9.6) captures the uncertainty associated with plume spread well enough. The joint distribution of A and B must be found in order to find the uncertainty on σ for any downwind distance x. Parameters A and B are not observable, but the plume spread σ at any given downwind distance is an observable quantity. Eight experts were therefore asked to quantify their uncertainty on $\sigma(x_i)$ for downwind distance $x_1, \ldots, x_5$. The experts were asked to express their uncertainty in terms of 5, 50 and 95% quantile points on plum spread at downwind distances of 500 m, 1 km, 3 km, 10 km and 30 km. The performance-based weighted combinations of the experts' distributions led to results presented in Table 9.1 (for more information about structured expert judgment for this example, see Cooke and Goossens (2000b)).

One can see in Kraan (2002) how the distributions of A and B are obtained using the PREJUDICE method. To simplify the presentation, we consider only two

Quantile	$\sigma(500\text{ m})$	$\sigma(1\text{ km})$	$\sigma(3\text{ km})$	$\sigma(10\text{ km})$	$\sigma(30\text{ km})$
5%	33	64.8	175	448	1100
50%	94.9	172	446	1220	2820
95%	195	346	1040	3370	8250

Table 9.1   Assessments of σ.

downwind distances here, that is, 500 m and 1 km, and show how this problem can be solved with iterative procedures PARFUM and IPF. Let us take A to be uniform on the interval $(0,1.5)$ and B to be uniform on the interval $(0.5,1.3)$. Now, by sampling N values of A and B, (a_i, b_i), $i = 1, \ldots, N$, we apply (9.6) to each sample for both downwind distances, say $y_{1i} = a_i 500^{b_i}$ and $y_{2i} = a_i 1000^{b_i}$. We assign each sample to a cell in a 4×4 matrix according to the inter-quantile intervals of Y_1, Y_2 catching the sample. Thus, if $y_{1i} \leq 33$ and $64.8 < y_{2i} \leq 172$, then Y_1 falls in the first inter-quantile interval and Y_2 in the second. Let c be the 4×4 matrix obtained in this way.

The distribution $p(0)$ is obtained from the matrix c by dividing each cell of c by the number of samples $N = 10,000$: $p(0) = \frac{1}{N}c$. Hence, we get

$$
p(0) = \begin{bmatrix} 0.1966 & 0.0006 & 0 & 0 \\ 0.0407 & 0.1642 & 0.0050 & 0 \\ 0 & 0.0094 & 0.1196 & 0.0155 \\ 0 & 0 & 0.0008 & 0.4476 \end{bmatrix}.
$$

It may be noticed that the cells $(4, 1)$, $(4, 2)$ are empty. Indeed, these cells are physically impossible, since a plume cannot be wider at 500 m than at 1000 m.

After 200 iterations, we get the following results using PARFUM and IPF respectively:

$$
p_{PARFUM} = \begin{bmatrix} 0.0439 & 0.0061 & 0 & 0 \\ 0.0061 & 0.4229 & 0.0210 & 0 \\ 0 & 0.0210 & 0.4233 & 0.0057 \\ 0 & 0 & 0.0057 & 0.0443 \end{bmatrix}
$$

$$
p_{IPF} = \begin{bmatrix} 0.0461 & 0.0039 & 0 & 0 \\ 0.0039 & 0.4253 & 0.0208 & 0 \\ 0 & 0.0208 & 0.4270 & 0.0022 \\ 0 & 0 & 0.0022 & 0.0478 \end{bmatrix}
$$

One can verify that both marginal distributions for p_{PARFUM} and p_{IPF} are equal to $q = [0.05, 0.45, 0.45, 0.05]$. The relative information of p_{PARFUM} and p_{IPF} with respect to $p(0)$ is 0.8219 and 0.8166, respectively. The vector of samples weights can be obtained from p_{PARFUM} or p_{IPF} by distributing probabilities in each cell of p_{PARFUM} or p_{IPF} uniformly over all samples that fall in this cell. The number of samples (10,000) is not unrealistic, and iterative algorithms are quite fast. There is, however, no guidance on how the initial distributions of A and B have to be chosen.

Initially, (A, B) had a joint distribution uniform on $[0, 1.5] \times [0.5, 1.3]$. Figure 9.1 shows how the joint distribution of (A, B) has changed after applying PARFUM algorithm. One can see that A and B are now highly negatively correlated.

The marginal distributions of $\sigma(500)$ and $\sigma(1000)$ are also changed to match expert's quantile information. Figure 9.2 shows the difference between the original marginal distributions for $\sigma(500)$ and $\sigma(1000)$ and marginal distributions after applying PARFUM algorithm.

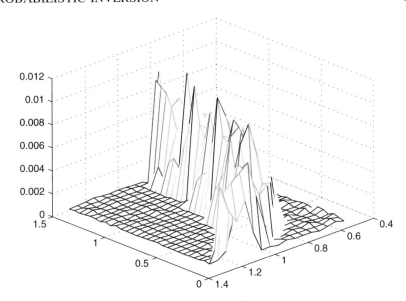

Figure 9.1   A joint distribution of (A, B) after applying PARFUM algorithm.

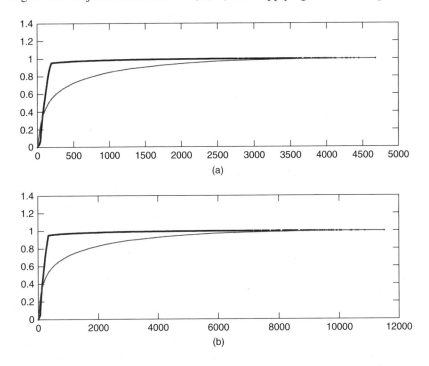

Figure 9.2   Marginal distributions of $\sigma(500)$ (a) and $\sigma(1000)$ (b) before (more diffuse) and after (less diffuse) applying PARFUM algorithm.

9.5.2 Chicken processing line

A model of a chicken processing line is presented in Chapter 1. For details see Van der Fels-Klerx et al. (2005). Suffice to say that for each of the phases, *scalding, evisceration, defeathering*, 5, 50 and 95% quantiles were determined for six predicted variables using structured expert judgment. Probabilistic inversion was used to pull this distribution back onto the parameter space of the model. The parameters in this case are transfer coefficients. The table in Figure 9.3 summarizes the results with IPF and with PARFUM. In each phase, 50,000 samples were generated from a diffuse distribution over the transfer coefficients, and for each sample, the values of the predicted variables were computed. In each case, 50 iterations were performed, beyond which no significant improvements were observed.

In all cases, there was discrepancy between the percentile aimed for and the percentile realized by the probabilistic inversion. This indicated that the inversion problem for the given samples is not feasible. Of course, one option would be to take more samples; however, after some effort in this direction, it was concluded that the expert assessments were not really compliant with the model to be fitted – a situation that arises not infrequently in practice. The table in Figure 9.3 thus illustrates the behaviour of these algorithms on infeasible problems. Note that in each case PARFUM has *lower* relative information with respect to the starting distribution, that is, the distribution assigning each sample the weight 1/50,000. Indeed, IPF converges to the I-projection of the starting measure on the set of feasible measures, but when the problem is infeasible, the set of feasible measures

Prediction variable	Percentile	Evisceration		Scalding		Defeathering	
		PARFUM	IPF	PARFUM	IPF	PARFUM	IPF
1	5%	0.050	0.050	0.060	0.063	0.053	0.014
	50%	0.497	0.500	0.541	0.618	0.424	0.175
	95%	0.950	0.950	0.957	0.962	0.871	0.719
2	5%	0.050	0.050	0.050	0.062	0.370	0.357
	50%	0.497	0.500	0.539	0.618	0.736	0.921
	95%	0.950	0.950	0.957	0.962	0.972	0.993
3	5%	0.090	0.161	0.043	0.050	0.053	0.014
	50%	0.557	0.559	0.464	0.498	0.424	0.175
	95%	0.956	0.956	0.940	0.954	0.871	0.719
4	5%	0.032	0.040	0.033	0.034	0.030	0.033
	50%	0.404	0.403	0.386	0.357	0.256	0.151
	95%	0.851	0.853	0.857	0.912	0.543	0.654
5	5%	0.121	0.116	0.137	0.068	0.184	0.158
	50%	0.532	0.540	0.572	0.531	0.501	0.702
	95%	0.942	0.884	0.959	0.943	0.839	0.969
6	5%	0.065	0.050	0.106	0.050	0.133	0.050
	50%	0.647	0.500	0.628	0.500	0.606	0.500
	95%	0.966	0.950	0.966	0.950	0.972	0.950
Relative information		6.0769949	6.604376	4.591191	6.796448	4.591191	8.42945

Figure 9.3 Comparison PARFUM and IPF – 50,000 samples, 50 iterations

is empty and very little is known about IPF's behaviour. It should be noted that IPF is stopped after cycling through all six variables, hence for the sixth variable, the constraints are always satisfied. Glancing at the other variables, each of whose constraints would be satisfied at an appropriate point in the cycle, we can appreciate that the IPF algorithm is far from converging. We note that in case of infeasibility, PARFUM's performance seems to degrade more gracefully than that of IPF, though no quantitative measures have been applied in this regard.

9.6 Convolution constraints with prescribed margins

The iterative re-weighting algorithms can be used to impose constraints on the joint distribution when these can be expressed as quantile constraints on functions of the margins. To illustrate this, assume that samples $(x_i, y_i), i = 1, \ldots, N$ from variables (X, Y) are given. The samples have to be re-weighted in such a way that the quantiles of the distribution of $X + Y$ will agree with those of the convolution $X * Y$, that is, with the distribution of the sum of independent variables with margins X and Y^2. We conjecture that imposing more constraints of this form will make the distributions of X and Y more independent, as suggested by the following (Girardin and Limnios (2001)) remark.

Remark 9.1 *Let (X, Y) be a random vector such that for all $a, b \in \mathbb{R}, aX + bY$ is distributed as the convolution of aX and bY; then X and Y are independent.*

We illustrate this with the following example:

Example 9.1 *Let U, V, W be three independent standard normal variables. Let $X = U + W$ and $Y = V + W$. Clearly, X and Y are not independent. Let us draw $N = 10,000$ samples from X and Y, say (x_i, y_i). We re-weight these samples such that X and Y satisfy 5, 50 and 95% quantiles for the exponential distribution with parameter $\lambda = 1$ (0.0513, 0.6931, 2.9957). Since, if X and Y are independent exponentials, $X + Y$ has a gamma distribution with scale 2 and shape 1, we require additionally that the sum of samples $x_i + y_i$ satisfies quantile constraints for the gamma distribution (0.3554, 1.6783, 4.7439).*

The PARFUM algorithm was used for this problem and the result is presented in Figure 9.4. We can see in Figure 9.4 how closely the sum of re-weighted samples resembles the gamma distribution. For this example we obtained a very good fit using only three quantiles.

In Figure 9.5 (a) 5000 samples from the original distribution of X and Y are shown. Since $X = U + W$ and $Y = V + W$ with U, V, W independent standard normals, the correlation between X and Y is approximately equal to 0.5.

$^2 X * Y$ is the random variable, the characteristic function of which is a product of the characteristic functions of X and Y.

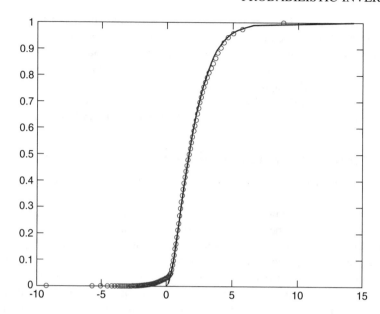

Figure 9.4 The result of re-weighting samples to match three quantiles for the exponential distribution and three quantiles for the gamma distribution for the sum of X and Y. The solid curve represents gamma distribution with scale 2 and shape 1; the 'circle' curve shows the distribution of the sum of re-weighted PARFUM algorithm samples.

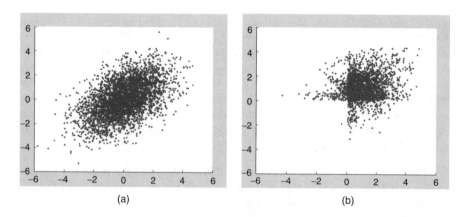

Figure 9.5 Scatter plots with 5000 samples from the original distribution of X and Y (correlation 0.4999) (a) and the distribution after re-weighting to match three quantiles of the exponential distribution and three quantiles of the gamma distribution for the sum of the re-weighted samples (correlation 0.2043) (b).

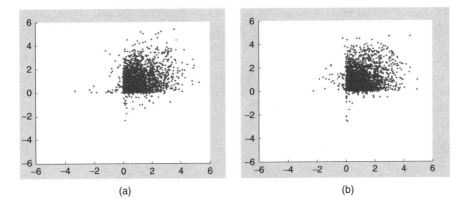

Figure 9.6 Scatter plots with 5000 samples from the distribution after re-weighting to match seven quantiles of the exponential distribution and seven quantiles of $X + Y$ (correlation 0.1449) (a) and the distribution after re-weighting to match seven quantiles of the exponential distribution and seven quantiles for $X + Y$, $X + 2Y$ and $X + 4Y$ (correlation 0.1328) (b).

Figure 9.5 (b) shows the scatter plot of the distribution after re-weighting samples to match three quantiles of the exponential distribution and three quantiles of the gamma distribution for the sum of the re-weighted samples. Since we fit only three quantiles, the probability that the samples will be negative is still quite high. In Figure 9.6 (a) we present the scatter plot for the same case when seven quantiles $(1\%, 5\%, 25\%, 50\%, 75\%, 95\%, 99\%)$ were used. One can see that increasing the number of quantiles leads to a better approximation of the independent exponential distributions. By adding one constraint concerning the sum of the re-weighted samples, we could reduce the correlation to 0.2043 in case of three quantiles and to 0.1449 in case of seven quantiles. Adding additional constraints on linear combinations of X and Y leads to greater reduction of correlation (Figure 9.6 (b)).

For the preceding examples we could easily stipulate quantile constraints on $aX + bY$ for $a > 0$ and $b > 0$ because it is known that if $a = b$, then $aX + bY$ has a gamma distribution, and if $a \neq b$, it has a generalized Erlang distribution with density $f_{aX+bY}(z) = \frac{1}{a-b}e^{-z/a} + \frac{1}{b-a}e^{-z/b}$. If we do not know the distribution of the sum of independent linear combinations of X and Y, then this distribution must first be simulated to extract the quantile constraints.

9.7   Conclusions

We see that iterative algorithms possess attractive features for solving probabilistic inversion problems. These algorithms do not require intelligent steering, other than the choice of the initial sample. Further, they do not pose size restrictions on the sample, so the sample may be large. Of course, large samples will increase run time.

We note that in practice probabilistic inversions are typically infeasible. In such cases, IPF is problematic, as little is known about its behaviour in the case of infeasibility. The iterative PARFUM algorithm, on the other hand, always converges to a distribution that minimizes the sum of relative information of the marginal distributions of the solution, relative to their arithmetic average.

9.8   Unicorn projects

Project 9.1

We consider Example 9.1. Create a case with three standard normal input variables U, V, W. Create UDFs

$$X : U + W$$

$$Y : V + W$$

$$Z : X + Y$$

Evidently, X and Y are not independent. Following Example 9.1, we wish to re-weight a sample file so that X and Y are independent and follow an exponential distribution.

Run this case with 100 runs, saving input and output samples. Name the case 'probinver'. When the simulation is complete, hit 'Prob. Inversion'. This satellite program is an EXCEL spreadsheet. The IPF and PARFUM algorithms are programmed in Visual Basic, and the aficionado is invited to peruse the code to see how it works. This can serve as a model for writing your own post-processing programs, which we hope you will share with us.

When samples are saved, UNICORN creates a *.sam file and an *.sae file, the latter in EXCEL-compliant format. The spreadsheet automatically imports the sae file. Note that you must have given this file a name, otherwise EXCEL has a run time error. On the prompt, select the output variables X, Y and Z for inclusion. The help file here is very summary. To specify the quantile constraints you wish to impose, hit '% Enter quantiles'. You are asked to specify the number of quantiles and their values. Enter '3' and then enter 0.05, 0.50 and 0.95. Next, you are asked to specify the quantile numbers; these are the numbers that will correspond to the 5, 50 and 95% quantiles. Enter the values from Exercise 9.1: For X and Y these are (0.0513, 0.6931, 2.9957) and for Z these are (0.3554, 1.6783, 4.7439). When this is done, hit 'build constraint matrix'. If you entered K quantiles, there are $K + 1$ inter-quantile intervals into which each sample might fall. The constraint matrix consists of $K + 1$ columns for each variable. For each sample a '1' is placed in the column corresponding to the interval in which the sample falls; other columns are assigned '0'. To view the constraint matrix, hit 'see constraint matrix'. You can

Nr.	Variable	Mean	Variance	5% Perc	50% Perc	95% Perc
14	x	9.84E – 001	9.15E – 001	5.90E – 002	6.93E – 001	2.89E + 000
15	y	1.00E + 000	9.78E – 001	3.28E – 002	6.97E – 001	3.02E + 000
16	z	1.99E + 000	2.42E + 000	3.67E – 001	1.69E + 000	4.76E + 000

Figure 9.7 Probabilistic inversion quantile output.

verify that the '0s' and '1s' have been placed correctly by viewing the data and comparing with the quantile numbers you have entered. The first sample should fall beneath the 5% quantile for all three variables, so the first column under each variable contains the '1'. You may have to widen the columns in the data worksheet to view the numbers.

You are now ready to calculate weights. Hit IPF, and choose 20 iterations. The average error on each iteration is shown, as well as the probability of mass falling beneath the specified quantile numbers. In this case, the quantile constraints are satisfied exactly after 20 iterations. For all three variables, the quantiles after re-weighting are 0.05, 0.5 and 0.95, exactly as specified.

You can now hit 'export prob.file'. This writes a file named 'probinver.prb' to the directory where the file probinver.unc lives.

You can now re-sample the sample file with the weights just calculated. To do this, first close UNICORN and re-launch. This closes the open *.sam and *.sae files. Create a new UNICORN file; instead of defining variables and assigning them distributions and dependencies, you will simply use the sample file 'probinver.sam'. To do this hit 'Batch' and choose 'Create from sample file...'. You are shown a list of all resident *.sam files. Choose 'probinver.sam'. You can choose the variables that you wish to import; choose all. You are told that a prb file is not specified; hit 'browse' to see the list of resident prb files. Choose 'probinver.prb', which you have just created.

You are now ready to re-sample. Go to Run/Simulate and choose 200 runs, saving input and output. Sampling takes a bit longer, as UNICORN must read the prb file. When sampling is finished, the report confirms that the quantile constraints have been satisfied up to sampling error (see Figure 9.7).

Go to graphics. The cumulative distribution function for Z should resemble that of Figure 9.4. The scatter plot of X and Y should resemble the one shown in Figure 9.8. To reproduce this scatter plot, you will have to use the sliders to adjust the vertical and horizontal axes. Although the quantile constraints are satisfied, there are samples of X and Y below zero, which of course is impossible for the exponential distribution. As indicated in the text of Example 9.1, you can impose additional constraints and concentrate more weight in the positive quadrant, and you can introduce another UDF $ZZ = X + 2Y$ to make X and Y less dependent.

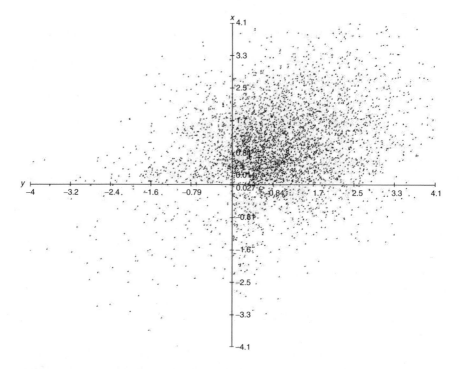

Figure 9.8 Probabilistic inversion scatter plot for X and Y.

9.9 Supplement

9.9.1 Proofs

Theorem 9.1

Let $\mathcal{Q}_1$, $\mathcal{Q}_2$ be closed convex subsets of $S^{K \times K}$. For $p(j) \in S^{*K \times K}$, let $q(j)^m$ be the I-projection of $p(j)$ on $\mathcal{Q}_m$, $m = 1, 2$. Let $p(j+1) = \frac{q(j)^1 + q(j)^2}{2}$. Then $I(p(j+1)|p(j)) \to 0$ as $j \to \infty$.

Proof. Since $\mathcal{Q}_m$ is closed and convex, $q(j)^m$ exists and is unique (Csiszar (1975), Theorem 2.1). Define $F(p(j)) = I(q(j)^1|p(j)) + I(q(j)^2|p(j)) \geq 0$. By Proposition 9.2 and the fact that $q(j+1)^m$ is the I-projection of $p(j+1)$ on $\mathcal{Q}_m$, we have

$$F(p(j)) = I(q(j)^1|p(j)) + I(q(j)^2|p(j)) \tag{9.7}$$

$$\geq I(q(j)^1|p(j+1)) + I(q(j)^2|p(j+1)) \tag{9.8}$$

$$\geq I(q(j+1)^1|p(j+1)) + I(q(j+1)^2|p(j+1)) = F(p(j+1)). \tag{9.9}$$

Equality holds if and only if $p(j) = p(j+1)$. Thus, $F(p(j))$ is decreasing in j and converges. To show that $I(p(j+1)|p(j))$ converges to zero, pick $\varepsilon > 0$

and $j \in \mathbf{N}$ such that $F(p(j)) - F(p(j+1)) < \varepsilon$. Then $\sum_{m=1}^{2} I(q(j)^m | p(j)) - \sum_{m=1}^{2} I(q(j)^m | p(j+1)) < \varepsilon$. Writing this inequality element-wise:

$$\sum_{m=1}^{2} \sum_{i,k=1}^{K} q(j)_{ik}^m (\ln(q(j)_{ik}^m / p(j)_{ik}) - \ln(q(j)_{ik}^m / p(j+1)_{ik})) < \varepsilon.$$

Reversing the order of summation and substituting $\sum_{m=1}^{2} q(j)_{ik}^m = 2p(j+1)_{ik}$, we find $2I(p(j+1) | p(j)) < \varepsilon$. $\square$

Theorem 9.2
Let p be a fixed point of the PARFUM algorithm with $p_{ij} > 0$ for all $i, j = 1, \ldots, K$ then p is feasible, that is, $p_{i,\cdot} = a_i, i = 1, \ldots, K$ and $p_{\cdot,j} = b_j, j = 1, \ldots, K$.

Proof. Since p is a fixed point we have:

$$2p_{ij} = p_{ij} \left(\frac{a_i}{p_{i,\cdot}} + \frac{b_j}{p_{\cdot,j}} \right). \qquad (9.10)$$

Since $p_{ij} > 0$, we may divide both sides by p_{ij} and obtain:

$$\left(2 - \frac{a_i}{p_{i,\cdot}} \right) p_{\cdot,j} = b_j.$$

Summing both sides over j we find

$$\left(2 - \frac{a_i}{p_{i,\cdot}} \right) = 1 = \frac{b_j}{p_{\cdot,j}}.$$

Similarly, we can show that $\frac{a_i}{p_{i,\cdot}} = 1$. $\square$

Theorem 9.3
Let $\mathcal{Q}_1 \cap \mathcal{Q}_2 = \mathcal{Q} \neq \emptyset$. If p is a fixed point of the PARFUM algorithm, then $p \in \mathcal{Q}$.

We have formulated this theorem only for the two-dimensional case. This case is easy to demonstrate and shows all steps of the proof. This theorem can be generalized to higher dimensions and the proof for higher dimensions is similar. Before we start the proof of Theorem 9.3, we introduce the necessary notation, definitions and facts used later on in the proof.

Let $r(0)$ denote a starting point of the PARFUM algorithm and $s \in \mathcal{A}$, hence s is a solution. For two-dimensional case, $r(0)$, p and s are $K \times K$ (K-number of inter-quantile intervals) matrices. It is easy to notice the following:

Proposition 9.3 *The problem will remain unchanged if we exchange row i_1 and i_2 of $p, r(0)$ and s, simultaneously. The same applies to columns.*

Definition 9.2 *We call two columns j^* and j^{**} of p related if there exists a sequence* (i_k, j_k), $k = 1, \ldots, t$ *satisfying*

$$j^* = j_1,$$
$$i_k = i_{k-1} \text{ when } k \text{ is even and } 1 < k < t,$$
$$j_k = j_{k-1} \text{ when } k \text{ is odd and } 1 < k < t,$$
$$j_t = j^{**},$$

where $p_{i_k, j_k} > 0$ for all $k = 1, 2, \ldots, t$.

In prose, the column j^* is related to j^{**} ($j^* \mathcal{R} j^{**}$) if there exists a 'route' through the non-zero cells of p from column j^* to column j^{**}.

Proposition 9.4 *If $j \mathcal{R} j^*$, then*

$$\frac{q_j}{p_{.,j}} = \frac{q_{j^*}}{p_{.,j^*}}$$

Proof.
Since p is a fixed point of PARFUM and $p_{i_k, j_k} > 0$ for all $k = 1, 2, \ldots, t$, we can divide both sides of the equation given below by p_{i_k, j_k}.

$$p_{i_k, j_k} = \frac{p_{i_k, j_k}}{2} \left(\frac{q_{i_k}}{p_{i_k, .}} + \frac{q_{j_k}}{p_{., j_k}} \right)$$

Applying this to all cells in the 'route' we get

$$2 = \left(\frac{q_{i_1}}{p_{i_1, .}} + \frac{q_{j_1}}{p_{., j_1}} \right) = \left(\frac{q_{i_1}}{p_{i_1, .}} + \frac{q_j}{p_{., j}} \right)$$

$$2 = \left(\frac{q_{i_2}}{p_{i_2, .}} + \frac{q_{j_2}}{p_{., j_2}} \right) = \left(\frac{q_{i_1}}{p_{i_1, .}} + \frac{q_{j_2}}{p_{., j_2}} \right)$$

$$2 = \left(\frac{q_{i_3}}{p_{i_3, .}} + \frac{q_{j_3}}{p_{., j_3}} \right) = \left(\frac{q_{i_3}}{p_{i_3, .}} + \frac{q_{j_2}}{p_{., j_2}} \right)$$

$$2 = \left(\frac{q_{i_4}}{p_{i_4, .}} + \frac{q_{j_4}}{p_{., j_4}} \right) = \left(\frac{q_{i_3}}{p_{i_3, .}} + \frac{q_{j_4}}{p_{., j_4}} \right)$$

$$\ldots$$

$$2 = \left(\frac{q_{i_{t-1}}}{p_{i_{t-1}, .}} + \frac{q_{j_{t-1}}}{p_{., j_{t-1}}} \right) = \left(\frac{q_{i_t}}{p_{i_t, .}} + \frac{q_{j_{t-1}}}{p_{., j_{t-1}}} \right)$$

$$2 = \left(\frac{q_{i_t}}{p_{i_t, .}} + \frac{q_{j_t}}{p_{., j_t}} \right) = \left(\frac{q_{i_t}}{p_{i_t, .}} + \frac{q_{j^*}}{p_{., j^*}} \right).$$

Thus,

$$\frac{q_j}{p_{.,j}} = \frac{q_{j_1}}{p_{.,j_1}} = \frac{q_{j_2}}{p_{.,j_2}} = \ldots = \frac{q_{j^*}}{p_{.,j^*}},$$

which concludes the proof. $\square$

We call $\frac{q_j}{p_{.,j}}$ the weight of the column j. Thus, the preceding proposition says that if two columns are related then they have the same weights.

Proof of Theorem 9.3.
We start the proof by rearranging columns and rows of $r(0)$, p and s, simultaneously.

Step 1. Choose one column of p, say column 1, and find all columns related to 1. Rearrange these columns to the left of p. We denote this changed distribution by $T_1 p$, where T_1 is a matrix operator in step 1. $T_1 p$ is a fixed point of PARFUM such that columns 1 to n_1 are related and not related with any other columns.

Step 2. Next, we rearrange rows of $T_1 p$. All rows such that $\sum_1^{n_1} T_1 p_{i,j} \neq 0$ are moved to the upper part of the matrix. Let us assume that we have found m_1 rows in this manner; hence, in this step we have obtained the matrix $T_2 T_1 p$ with the 'lower-left' sub-matrix consisting of 0s (it can of course happen that $m_1 = K$).

If $m_1 < K$, then we claim that upper-right sub-matrix of $T_2 T_1 p$ consists of 0s. Otherwise, there exists $T_2 T_1 p_{i^*, j^*} > 0$ where $i^* \leq m_1$, $j^* > n_1$. By the construction, there exists $j^{**} \leq n_1$ satisfying $T_2 T_1 p_{i^*, j^{**}} > 0$; thus, columns j^* and j^{**} are related, which is impossible because we have chosen all n_1 related columns already.

Step 3. By induction we can rearrange the whole p, finding all related columns, hence the matrix transformation T_3 of this step can be written as

$$T_3 p = T_2^l T_1^l T_2^{l-1} T_1^{l-1} \ldots T_2^1 T_1^1 p$$

and the obtained matrix can be written as

$$T_3 p = \begin{bmatrix} \Gamma_1 & & & \\ & \Gamma_2 & & \\ & & \ldots & \\ & & & \Gamma_l \end{bmatrix}$$

Step 4. In this step we rearrange Γ_i such that corresponding to columns of Γ_i weights are in increasing order. We also combine Γ_i and Γ_j if their corresponding columns weights are equal. After this step we obtain the following matrix.

$$T_4 T_3 p = \begin{bmatrix} \Omega_1 & & & \\ & \Omega_2 & & \\ & & \ldots & \\ & & & \Omega_r \end{bmatrix}$$

such that weights w_i corresponding to columns of sub-matrices Ω_i satisfy $w_1 < w_2 < \ldots < w_r, r \leq l$.

All matrix operations were performed on $p, r(0)$ and s. From now on, when we refer to $p, r(0)$ or s, we mean $T_4 T_3 p, T_4 T_3 r(0)$ or $T_4 T_3 s$, respectively.

Since the problem is feasible, there exists a solution s satisfying the marginal constraints q that is absolutely continuous with respect to the starting point $r(0)$. After transformations in Steps 1–4, we get that

$$p = \begin{bmatrix} \Omega_1 & 0_1 \\ 0_2 & \Omega_{others} \end{bmatrix},$$

where 0_1 and 0_2 are matrices of zeros with respective sizes and

$$\Omega_{others} = \begin{bmatrix} \Omega_2 & & & \\ & \Omega_3 & & \\ & & \cdots & \\ & & & \Omega_r \end{bmatrix}.$$

The rearranged solution is:

$$s = \begin{bmatrix} \Delta_1 & \Delta_2 \\ \Delta_3 & \Delta_4 \end{bmatrix},$$

where the sizes of respective matrices in p and s are the same. Let Ω_1 and Δ_1 be $m_1 \times n_1$ matrices. In order to prove our theorem, we must show that $r = 1$ and hence that $p = \Omega_1$. Suppose on the contrary that $r > 1$, then w_1 has to be smaller than, 1 since $w_2 > w_1$. By the definition of weight, we get

$$\frac{q_j}{p_{.,j}} = w_1, j = 1, 2, \ldots, n_1,$$

$$\frac{q_i}{p_{i,.}} = 2 - w_1, i = 1, 2, \ldots, m_1.$$

If we denote the sum of all elements of the matrix X by $\| X \|$, then we get

$$\sum_{j=1}^{n_1} q_j = w_1 \sum_{j=1}^{n_1} p_{.,j} = w_1 \| \Omega_1 + 0_2 \| = w_1 \| \Omega_1 + 0_1 \|$$

$$= w_1 \sum_{i=1}^{m_1} p_{i,.} = \frac{w_1}{2 - w_1} \sum_{i=1}^{m_1} q_i.$$

Since $w_1 < 1$, $\frac{w_1}{2 - w_1} < 1$, so from the above we get that $\sum_{j=1}^{n_1} q_j < \sum_{i=1}^{m_1} q_i$.

The solution s satisfies the marginal constraints q, hence from the above we get

$$\| \Delta_1 + \Delta_3 \| = \sum_{j=1}^{n_1} q_j < \sum_{i=1}^{m_1} q_i = \| \Delta_1 + \Delta_2 \|,$$

which implies that

$$\| \Delta_2 \| > \| \Delta_3 \| \geq 0.$$

From the above, it is seen that there exists at least one cell in Δ_2 that is non-zero. Let $s_{i^*, j^*} > 0$, $i^* \leq m_1$, $j^* > n_1$. s is absolutely continuous with respect to $r(0)$, hence $r(0)_{i^*, j^*} > 0$. Notice also that $p_{i^*, j^*} = 0$, since it is one of the elements of 0_1.

We have a situation where a non-zero cell in the initial distribution $r(0)$ converges to zero by the PARFUM algorithm. Let us consider the following.

$$\frac{1}{2} \left(\frac{q_{i^*}}{r(n)_{i^*, .}} + \frac{q_{j^*}}{r(n)_{., j^*}} \right) \to \frac{1}{2} \left(\frac{q_{i^*}}{p_{i^*, .}} + \frac{q_{j^*}}{p_{., j^*}} \right) = \lambda,$$

where $r(n)$ denotes nth step of the PARFUM algorithm. The $(n + 1)$th step of PARFUM algorithm is given

$$r(n + 1)_{i^*, j^*} = \frac{r(n)_{i^*, j^*}}{2} \left(\frac{q_{i^*}}{r(n)_{i^*, .}} + \frac{q_{j^*}}{r(n)_{., j^*}} \right).$$

Since $r(n)_{i^*, j^*} \to 0$ as $n \to \infty$, there exists a monotonically convergent subsequence of $r(n)_{i^*, j^*}$ that converges to zero. Hence

$$\lambda \leq 1. \tag{9.11}$$

Since $i^* \leq m_1$, $\frac{q_{i^*}}{p_{i^*, .}} = 2 - w_1$. We also know that $\frac{q_{j^*}}{p_{., j^*}} = w_f$, where $1 < f < r$ and $w_1 < w_f$. Hence

$$\lambda = \frac{1}{2} \left(\frac{q_{i^*}}{p_{i^*, .}} + \frac{q_{j^*}}{p_{., j^*}} \right) = \frac{1}{2}(2 - w_1 + w_f) > \frac{1}{2}2 = 1.$$

This contradicts (9.11); hence $r = 1$ that implies that p has both margins equal to q and hence p is a solution. $\square$

Remark 9.2 *Note that in the proof of Theorem 9.3, both margins are equal to q. The proof will be valid if the margins are different.*

9.9.2 IPF and PARFUM

Relationship between IPF and PARFUM IPF and PARFUM algorithms are both easy to apply. In case of a feasible problem, IPF converges quite fast to the *I*-projection of a starting distribution onto a set of distributions with given margins. Some remarks about the 'speed' of convergence can be found in Haberman (1974). For infeasible problems in two dimensions, IPF oscillates between two distributions (Csiszar and Tusnady (1984)). The PARFUM algorithm is shown to have fixed points minimizing an information functional, even if the problem is not feasible, and is shown to have only feasible fixed points if the problem is feasible. The algorithm is not shown to converge, but the relative information of successive iterates is shown to converge to zero. The PARFUM algorithm is slower then IPF. We show now that for two-dimensional problem there is a relationship between fixed points of PARFUM and oscillation point of IPF.

Proposition 9.5 *Let r be a fixed point of PARFUM and let p, q be I-projections of r onto $\mathcal{Q}_1$ and $\mathcal{Q}_2$, respectively, then q is the I-projection of p on $\mathcal{Q}_2$ and p is the I-projection of q on $\mathcal{Q}_1$.*

Proof. We get

$$p_{ij} = r_{ij} \frac{a_i}{r_{i,\cdot}} \tag{9.12}$$

and

$$q_{ij} = r_{ij} \frac{b_j}{r_{\cdot,j}}. \tag{9.13}$$

Since r is a fixed point of PARFUM algorithm, $r = \frac{1}{2}(p + q)$, hence from (9.12)

$$p_{ij} = \frac{\frac{1}{2}(p_{ij} + q_{ij})a_i}{\frac{1}{2}(p_{i,\cdot} + q_{i,\cdot})}.$$

Since $p \in \mathcal{Q}_1$, $p_{i,\cdot} = a_i$, and from the above we get

$$p_{ij}(a_i + q_{i,\cdot}) = (p_{ij} + q_{ij})a_i.$$

Hence

$$\frac{q_{ij}}{p_{ij}} = \frac{q_{i,\cdot}}{a_i}.$$

Similarly, from (9.13)

$$\frac{q_{ij}}{p_{ij}} = \frac{b_j}{p_{\cdot,j}}.$$

Now, assuming that z is an I-projection of p onto $\mathcal{Q}_2$ we get that $z_{ij} = p_{ij} \frac{b_j}{p_{\cdot,j}}$, and from the above

$$\frac{z_{ij}}{p_{ij}} = \frac{b_j}{p_{\cdot,j}} = \frac{q_{ij}}{p_{ij}}.$$

Hence $z = q$. Similarly, we can show that p is an I-projection of q onto $\mathcal{Q}_1$. $\square$

We show now that the average of the oscillating distributions of IPF is a fixed point of PARFUM.

Proposition 9.6 *Let $p \in \mathcal{Q}_1$ and $q \in \mathcal{Q}_2$ be such that q is the I-projection of p on $\mathcal{Q}_2$ and p is the I-projection of q on $\mathcal{Q}_1$, then $r = \frac{1}{2}(p + q)$ is a fixed point of the PARFUM algorithm.*

Proof. We notice that for all $i = 1, \ldots, K$, $j = 1, \ldots, M$, $p_{ij} = 0$ if and only if $q_{ij} = 0$. Hence even $r_{ij} = 0$ if and only if $p_{ij} = 0$. Since

$$p_{ij} = q_{ij} \frac{a_i}{q_{i,\cdot}} \quad q_{ij} = p_{ij} \frac{b_j}{p_{\cdot,j}},$$

for all i, j such that $p_{ij} \neq 0$

$$\frac{a_i}{q_{i,\cdot}} \frac{b_j}{p_{\cdot,j}} = 1. \tag{9.14}$$

To show that r is a fixed point of PARFUM is enough to show that

$$\left(\frac{a_i}{r_{i,\cdot}} + \frac{b_j}{r_{\cdot,j}} \right) = 2.$$

We get

$$\frac{a_i}{r_{i,\cdot}} = \frac{a_i}{0.5(a_i + q_{i,\cdot})} = \frac{1}{0.5 \left(1 + \frac{q_{i,\cdot}}{a_i} \right)}$$

$$\frac{b_j}{r_{\cdot,j}} = \frac{b_i}{0.5(p_{\cdot,j} + b_j)} = \frac{1}{0.5 \left(1 + \frac{p_{\cdot,j}}{b_j} \right)}.$$

From the above and using (9.14) we obtain

$$\frac{a_i}{r_{i,\cdot}} + \frac{b_j}{r_{\cdot,j}} = \left(\frac{1}{0.5 \left(1 + \frac{q_{i,\cdot}}{a_i} \right)} + \frac{1}{0.5 \left(1 + \frac{p_{\cdot,j}}{b_j} \right)} \right) = 2. \;\; \square$$

Remark 9.3 *Notice that from Propositions 9.5 and 9.6 it does not follow that the fixed point of the PARFUM algorithm is equal to an average of oscillating distributions of the IPF algorithm.*

In the example below, we illustrate the performance of IPF and PARFUM in case of an infeasible problem. You can see that the average of oscillating distributions of the IPF algorithm is different from the fixed point of PARFUM.

Example 9.2 *Let us consider a distribution:*

$$p = \begin{bmatrix} 0.1 & 0 & 0.1 & 0.1 \\ 0 & 0 & 0.1 & 0.1 \\ 0.1 & 0 & 0 & 0 \\ 0.1 & 0.1 & 0.1 & 0.1 \end{bmatrix}.$$

We would like to find the closest distribution to p, in a sense of relative information, with both margins equal to $a = b = [0.05, 0.45, 0.45, 0.05]$.

Clearly, this problem is infeasible. The PARFUM algorithm leads to the following result (marginal distributions printed in bold).

0.05	0.0000	0	0.0491	0.0009
0.45	0	0	0.4009	0.0491
0.25	0.2500	0	0	0
0.25	0.0000	0.2500	0.0000	0.0000
	0.25	**0.25**	**0.45**	**0.05**

The IPF algorithm cycles between following two distributions.

0.05	0.0000	0	0.0450	0.0050
0.45	0	0	0.4050	0.0450
0.05	0.0500	0	0	0
0.45	0.0000	0.4500	0.0000	0.0000
	0.05	**0.45**	**0.45**	**0.05**

0.05	0.0000	0	0.0450	0.0050
0.45	0	0	0.4050	0.045
0.45	0.4500	0	0	0
0.05	0.0000	0.0500	0.0000	0.0000
	0.45	**0.05**	**0.45**	**0.05**

The average of oscillating distributions of IPF is:

0.05	0.0000	0	0.0450	0.0050
0.45	0	0	0.4050	0.0450
0.25	0.2500	0	0	0
0.25	0.0000	0.2500	0.0000	0.0000
	0.25	**0.25**	**0.45**	**0.05**

Let us denote the PARFUM result by p_{PAR} and the oscillating distributions of IPF by p_{IPF}^1 and p_{IPF}^2. p_{IPF}^1 and p_{IPF}^2 have column and row margins equal to [0.05 0.45 0.45 0.05], respectively. Moreover, let us denote the average of oscillating distributions of IPF as p_{IPF}^{avr}. Notice that the marginal probabilities in the first and second rows and the fourth and third columns of p_{PAR} and both p_{IPF}^1 and p_{IPF}^1, as well as p_{IPF}^{avr}, are 0.05 and 0.45, hence cells in these rows and columns will be invariant, under projections. Projecting p_{PAR} onto $\mathcal{Q}_1$ sets $p_{PAR}^1(3,1) = 0.05$ and $p_{PAR}^1(4,2) = 0.45$; similarly, taking the projection of p_{PAR} onto $\mathcal{Q}_2$ makes $p_{PAR}^2(3,1) = 0.45$ and $p_{PAR}^2(4,2) = 0.05$. Now it can be easily noticed that the projection of p_{PAR}^2 onto $\mathcal{Q}_1$ is equal to p_{PAR}^1 and the projection of p_{PAR}^1 onto $\mathcal{Q}_2$ is equal to p_{PAR}^2, as assured in Proposition 9.5. Moreover, p_{IPF}^{avr} is a fixed point of the PARFUM algorithm. However, the average of oscillating distributions of IPF is not equal to the PARFUM result.

p_{PAR} and p_{IPF}^{avr} have the same marginal distributions, hence the sum of relative information obtained with respect to required margins is the same and equal to 0.3681. However, the relative information of p_{PAR} and p_{IPF}^{avr} with respect to the starting distribution p are 0.9407 and 0.9378, respectively. From the above, we conclude that the best distribution for this problem is p_{IPF}^{avr}.

Results of this subsection, however, do not generalize to higher dimensions. It is not clear that in case of infeasibility, in higher dimensions, the IPF algorithm will oscillate. Moreover, using arguments similar to those of Proposition 9.5, we can easily see that already for the three-dimensional problem, projections of a fixed point of PARFUM, do not give an oscillating triple for IPF.

Generalizations of IPF and PARFUM For easier presentation of both the IPF and the PARFUM algorithms we have assumed that both required margins are from $\mathcal{S}^K$. This can be trivially extended. Moreover, the results presented in Section 9.3 can be generalized to higher dimensions ($M > 2$). Iterative algorithms can easily be adopted to satisfy joint and marginal constraints (Kullback (1971)). Generalizations of the IPF to the continuous case have been introduced in Ireland and Kullback (1968); Kullback (1968). However, the convergence of the IPF in the continuous case, under certain regularity conditions, was proven much later in Rüschendorf (1995). We present here a general formulation of IPF and PARFUM following a very compact notation used in Whittaker (1990).

Let $N = \{1, 2, \ldots, n\}$ and $X_N = (X_1, \ldots, X_n)$. Suppose that X_N is portioned into two sub-vectors $X_N = (X_a, X_b)$. Let the starting distribution be denoted by g_{ab}^0 and the required margins as $[f_a, f_b]$. The IPF procedure scales the rows according to

$$g_{ab}^{2k+1} = g_{ab}^{2k} \frac{f_a}{g_a^{2k}} = g_{b|a}^{2k} f_a, \quad k = 0, 1, \ldots \tag{9.15}$$

and the columns according to

$$g_{ab}^{2k+2} = g_{ab}^{2k+1} \frac{f_b}{g_a^{2k+1}} = g_{a|b}^{2k+1} f_b, \quad k = 0, 1, \ldots \tag{9.16}$$

until some convergence criterion is met. In this case the PARFUM algorithm scales the rows according to (9.15) and the columns according to (9.16) and averages them to produce the next iteration.

$$g_{ab}^k = g_{ab}^{k-1} \left(\frac{f_a}{g_a^{k-1}} + \frac{f_b}{g_b^{k-1}} \right), \quad k = 1, 2, \ldots \tag{9.17}$$

Proposition 9.2 extends easily to this case. We must show that for all distributions P

$$I(g_{b|a}^{k-1} f_a | P) + I(g_{a|b}^{k-1} f_b | P) \geq I(g_{b|a}^{k-1} f_a \mid g_{ab}^k) + I(g_{a|b}^{k-1} f_b \mid g_{ab}^k).$$

Hence

$$I(g_{b|a}^{k-1} f_a \mid P) - I(g_{b|a}^{k-1} f_a \mid g_{ab}^k) + I(g_{a|b}^{k-1} f_b \mid P) - I(g_{a|b}^{k-1} f_b \mid g_{ab}^k) \geq 0.$$

From the definition of relative information and (9.17), we get that the above inequality is equivalent to

$$2I(p_{ab}^{k-1} \mid P) \geq 0,$$

which always holds (Kullback (1959)). Using the generalized version of Proposition 9.4 in the preceding text, the proof of Theorem 9.1 goes through in this case. Using the notation in the preceding text, we must show that if the PARFUM process is given by (9.17), then

$$I(g_{ab}^{k+1}) \mid g_{ab}^k) \to 0 \text{ as } k \to \infty.$$

As in the proof of Theorem 9.1, we define

$$F(g^k_{ab}) = I(g^k_{a|b} f_b \mid g^k_{ab}) + I(g^k_{b|a} f_a \mid g^k_{ab})$$

By the generalized version of Proposition 9.2 in the preceding text and because $g^{k+1}_{a|b} f_b$, $g^{k+1}_{b|a} f_a$ are projections of g^{k+1}_{ab} onto sets of distributions with margins equal to f_b and f_a, respectively, we get

$$F(g^k_{ab}) \geq I(g^k_{a|b} f_b \mid g^{k+1}_{ab}) + I(g^k_{b|a} f_a \mid g^{k+1}_{ab})$$
$$\geq I(g^{k+1}_{a|b} f_b \mid g^{k+1}_{ab}) + I(g^{k+1}_{b|a} f_a \mid g^{k+1}_{ab}) = F(g^{k+1}_{ab}).$$

Hence, $F(g^k_{ab})$ decreases in k and converges. To see that $I(g^{k+1}_{ab} \mid g^k_{ab})$ converges to zero, pick ϵ and $k \in \mathbf{N}$ such that

$$F(g^k_{ab}) - F(g^{k+1}_{ab}) < \epsilon.$$

We get

$$F(g^k_{ab}) - F(g^{k+1}_{ab})$$
$$\geq I(g^k_{a|b} f_b \mid g^k_{ab}) + I(g^k_{b|a} f_a \mid g^k_{ab}) - I(g^k_{a|b} f_b \mid g^{k+1}_{ab}) - I(g^k_{b|a} f_a \mid g^{k+1}_{ab})$$
$$= 2I(g^{k+1}_{ab} \mid g^k_{ab}).$$

Hence, $I(g^{k+1}_{ab} \mid g^k_{ab})$ converges to zero.

The above can be presented in a more general set-up by considering a set $\{a_1, \ldots, a_m\}$ of subsets of N such that no a_i is contained in a_j and $N = \cup a_i$. The starting distribution is g^0_N and the required margins are the those of the distribution f_N. The IPF algorithm proceeds by cycling through the subsets

$$a = a_i, \quad i = 1, \ldots, m$$

and adjusting each margin in turn by

$$g^{k+1}_{ab} = g^k_{b|a} f_a,$$

where $b = N \setminus a_i$.

The PARFUM algorithm adjusts each margin and averages the results to obtain the next iteration. Denoting $a^c_i = N \setminus a_i$, we obtain

$$g^{k+1}_{ab} = \frac{1}{m} \sum_{i=1}^{m} g^k_{a^c_i|a_i} f_{a_i}.$$

10

Epilogue: Uncertainty and the United Nation Compensation Commission

10.1 Introduction

Continents do not collide with a bang. Systems of reasoning in science and law are two continents colliding with excruciating languor. Each has its own rules for dealing with uncertainty, and these rules are very different. In jurisprudence, the traditional scientific expert simply reports on a scientific consensus, and the individuality of the expert plays no role, provided he can competently report what any competent expert would report. Think of the testimony of a ballistics expert. There is no 'expert uncertainty'.

However, expert input in legal proceedings is often sought in cases where the experts are not certain, and where all experts do not say the same thing. Differences in scientific and legal methods of reasoning under uncertainty were brought sharply into focus in a recent sitting of the United Nations Compensation Commission (UNCC) UNCC. A short summary of these events forms a fitting conclusion to this book.

The Governing Council of the UNCC recently completed 12 years of claims processing at the UNCC for damage claims against Iraq that resulted from the 1990–1991 Gulf War. Since the appointment of the first panels in 1993, over 2.68 million claims, seeking approximately USD 354 billion in compensation, have been resolved by the various panels of Commissioners. Awards of approximately USD 52.5 billion have been approved, representing roughly 14.8% of the amount claimed.

Uncertainty Analysis with High Dimensional Dependence Modelling   D. Kurowicka and R. Cooke
© 2006 John Wiley & Sons, Ltd

10.2   Claims based on uncertainty

The final sitting included claims from Kuwait for damage to public health resulting from the oil fires. These claims were based on modelling the dispersion of particulate matter from the oil fires, estimating the public exposure and estimating the health consequences. Although most experts agree that exposure to particulate matter increases mortality, especially because of heart attack, the actual dose–response relation is uncertain. The Iraqi lawyers accordingly noted:

In all domestic or international tort law systems, in order to succeed with a claim for damages, it has to be proven with certainty that damage or harm to a legally protected interest, for example to health, life or property, has actually occurred. Statistical evidence that damage might have occurred is not sufficient in any private law system. (Transcripts, UNCC, Wednesday, 15th September 2004, p. 28, lines 9–15)

The Kuwaiti claim was based on recognition of the uncertainty in the dose–response relation. Indeed, they endeavoured to quantify this uncertainty in the best scientific manner. Figure 10.1 shows the uncertainty of six pre-eminent experts in this area about the percentage increase in mortality among Kuwaiti nationals in Kuwait at the time of the fires and the number of excess deaths attributable to the fires. Evidently, the uncertainties are large.

Equally evident, all experts agreed that there will most likely be excess deaths due to the oil fires; no one believes that the fires have no effect. Quantifying the uncertainty in attributable deaths for the Commission means somehow combining each expert's individual uncertainties. Figure 10.1 also shows the results of combining the experts with equal weighting and with 'performance-based weighting'. The latter is sketched very briefly in Chapter 2, in which further references are given[1]. The Kuwaiti lawyers thus acknowledged the uncertainty and based a claim on the performance-based decision maker's 35 expected deaths due to the oil fires.

In adjudicating these claims, the UNCC was free to base its decisions on such precedents that it may choose from national tort law. The Iraqi lawyers pointed out:

What is the role of statistical evidence in a situation of uncertainty, such as the present one? In some legal systems, for example the French, Belgian, German or Austrian systems, in a situation of different potential sources of harm the plaintiff needs to prove with a probability close to certainty that a given source – in our case burning oil wells – caused him harm. In others, mostly common law systems, notably the laws of the US and England, in order to presume causation the plaintiff needs to show that it is more probable than not – that means that there is a 51 per cent probability at least – that a certain cause, for example toxic emissions, caused his

---

[1] Suffice to say that the experts also assessed uncertainty for changes in mortality in London and Athens during periods of heightened pollution from particulate matter, as well as fluctuations in concentration peaks. Performance on these assessment tasks formed the basis for the performance-based combination.

```
Kuwait percentage increase in mortality
Experts
   1 [--<---*---------->------------ ]
   2 <*> ]
   3 > ]
   4 [ ----<------*>------------------------------------------------------------------------]
   5 [ --<-*--->--]
   6 [ -<-*->-]
per [<*=>============================= ]
equ [<===*=====>=================================== ]
   ~~~~~~~~~~~~~~~~~~~~~~~~~~~~~~~~~~~~~~~~~~~~~~~~~~~~~~~~
       0.000338                                                      11.25

Kuwait: number or attributable deaths
Experts
   1 [---------<------*------------------>----------------------------------------------------]
   2 |
   3 >- ]
   4 <>--- ]
   5 * ]
   6 |
per >=== ]
equ *>=========================================== ]
   ~~~~~~~~~~~~~~~~~~~~~~~~~~~~~~~~~~~~~~~~~~~~~~~~~~~~~~~~~
       0.5                                                          1.15E004
```

Figure 10.1 Increase in mortality among the Kuwaiti national population exposed to oil fires (above) and the attributable deaths (below). The 90% central confidence bands are indicated by '[,] ', 50% central confidence bands by '<, >' and median's by '#'.

disease. A probability of causation below this threshold is not sufficient to establish a presumption of causation. (Transcripts, UNCC, Wednesday, 15th September 2004, p. 31, lines 10–25)

The continents collide when the Iraqi lawyers quip:

The authors [of the Kuwaiti risk assessment] themselves acknowledge that there is a considerable uncertainty that may lead to estimates of excess death, ranging from zero to perhaps several hundreds. This shows, also, that the authors were not very certain of the results. (Transcripts, UNCC, Wednesday, 15th September 2004, p. 13, lines 3–7)

The authors were quite certain about the uncertainty.

10.3 Who pays for uncertainty

The state of Kuwait did not argue that specific deaths were caused by pollution from the oil fires. We never determine 'the' cause of a heart attack. Acknowledging the uncertainty, Kuwait argued that this same uncertainty entailed 35 excess expected deaths. Does this warrant compensation? The UNCC's decision was negative:

However, the evidence submitted by Kuwait is not sufficient to demonstrate either that 35 premature deaths actually occurred or that any such premature deaths were the direct result of the invasion and occupation. In particular, Kuwait provides no information on the specific circumstances of actual deaths that would enable the Panel to determine whether such premature deaths could reasonably be attributed, wholly or partially, to factors resulting from Iraq's invasion and occupation. Consequently, Kuwait has failed to meet the evidentiary requirements for compensation as specified in article 35(3) of the Rules. Accordingly, the Panel recommends no compensation for this claim unit. (UNCC Decision, June 30, 2005, p. 89)

Significantly, the Commission did not dispute the assessment of uncertainty leading to 35 expected deaths, but they required that the 'specific circumstances of actual deaths' be reasonably linked to the oil fires. Of course, this places a proof burden on victims that far exceeds what is scientifically achievable. Statistical deaths are thus not actionable.

Pithily put, probable guilt is not guilt. This principle has much to recommend it. We do not wish to return to the jurisprudence of the Inquisition, in which the severity of sentence for witchcraft or heresy was proportional to the probability of guilt. Denying the existence of witches was already good grounds for suspicion, as this would entail belief that the Church was torturing and burning innocent women, a manifest heresy.

The Inquisitors also invoked experts to remove uncertainty. The typical sentence read: 'And wishing to conclude your trial in a manner beyond all doubt, we convened in solemn council men learned in the Theological Faculty and in the Canon and Civil Laws, and having diligently examined and discussed each circumstance of the process and maturely and carefully considered with the said learned men everything which has be said and done in this present case, we find that you have been legally convicted of having been infected with the sin of heresy...'. But mercy tempers justice: 'But if it should happen that after the sentence, and when the prisoner is already at the place where he is to be burned, he should say that he wishes to confess the truth ... although it may be presumed that he does this rather from fear of death than for love of the truth, yet I should be of the opinion that he may in mercy be received as a penitent heretic and be imprisoned for life' (Kramer and Sprenger (1971, first published in 1486), p. 212, p. 261).

Unlike the present case, the learned Doctors of the Theological Faculty never faltered in the removal of doubt. Continents collide slowly.

Bibliography

Bedford T and Cooke R 2001a *Probabilistic Risk Analysis; Foundations and Methods*. Cambridge University Press, Cambridge, United Kingdom.

Bedford T and Cooke R 2001b Probability density decomposition for conditionally dependent random variables modeled by vines. *Annals of Mathematics and Artificial Intelligence* **32**, 245–268.

Bedford T and Cooke R 2002 Vines – a new graphical model for dependent random variables. *Annals Of Statistics* **30**(4), 1031–1068.

Bier V 1983 *A Measure of Uncertainty Importance for Components in Fault Trees*. Phd thesis, Laboratory for Information and Decision systems, MIT Boston.

Bishop Y 1967 *Multidimensional Contingency Tables: Cell Estimates*. Phd dissertation, Harvard University.

Bojarski J 2002 A new class of band copulas – distributions with uniform marginals. *Journal of Mathematical Sciences* **111**(3), 3520–3523.

Brown D 1959 A note on approximations to discrete probability distributions. *Information and Control* **2**, 386–392.

Budnitz R, Apostolakis G, Boore D, Cluff L, Coppersmith K, Cornell C and Morris P 1997 Recommendations for probabilistic seismic hazard analysis: guidance on uncertainty and use of experts. Technical Report NUREG/CR-6372; UCRL-ID-122160 Vol. 1, Lawrence Livermore National Laboratory.

Callies U, Kurowicka D and Cooke R 2003 Graphical models for the evaluation of multisite temperature forecasts: comparison of vines and independence graphs. *Proceedings of ESREL 2003, Safety and Reliability* **1**, 363–371.

Cario M and Nelsen B 1997 Modeling and generating random vectors with arbitrary marginal distributions and correlation matrix. Technical report, Department of Industrial Engineering and Menagement Sciences, Northwestern University, Evanston.

Cheng J, Bell D and Liu W 1997 Learning bayesian networks from data: and efficient approach based on information theory. *Proceedings of the 6th ACM International Conference on Information and Knowledge Management*, Las Vegas, USA.

Clemen R and Winkler R 1995 Screening probability forecasts: contrasts between choosing and combining. *International Journal of Forecasting* **11**, 133–146.

Clemen R and Winkler R 1999 Combining probabiity distributions from experts in risk analysis. *Risk Analysis* **19**, 187–302.

Clemen R, DeWispelare A and Herren L 1995 The use of probability elicitation in the high-level waste regulation program. *International Journal of Forecasting* **11**, 5–24.

Cleveland W 1993 *Vizualizing Data.* AT&T Bell Laboratories, Murray Hill, New Jersey.

Cooke R 1991 *Experts in Uncertainty.* Oxford University Press, New York.

Cooke R 1994 Parameter fitting for uncertain models: modelling uncertainty in small models. *Reliability Engineering and System Safety* **44**, 89–102.

Cooke R 1997a Markov and entropy properties of tree and vines-dependent variables *Proceedings of the ASA Section of Bayesian Statistical Science*, Anaheim, USA.

Cooke R 1997b Uncertainty modeling: examples and issues. *Safety Science* **26**(1/2), 49–60.

Cooke R 2004 The anatomy of the squizzel, the role of operational definitions in science. *Reliability Engineering and System Safety* **85**(1/3), 313–321.

Cooke R and Goossens L 2000a Procedures guide for structured expert judgment. Technical Report EUR 18820 EN, European Commission, Directorate-General for Research, Nuclear Science and Technology, Brussels.

Cooke R and Goossens L 2000b Procedures guide for structured expert judgment in accident consequence modelling. *Radiation Protection Dosimetry* **90**(3), 303–309.

Cooke R and Noortwijk JV 1998 Local probabilistic sensitivity measures for comparing form and monte carlo calculations illustrated with dike ring reliability calculations. *Computer Physics Communications* **117**, 86–98.

Cooke R and Noortwijk Jv 2000 *Graphical Methods for Uncertainty and Sensitivity Analysis* in Saltelli, Chan and Scott, Wiley, New York.

Cooke R and Waij R 1986 Monte carlo sampling for generalized knowledge dependence with application to human reliability. *Risk Analysis* **6**, 335–343.

Cooke R, Kurowicka D and Meilijson I 2003 Linearization of local probabilistic sensitivity via sample re-weighting. *Reliability Engineering and System Safety* **79**, 131–137.

Cooke R, Nauta M, Havelaar A and van der Fels I 2006 Probabilistic invesion for chicken processing lines. *Appearing in Reliability Engineering and System Safety.*

Cowell R, Dawid A, Lauritzen S and Spiegelhalter D 1999 *Probabilistic Networks and Expert Systems* Statistics for Engineering and Information Sciences. Springer-Verlag, New York.

Crick J, Hofer E, Johnes J and Haywood S 1988 Uncertainty analysis of the foodchain and atmospheric dispersion modules of marc. Technical Report 184, National Radiological Protection Board.

Csiszar I 1975 I-divergence geometry of probability distributions and minimization problems. *Annals of Probability* **3**, 146–158.

Csiszar I and Tusnady G 1984 Information geometry and alternating minimization procedures. *Statistics & Decisions* **1**, 205–237.

Dall'Aglio G, Kotz S and Salinetti G 1991 *Probability Distributions with Given Marginals; Beyond the Copulas.* Kulwer Academic Publishers.

Doruet Mari D and Kotz S 2001 *Correlation and Dependence.* Imperial College Press, London.

Duintjer-Tebbens R 2006 *Quantitative Analysis of Trade-offs and Input Sensitivities in Public Health.* Phd thesis, TU Delft Department of Mathematics, Delft University of Technology.

Edmunds W, Medley G and Nokes D 1999 Evaluating the cost-effectiveness of vaccination programmes: a dynamic perspective. *Statistics in Medicine* **18**(23), 3263–3268.

Embrechts P, McNeil A and Straumann D 2002 Correlation and dependence in risk management: properties and pitfalls. *In Risk Management: Value at Risk and Beyond* ed. M.A.H. Dempster, pp. 176–223.

EXCEL 1995 for windows 95. *Microsoft Corporation*.

Fels-Klerx Hvd, Cooke R, Nauta M, Goossens L and Havelaar A 2005 A structured expert judgment study for a model of campylobacter contamination during broiler chicken processing. *Risk Analysis* **25**, 109–124.

Ferguson T 1995 A class of symmetric bivariate uniform distributions. *Statistical Papers* **35**(1), 31–40.

Fienberg S 1970 An iterative procedure for estimation in contingency tables. *Annals of Mathematical Statistics* **41**, 907–917.

Fischer F, Ehrhardt J and Hasemann I 1990 Uncertainty and sensitivity analyses of the complete program system ufomod and selected submodels. Technical Report 4627, Kernforschungzentrum Karlsruhe.

Fisher R 1935 *Design of Experiments*. Oliver and Boyd.

Fox B 1986 Implementation and relative efficiency of quasirandom sequence generators. *ACM Transactions on Mathematical Software* **12**(4), 362–376.

Frank M 1979 On the simultaneous associativity of f(x,y) and x+y-f(x,y). *Aequationes Mathematicae* **19**, 194–226.

Fréchet M 1951 Sur les tableaux de corrélation dont les marges sont données. *Annales de l'Université de Lyon* **Sec A, Ser.3,14**, 53–77.

Garthwaite P, Kadane J and O'Hagan A 2005 Elicitation. *Journal of the American Statistical Association* **100**(407), 680–701.

Genest C and MacKay J 1986 The joy of copulas: bivariate distributions with uniform marginals. *American Statistician* **40**, 280–283.

Ghosh S and Henderson S 2002 Properties of the notra method in higher dimensions. *Proceedings of the 2002 Winter Simulation Conference*, San Diego, USA, pp. 263–269.

Girardin V and Limnios N 2001 *Probabilités en vue des Applications*. Vuibert, Paris.

Glivenko V 1933 Sulla determinazione empirica della legge di probabilita. *Gior. Ist. Ital. Attuari* **4**, 92–99.

Goossens L, Cooke R and Kraan B 1998 *Evaluation Of Weighting Schemes For Expert Judgment Studies*, vol. PSAM 4 Proceedings, eds. A Moselh and RA Bari. Springer, pp. 1937–1942.

Goossens L, Harrison J, Kraan B, Cooke R, Harper F and Hora S 1997 Probabilistic accident consequence uncertainty analysis: uncertainty assessment for internal dosimetry. Technical Report, NUREG/CR-6571, EUR 16773, SAND98-0119, vol. 1, Brussels.

Gould S 1981 *The Mismeasure of Man*. W.W. Norton and Company.

Granger Morgan M and Henrion M 1990 *Uncertainty: A Guide to Dealing with Uncertainty in Quantitative Risk and Policy Analysis*. Cambridge University Press, London.

Greenberger M 1961 An a priori determinantion of serial correlation in computer generated random numbers. *Mathematics of Computation* **15**, 383–389.

Haberman S 1974 *An Analysis of Frequency Data*. University Chicago Press.

Haberman S 1984 Adjustment by minimum discriminant information. *Annals of Statistics* **12**, 971–988.

Hackerman D 1998 *Learning in Graphical Models* Kluwer chapter A tutorial on learning with Bayesian networks.

Halton J 1960 On the efficiency of certain quasi-random sequences of points in evaluating multidimensional integrals. *Numerische Mathematik* **2**, 84–90.

Hamby D 1994 A review of techniques for parameter sensitivity analysis of environmental models. *Environmental Monitoring and Assesment* **32**(2), 135–154.

Hardin C 1982 On the linearity of regression. *Zeitschrift Fur Wahrscheinlichkeitstheorie Und Verwandte Gebiete* **61**, 293–302.

Harper F, Goossens L, Cooke R, Hora S, Young M, Pasler-Ssauer J, Miller L, Kraan B, Lui C, McKay M, Helton J and Jones A 1994 Joint usnrc cec consequence uncertainty study: summary of objectives, approach, application, and results for the dispersion and deposition uncertainty assessment. Technical Report VOL. III, NUREG/CR-6244, EUR 15755 EN, SAND94-1453, Sandia National Laboratories, Albuquerque, New Mexico, USA.

Helton J 1993 Uncertainty and sensitivity analysis techniques for use in performance assesment for radioactive waste disposal. *Reliability Engineering and System Safet* **42**(2-3), 327–367.

Hoeffding W 1940 Masstabinvariante korrelationstheorie. *Schrijtfen des Mathematischen Instituts und des Instituts für Angewandte Mathematik der Universität Berlin* **5**, 179–233.

Hogarth R 1987 *Judgment and Choice*. Wiley, New York.

Homma T and Saltelli A 1996 Importance measures in global sensitivity analisis of nonlinear models. *Reliability Engineering and System Safety* **52**, 1–17.

Hora S and Iman R 1989 Expert opinion in risk analysis: the nureg-1150 methodology. *Nuclear Science and Engineering* **102**, pp. 102–323.

Iman R and Conver W 1982 A distribution-free approach to inducing rank correlation among input variables. *Communications in Statistics-Simulation and Computation* **11**(3), 311–334.

Iman R and Davenport J 1982 Rank correlation plots for use with correlated input variables. *Communications in Statistics-Simulation and Computation* **11**(3), 335–360.

Iman R and Helton J 1985 A comparison of uncertainty and sensitivity analysis techniques for computer models. Technical Report, NUREG/CR-3904 SAND84-1461 RG, Sandia National Laboratories, Albuquerque, New Mexico, USA.

Iman R and Helton J 1988 Investigation of uncertainty and sensitivity analysis techniques for computer models. *Risk Analysis* **8**, 71.

Iman R and Shortencarier M 1984 A fortran 77 program and user's guide for the generation of latin hypercube and random samples for use with computer models. Technical Report NUREG/CR-3624, Sandia National Laboratories, Albuquerque.

Iman R, Helton J and Campbell J 1981 An approach to sensitivity analysis of computer models. *Journal of Quality Technology*.

Ireland C and Kullback S 1968 Contingency tables with given marginals. *Biometrika* **55**, 179–188.

Ishigami T and Homma T 1990 An importance quantification technique in uncertainty analysis for computer models. *Proceedings of the ISUMA '90 First International Symposium on Uncertainty Modeling and Analysis*. University of Maryland, pp. 398–403.

Jensen F 1996 *An Introduction to Bayesian Networks*. Taylor and Francis, London.

Jensen F 2001 *Bayesian Networks and Decision Graphs*. Springer-Verlag, New York.

Joe H 1997 *Multivariate Models and Dependence Concepts*. Chapman & Hall, London.

Kahn H and Wiener A 1967 *The Year 2000, A Framework for Speculation*. Macmillan Publishing, New York.

Kamin L 1974 *The Science and Politics of I.Q.* Lawrence Erlbaum Associates, Potomac, Marhyland.

Kendall M 1938 A new measure of rank correlation. *Biometrika* **30**, 81–93.

Kendall M and Stuart A 1961 *The Advenced Theory of Statistics*. Charles Griffin & Company Limited, London.

Kiiveri H and Speed T 1984 *Sociological Methodology* Jossey Bass San Francisco chapter Structural analysis of multivariate data.

Kleijnen J and Helton J 1999 Statistical analyses of scatterplots to identify important factors in large-scale simulations. Technical Report SAND98-2202, Sandia National Laboratories, Albuquerque, New Mexico, USA.

Kocis L and Whiten W 1997 Computational investigations of low-discrepancy sequences. *ACM Transactions on Mathematical Software* **23**(2), 266–294.

Koster A 1995 Dualnet een Grafische Interface Voor het Oplossen van Netwerk Problemen. Masters Thesis, Department of Mathematics, TU Delft.

Kraan B 2002 *Probabilistic Inversion in Uncertainty Analysis and Related Topics*. Isbn 90-9015710-7, PhD dissertation, Delft University of Technology, Delft, the Netherlands.

Kraan B and Bedford T 2005 Probabilistic inversion of expert judgments in the quantification of model uncertainty. *Management Science* **51**(6), 995–1006.

Kraan B and Cooke R 1997 Post processing techniques for the joint cec/usnrc uncertainty analysis of accident consequence codes. *Journal of Statistical Computation and Simulation* **57**, 243–259.

Kraan B and Cooke R 2000a Processing expert judgments in accident consequence modeling. *Radiation Protection Dosimetry* **90**(3), 311–315.

Kraan B and Cooke R 2000b Uncertainty in compartmental models for hazardous materials – a case study. *Journal of Hazardous Materials* **71**, 253–268.

Kramer H and Sprenger J 1971, first published in 1486 *The Malleus Maleficarum*. Dover Publications, New York.

Kruithof J 1937 Telefoonverkeersrekening. *De Ingenieur* **52**(8), E15–E25.

Kullback S 1959 *Information Theory and Statistics*. John Wiley & Sons, New York.

Kullback S 1967 A lower bound for discrimination information in tetms of variation. *IEEE Transactions on Information Theory* **IT-13**, 126–127.

Kullback S 1968 Probability densities with given marginals. *The Annals of Mathematical Statistics* **39**(4), 1236–1243.

Kullback S 1971 Marginal homogeneity of multidimensional contingency tables. *The Annals of Methematical Statistics* **42**(2), 594–606.

Kurowicka D 2001 *Techniques in Representing high Dimensional Distributions*. Phd thesis, Delft University of Technology. ISBN 9040721475.

Kurowicka D and Cooke R 2000 Conditional and partial correlation for graphical uncertainty models. *Recent Advances in Reldiability Theory*. Birkhauser, Boston, Massachusetts, pp. 259–276.

Kurowicka D and Cooke R 2001 Conditional, partial and rank correlation for elliptical copula; dependence modeling in uncertainty analysis. *Proceedings of ESREL 2001* Turyn, Italy.

Kurowicka D and Cooke R 2003 A parametrization of positive definite matrices in terms of partial correlation vines. *Linear Algebra and its Applications* **372**, 225–251.

Kurowicka D, Misiewicz J and Cooke R 2000 Elliptical copulae. *Proceedings of the International Conference on Monte Carlo Simulation – Monte Carlo* pp. 209–214.

Lauritzen S and Spiegelhalter D 1998 Local computations with probabilities on graphical structures and their application to expert systems. *Journal of the Royal Statistical Society* **50**(B), 157–224.

Lewandowski D 2004 Generalized diagonal band copulae – applications to the vine–copula method. Presented at the *DeMoSTAFI 2004 Conference in Quebec*, Canada.

Marshall A and Olkin I 1979 *Inequalities: Theory of Majorization and its Applications*. Academic Press, San Diego, California.

McKay M, Beckman R and Conover W 1979 A comparison of three methods for selecting values of input variables in the analysis of output from a computer code. *Technometrics* **21**(2), 239–245.

Meeuwissen A 1993 *Dependent Random Variables in Uncertainty Analysis*. PhD thesis, Delft University of Technology, Delft, Netherlands.

Meeuwissen A and Bedford T 1997 Minimally informative distributions with given rank correlation for use in uncertainty analysis. *Journal of Statistical Computation and Simulation* **57**(1-4), 143–175.

Meeuwissen A and Cooke R 1994 Tree dependent random variables. Technical Report 94-28, Department of Mathematics, Delft University of Technology.

Misiewicz J 1996 Substable and pseudo-isotropic processes. connections with the geometry of subspaces of l_α-spaces. *Dissertationes Mathematicae*.

Morris M 1991 Factorial sampling plans for preliminary computational experiments. *Technometrics* **33**(2), 161–174.

Nauta M, Fels-Klerx Hvd and Havelaar A 2004 A poultry processing model for quantitative microbiological risk assessment. *Risk Analysis*, **25**, 85–98.

Nelsen R 1999 *An Introduction to Copulas*. Springer, New York.

Niederreiter H 1988 Low-discrepancy and low-dispersion sequences. *Journal Number Theory* **30**, 51–70.

Niederreither H 1992 Random number generation and quasi-monte carlo methods. *CBMS-NSF Regional Conference Series in Applied Mathematics 63*. SIAM, Philadelphia, Pennsylvania.

Nowosad P 1966 On the integral equation kf=1/f arising in a problem comunication. *Journal of Mathematical Analysis and Applications* **14**, 484–492.

O'Hagan A and Oakley J 2004 Probability is perfect, but we can't elicit it perfectly. *Reliability Engineering and System Safety* **85**, 239–248.

van Overbeek F 1999 *Financial Experts in Uncertainty*. Master's thesis, Delft University of Technology.

Papefthymiou G, Tsanakas A, Schavemaker P and Sluis vdL 2004 Design of 'distributed' energy systems based on probabilistic analysis. *8th International Conference on Probability Methods Applied to Power Systems (PMAPS)*. Iowa State University, Ames, Iowa.

Papefthymiou G, Tsanakas A, Schavemaker P and van der Sluis L 2005 Stochastic bounds for power systems uncertainty analysis, part i: theory. *IEEE Transactions on Power Systems* (in press).

Pearl J 1988 *Probabilistic Reasoning in Intelligent Systems: Networks of Plausible Inference.* Morgan Kaufman Publishers, San Mateo, California.

Pearson K 1904 Mathematical contributions to the theory of evolution. *Biometric* **Series I**, 1–34.

Press W, Flannery B, Teukolsky S and Vetterling W 1992 *Numerical Recipes in C : The Art of Scientific Computing*, 2nd edn. Cambridge University Press.

Rao C 1973 *Linear statistical Inference and its Applications.* Wiley, New York.

Ripley B 1987 *Stochastic Simulation.* Wiley, New York.

Roelen A, Wever R, Hale A, Goossenes L, Cooke R, Lopuhaa R, Simons M and Valk P 2003 Casual modeling for integrated safety at airport. *Proceedings of ESREL 2003, Safety and Reliability* **2**, 1321–1327.

Rosenblat M 1952 Remarks on a multivariate transformation. *Annals of Mathematical Statistics* **27**, 470–472.

Rubinstein R 1981 *Simulation and the Monte Carlo Method.* John Wiley & Sons, New York.

Rüschendorf L 1995 Convergence of the iterative proportional fitting procedure. *The Annals of Statistics* **23**(4), 1160–1174.

Saltelli A, Chan K and Schott E 2000 *Sensitivity Analysis.* Wiley, New York.

Saltelli A, Tarantola S, Compolongo F and Ratto M 2004 *Sensitivity Analsis in Practice. A guide to Assessing Scientific Models.* Wiley, Chichester.

Scott M and Saltelli A 1997 Special on theory and applications of sensitivity analysis of model output in computer simulation. *Journal of Statistical Computation and Simulation.*

Shachter R and Kenley C 1989 Gaussian influence diagrams. *Menagement Science* **35**(5), 527–550.

Shepard O 1930 *The Lore of the Unicorn.* Avenel Books, New York.

Sklar A 1959 Fonctions de réparation à n dimensions et leurs marges. *Publications de l'Institut de Statistique de l'Université de Paris* **8**, 229–231.

Sobol I 1967 The distribution of points in a cube and the approximate evaluation of integrals. *USSR Computational Mathematics and Mathematical Physics* **7**(4), 86–112.

Sobol I 1993 Sensitivity analisis for nonlinear mathematical models. *Mathematical Modeling & Computational Experiment* **1**, 407–414.

Spearman C 1904 The proof and measurment of association between two things. *American Journal of Psychology* **15**, 72–101.

SPSS 1997 for windows. *8.00.*

Trenberth K and Paolino D 1980 The northern hemisphere slp-database: trends, errors and discontinuities. *Monthly Weather Review* **108**, 855–872.

Tucker H 1967 *A Graduate Course in Probability.* Academic Press, New York.

Wand M and Jones M 1995 *Kernel Smoothing.* Chapman & Hall, London.

Wegman E 1990 Hyperdimensional data analysis using parallel coordinates. *Journal of the American Statistical Association* **90**(411), 664–675.

Whittaker J 1990 *Graphical Models in Applied Multivariate Statistics.* John Wiley & Sons, Chichester.

Whittle P 1992 *Probability via Expectation.* Springer-Verlag, New York.

Xu Q 2002 *Risk Analysis on Real Estate Investment Decision-Making.* PhD thesis, Delft University of Thechnology.

Yule G and Kendall M 1965 *An Introduction to the Theory of Statistics*, 14th edn. Charles Griffin & Company, Belmont, California.

Index

Bayesian belief nets (bbn's), 11,
  135–145, 146, 148, 159,
  160
  continuous, 11, 137
  discrete, 136
  non-parametric continuous,
    139, 178
bootstrap, 151

calibration, 14, 15
cobweb plot, **8**, 9, 12, 92, 101, 114,
  116, 117, 188–190, 195,
  199–201, 203, 205–207,
  234, 237
completion problem, **52**, **87–111**,
  111
copula, 12, 20, 21, 26, 32, **34**, 36,
  54, 55, 62, 70, 88, 91,
  98–100, 106, 114, 139,
  140, 154, 162–164, 172,
  173, 178, 180, 184
  archimedean, **45–46**
  diagonal band, 20, 22, **37–40**,
    68, 100, 105–107, 172,
    174, 176, 188
  elliptical, 10, 22, **42–45**, 49,
    62, 68, 104–106, 117, 188
  Fréchet, **36**
  Frank's, 21, 22, **45**, 49, 62,
    100, 105–107, 114, 117,
    158, 180, 188
  generalized diagonal band,
    41–42, 49

minimum information, 22,
  47–49, 49, 60, 62, 100,
  114, 173
  multivariate, 54
  normal, **34**
correlation, 7, 8, 15, 25, 26, 34–38,
  41–43, 46, 48, 50, 52–54,
  56, 60–62, 68, 92,
  99, 100, 102, 109, 110,
  114, 117, 170, 171, 174,
  181, 188, 191, 193, 199,
  200, 206, 218, 222, 223,
  229, 232, 257–259
  conditional, 26, **33**, 44, 55,
    87, 100, 102, 103,
    139, 141, 154, 188, 197,
    229
  Kendall's tau, 26, **32**, 35, 51,
    54
  matrix, 52, 54, 55, 61, 67, 84,
    86, 87, 92, 102, 105,
    107–109, 111, 113, 117,
    147, 150, 151, 171
  multiple, **34**
  partial, 26, **32–33**, 44, 54, 55,
    87, 101–105, 108, 109,
    111, 112, 148–152, 154,
    181, 218
  product moment, 12, 25,
    26–30, 51, 52, 54, 62, 67,
    85, 165, 200, 203, 218,
    233
  matrix, 11

correlation (*continued*)
    rank, 8, 12, 13, 17–19, 25, 26,
        30–32, 35, 40, 41, 47, 49,
        51, 55, 56, 60–62, 85,
        90–92, 99–101, 103–105,
        113, 114, 117, 118,
        139–142, 158, 159,
        172–174, 176, 180–182,
        187, 190, 191, 193, 194,
        196, 197, 218, 236
    matrix, 7, 8, 12
correlation ratio, 218, **219**, 220, 223,
        232, 233
    computing, 220–223
        kernel estimation, 221
        max square correlation, 222
        pedestrian method, 220

dependence, 3, 5, 7, 8, 10–13, 17,
        20–22, 25, 26, 34, 36, 51,
        54–57, 60, 83–92, 94, 98,
        99, 113, 114, 117, 139,
        149, 158, 159, 163, 164,
        170, 172, 184, 186, 187,
        195, 206
    linear, 27
Design of experiments, **212–217**
    fractional factorial design,
        216
    full factorial design, 214
determinant, 33, 34, 54, 67, 102,
        103, 149–151
    distribution, 151
    sampling distribution, 151
Discrepancy, **166**
distribution
    beta, 57, 115, 157
    chi-square, 151
    constant, 57
    exponential, 195, 206, 257,
        259–261
    gamma, 163, 257, 259
    generalized Erlang, 259
    lognormal, 184

loguniform, 57, 58
    normal, 11, 12, 29, 31–33, 44,
        54, 60, 84, 85, 87, 103,
        137, 138, 148, 149, 151,
        154, 163, 171, 187, 188,
        203, 228, 229, 237, 257
    bivariate, 21, 22, 34, **50–51**,
        55
    multivariate, 53, 54, **54–55**,
        138
    uniform, 20, 60, 68, 92, 113,
        157, 159, 172, 200, 205,
        244, 254
    Weibull, 186

expert judgment, 1–3, 5, 10, 13–15,
        83, 137, 139, 142, 146,
        203, 243, 253, 256

independence, **26**, 27, 36, 45, 50,
        55, 61, 62, 100, 139, 140,
        151, 162, 163, 244
    conditional, 12, 100, 135, 136,
        139, 140, 145, 146, 148,
        154, 162
independence graphs, 11, 135, **145**,
        146, 152, 154
independent, 3–5, 12, 17, 18, 26,
        26, 27, 30–33, 35–37,
        40–42, 45, 46, 49, 51, 56,
        60–62, 71, 72, 83, 85, 87,
        90, 91, 100, 102, 106,
        111, 113, 115, 117, 136,
        138–140, 145, 146, 149,
        151, 161, 162, 164,
        168–174, 179, 180, 182,
        184, 187, 190, 193–196,
        203, 205, 207, 212, 220,
        222–224, 226, 228, 229,
        236, 237, 245, 257, 259,
        260
    conditionally, 33, 90, 100, 101,
        117, 136, 139, 146, 148
information, 13–16, 18, 19, **47**, 49,
        84, 91, 100, 147, 148,

231, 244, 246, 248, 249,
    252, 254, 256, 260, 267,
    269–271
Inquisition, 276
iterative proportional fitting (IPF),
    87, 149, 150, 154, 244,
    247, **248**, 249, 253, 254,
    256, 260

majorization, **150**
min cost flow networks , 154
Morris'method, **209–212**
multiplicative congruential method,
    164

NETICA, 181–184
normal transform, **84–87**, 105–107,
    113

OAT, 210

PARFUM, 244, 246, 247, **249**, 250,
    253, 254, 256, 257, 260
positive definiteness, 11, 12, 17, 52,
    52, 53, 56, 84, 85, 87,
    102, 107–109, 111, 148,
    149
PREJUDICE, 247, 253
probabilistic dissonance, 202,
    205
probabilistic inversion, 6, 7, 9–12,
    16, 20, **243**, 244, 247,
    250–253, 256, 259–261

radar graphs, 189–192, 203–205

sampling
    continuous bbn's, **178–181**
    Latin hypercube, 163, **170–171**
    pseudo random, 163, **164–165**
    quasi-random, 163, **165–168**
    regular vines, 11, **173–178**
    stratified, 163, **168–170**
    trees, **172–173**

scatter plot, 21, 22, 189, 192–194,
    205, 259, 261, 262
Schur convex, **150**, 151
sensitivity analysis, 11, 189, 191,
    209
sensitivity measures, 203, 205
    global, 191, **218–226**
    local, 197, 199, **226–231**
    FORM, 226, **226–227**
    LPSM, 203, 226, **227–231**
Sobol indices, 218, **223–225**
social networks, 158

tornado graphs, 189–191,
    205
Total effect indices, 218, **225–226**
trees, 11, 20, 56, 60, **88–92**, 94–96,
    98, 100, 111–114, 117,
    148, 152, 172, 173, 177,
    178, 184, 186, 187
    bivariate, **89**
    copula, 84, **89**, 99, 113

UNICORN, 10, 11, **11**, 18, 20, 21,
    24, 56–58, 92, 109, 113,
    115, 154, 155, 159, 160,
    184, 185, 188, 193, 195,
    205, 223, 229, 232, 240,
    260, 261
unicorns, 1
United Nations Compensation
    Commission, 273

vines, 11, 12, 17, 20, 54, 84, 87, 91,
    94–98, 102, 105, 107,
    109, 111, 113, 137, 139,
    146, 172–174, 178,
    182–184, 186, 187
    bivariate, **98**
    C-vine, **95**, 96, 104, 105, 108,
    109, 112, 173, 174, 187,
    188
    copula, **98**, 103, 105, 107, 113,
    135

vines (*continued*)
    D-vine, **95**, 96, 100, 112, 114,
        174–176, 179, 188, 232
    non-regular, 94
    normal, **103**, 181

partial correlation, **102–118**
rank correlation, 103, 107, 181
regular, 94, 95, 98–100, 102,
    103, 105, 107, 111, 173,
    177, 188

WILEY SERIES IN PROBABILITY AND STATISTICS

ESTABLISHED BY WALTER A. SHEWHART AND SAMUEL S. WILKS

Editors
*David J. Balding, Peter Bloomfield, Noel A. C. Cressie, Nicholas I. Fisher,
Iain M. Johnstone, J. B. Kadane, Geert Molenberghs, Louise M. Ryan,
David W. Scott, Adrian F. M. Smith, Jozef L. Teugels*
Editors Emeriti
Vic Barnett, J. Stuart Hunter, David G. Kendall

The *Wiley Series in Probability and Statistics* is well established and authoritative. It covers many topics of current research interest in both pure and applied statistics and probability theory. Written by leading statisticians and institutions, the titles span both state-of-the-art developments in the field and classical methods.

Reflecting the wide range of current research in statistics, the series encompasses applied, methodological and theoretical statistics, ranging from applications and new techniques made possible by advances in computerized practice to rigorous treatment of theoretical approaches.

This series provides essential and invaluable reading for all statisticians, whether in academia, industry, government, or research.

ABRAHAM and LEDOLTER · Statistical Methods for Forecasting
AGRESTI · Analysis of Ordinal Categorical Data
AGRESTI · An Introduction to Categorical Data Analysis
AGRESTI · Categorical Data Analysis, *Second Edition*
ALTMAN, GILL, and McDONALD · Numerical Issues in Statistical Computing for the Social Scientist
AMARATUNGA and CABRERA · Exploration and Analysis of DNA Microarray and Protein Array Data
ANDĚL · Mathematics of Chance
ANDERSON · An Introduction to Multivariate Statistical Analysis, *Third Edition*
*ANDERSON · The Statistical Analysis of Time Series
ANDERSON, AUQUIER, HAUCK, OAKES, VANDAELE, and WEISBERG · Statistical Methods for Comparative Studies
ANDERSON and LOYNES · The Teaching of Practical Statistics
ARMITAGE and DAVID (editors) · Advances in Biometry
ARNOLD, BALAKRISHNAN, and NAGARAJA · Records
*ARTHANARI and DODGE · Mathematical Programming in Statistics
*BAILEY · The Elements of Stochastic Processes with Applications to the Natural Sciences
BALAKRISHNAN and KOUTRAS · Runs and Scans with Applications
BARNETT · Comparative Statistical Inference, *Third Edition*
BARNETT · Environmental Statistics: Methods & Applications
BARNETT and LEWIS · Outliers in Statistical Data, *Third Edition*
BARTOSZYNSKI and NIEWIADOMSKA-BUGAJ · Probability and Statistical Inference
BASILEVSKY · Statistical Factor Analysis and Related Methods: Theory and Applications
BASU and RIGDON · Statistical Methods for the Reliability of Repairable Systems
BATES and WATTS · Nonlinear Regression Analysis and Its Applications
BECHHOFER, SANTNER, and GOLDSMAN · Design and Analysis of Experiments for Statistical Selection, Screening, and Multiple Comparisons
BELSLEY · Conditioning Diagnostics: Collinearity and Weak Data in Regression

*Now available in a lower priced paperback edition in the Wiley Classics Library.

BELSLEY, KUH, and WELSCH · Regression Diagnostics: Identifying Influential Data and Sources of Collinearity

BENDAT and PIERSOL · Random Data: Analysis and Measurement Procedures, *Third Edition*

BERNARDO and SMITH · Bayesian Theory

BERRY, CHALONER, and GEWEKE · Bayesian Analysis in Statistics and Econometrics: Essays in Honor of Arnold Zellner

BHAT and MILLER · Elements of Applied Stochastic Processes, *Third Edition*

BHATTACHARYA and JOHNSON · Statistical Concepts and Methods

BHATTACHARYA and WAYMIRE · Stochastic Processes with Applications

BIEMER, GROVES, LYBERG, MATHIOWETZ, and SUDMAN · Measurement Errors in Surveys

BILLINGSLEY · Convergence of Probability Measures, *Second Edition*

BILLINGSLEY · Probability and Measure, *Third Edition*

BIRKES and DODGE · Alternative Methods of Regression

BLISCHKE and MURTHY (editors) · Case Studies in Reliability and Maintenance

BLISCHKE and MURTHY · Reliability: Modeling, Prediction, and Optimization

BLOOMFIELD · Fourier Analysis of Time Series: An Introduction, *Second Edition*

BOLLEN · Structural Equations with Latent Variables

BOLLEN and CURRAN · Latent Curve Models: A Structural Equation Perspective

BOROVKOV · Ergodicity and Stability of Stochastic Processes

BOULEAU · Numerical Methods for Stochastic Processes

*BOX · Bayesian Inference in Statistical Analysis

BOX · R. A. Fisher, the Life of a Scientist

BOX and DRAPER · Empirical Model-Building and Response Surfaces

BOX and DRAPER · Evolutionary Operation: A Statistical Metl

BOROVKOV · Ergodicity and Stability of Stochastic Processes

BOULEAU · Numerical Methods for Stochastic Processes

BOX · Bayesian Inference in Statistical Analysis

BOX · R. A. Fisher, the Life of a Scientist

BOX and DRAPER · Empirical Model-Building and Response Surfaces

BOX and DRAPER · Evolutionary Operation: A Statistical Method for Process Improvement

BOX, HUNTER, and HUNTER · Statistics for Experimenters: Design, Innovation and Discovery, *Second Edition*

BOX and LUCEÑO · Statistical Control by Monitoring and Feedback Adjustment

BRANDIMARTE · Numerical Methods in Finance: A MATLAB-Based Introduction

BROWN and HOLLANDER · Statistics: A Biomedical Introduction

BRUNNER, DOMHOF, and LANGER · Nonparametric Analysis of Longitudinal Data in Factorial Experiments

BUCKLEW · Large Deviation Techniques in Decision, Simulation, and Estimation

CAIROLI and DALANG · Sequential Stochastic Optimization

CASTILLO, HADI, BALAKRISHNAN, and SARABIA · Extreme Value and Related Models with Applications in Engineering and Science

CHAN · Time Series: Applications to Finance

CHATTERJEE and HADI · Sensitivity Analysis in Linear Regression

CHATTERJEE and PRICE · Regression Analysis by Example, *Third Edition*

CHERNICK · Bootstrap Methods: A Practitioner's Guide

CHERNICK and FRIIS · Introductory Biostatistics for the Health Sciences

CHILÈS and DELFINER · Geostatistics: Modeling Spatial Uncertainty

CHOW and LIU · Design and Analysis of Clinical Trials: Concepts and Methodologies, *Second Edition*

CLARKE and DISNEY · Probability and Random Processes: A First Course with Applications, *Second Edition*

*COCHRAN and COX · Experimental Designs, *Second Edition*

CONGDON · Applied Bayesian Modelling

CONGDON · Bayesian Statistical Modelling

CONGDON · Bayesian Models for Categorical Data

CONOVER · Practical Nonparametric Statistics, *Second Edition*

COOK · Regression Graphics

COOK and WEISBERG · Applied Regression Including Computing and Graphics

COOK and WEISBERG · An Introduction to Regression Graphics

CORNELL · Experiments with Mixtures, Designs, Models, and the Analysis of Mixture Data, *Third Edition*

Now available in a lower priced paperback edition in the Wiley Classics Library.

COVER and THOMAS · Elements of Information Theory

COX · A Handbook of Introductory Statistical Methods

*COX · Planning of Experiments

CRESSIE · Statistics for Spatial Data, *Revised Edition*

CSÖRGÖ and HORVÁTH · Limit Theorems in Change Point Analysis

DANIEL · Applications of Statistics to Industrial Experimentation

DANIEL · Biostatistics: A Foundation for Analysis in the Health Sciences, *Sixth Edition*

*DANIEL · Fitting Equations to Data: Computer Analysis of Multifactor Data, *Second Edition*

DASU and JOHNSON · Exploratory Data Mining and Data Cleaning

DAVID and NAGARAJA · Order Statistics, *Third Edition*

*DEGROOT, FIENBERG, and KADANE · Statistics and the Law

DEL CASTILLO · Statistical Process Adjustment for Quality Control

DeMARIS · Regression with Social Data: Modeling Continuous and Limited Response Variables

DEMIDENKO · Mixed Models: Theory and Applications

DENISON, HOLMES, MALLICK, and SMITH · Bayesian Methods for Nonlinear Classification and Regression

DETTE and STUDDEN · The Theory of Canonical Moments with Applications in Statistics, Probability, and Analysis

DEY and MUKERJEE · Fractional Factorial Plans

DILLON and GOLDSTEIN · Multivariate Analysis: Methods and Applications

DODGE · Alternative Methods of Regression

*DODGE and ROMIG · Sampling Inspection Tables, *Second Edition*

*DOOB · Stochastic Processes

DOWDY and WEARDEN, and CHILKO · Statistics for Research, *Third Edition*

DRAPER and SMITH · Applied Regression Analysis, *Third Edition*

DRYDEN and MARDIA · Statistical Shape Analysis

DUDEWICZ and MISHRA · Modern Mathematical Statistics

DUNN and CLARK · Applied Statistics: Analysis of Variance and Regression, *Second Edition*

DUNN and CLARK · Basic Statistics: A Primer for the Biomedical Sciences, *Third Edition*

DUPUIS and ELLIS · A Weak Convergence Approach to the Theory of Large Deviations

EDLER and KITSOS (editors) · Recent Advances in Quantitative Methods in Cancer and Human Health Risk Assessment

*ELANDT-JOHNSON and JOHNSON · Survival Models and Data Analysis

ENDERS · Applied Econometric Time Series

ETHIER and KURTZ · Markov Processes: Characterization and Convergence

EVANS, HASTINGS, and PEACOCK · Statistical Distributions, *Third Edition*

FELLER · An Introduction to Probability Theory and Its Applications, Volume I, *Third Edition, Revised*; Volume II, *Second Edition*

FISHER and VAN BELLE · Biostatistics: A Methodology for the Health Sciences

FITZMAURICE, LAIRD, and WARE · Applied Longitudinal Analysis

*FLEISS · The Design and Analysis of Clinical Experiments

FLEISS · Statistical Methods for Rates and Proportions, *Second Edition*

FLEMING and HARRINGTON · Counting Processes and Survival Analysis

FULLER · Introduction to Statistical Time Series, *Second Edition*

FULLER · Measurement Error Models

GALLANT · Nonlinear Statistical Models

GELMAN and MENG (editors): Applied Bayesian Modeling and Casual Inference from Incomplete-data Perspectives

GEWEKE · Contemporary Bayesian Econometrics and Statistics

GHOSH, MUKHOPADHYAY, and SEN · Sequential Estimation

GIESBRECHT and GUMPERTZ · Planning, Construction, and Statistical Analysis of Comparative Experiments

GIFI · Nonlinear Multivariate Analysis

GIVENS and HOETING · Computational Statistics

GLASSERMAN and YAO · Monotone Structure in Discrete-Event Systems

GNANADESIKAN · Methods for Statistical Data Analysis of Multivariate Observations, *Second Edition*

*Now available in a lower priced paperback edition in the Wiley Classics Library.

GOLDSTEIN and LEWIS · Assessment: Problems, Development, and Statistical Issues
GREENWOOD and NIKULIN · A Guide to Chi-Squared Testing
GROSS and HARRIS · Fundamentals of Queueing Theory, *Third Edition*
*HAHN and SHAPIRO · Statistical Models in Engineering
HAHN and MEEKER · Statistical Intervals: A Guide for Practitioners
HALD · A History of Probability and Statistics and their Applications Before 1750
HALD · A History of Mathematical Statistics from 1750 to 1930
HAMPEL · Robust Statistics: The Approach Based on Influence Functions
HANNAN and DEISTLER · The Statistical Theory of Linear Systems
HEIBERGER · Computation for the Analysis of Designed Experiments
HEDAYAT and SINHA · Design and Inference in Finite Population Sampling
HELLER · MACSYMA for Statisticians
HINKELMAN and KEMPTHORNE: · Design and Analysis of Experiments, Volume 1:
    Introduction to Experimental Design
HINKELMANN and KEMPTHORNE · Design and Analysis of Experiments, Volume 2:
    Advanced Experimental Design
HOAGLIN, MOSTELLER, and TUKEY · Exploratory Approach to Analysis of Variance
HOAGLIN, MOSTELLER, and TUKEY · Exploring Data Tables, Trends and Shapes
*HOAGLIN, MOSTELLER, and TUKEY · Understanding Robust and Exploratory Data Analysis
HOCHBERG and TAMHANE · Multiple Comparison Procedures
HOCKING · Methods and Applications of Linear Models: Regression and the Analysis of Variance,
    Second Edition
HOEL · Introduction to Mathematical Statistics, *Fifth Edition*
HOGG and KLUGMAN · Loss Distributions
HOLLANDER and WOLFE · Nonparametric Statistical Methods, *Second Edition*
HOSMER and LEMESHOW · Applied Logistic Regression, *Second Edition*
HOSMER and LEMESHOW · Applied Survival Analysis: Regression Modeling of
    Time to Event Data
HUBER · Robust Statistics
HUBERTY · Applied Discriminant Analysis
HUNT and KENNEDY · Financial Derivatives in Theory and Practice, *Revised Edition*
HUSKOVA, BERAN, and DUPAC · Collected Works of Jaroslav Hajek—with
    Commentary
HUZURBAZAR · Flowgraph Models for Multistate Time-to-Event Data
IMAN and CONOVER · A Modern Approach to Statistics
JACKSON · A User's Guide to Principle Components
JOHN · Statistical Methods in Engineering and Quality Assurance
JOHNSON · Multivariate Statistical Simulation
JOHNSON and BALAKRISHNAN · Advances in the Theory and Practice of Statistics: A
    Volume in Honor of Samuel Kotz
JOHNSON and BHATTACHARYYA · Statistics: Principles and Methods, *Fifth Edition*
JUDGE, GRIFFITHS, HILL, LÜTKEPOHL, and LEE · The Theory and Practice of
    Econometrics, *Second Edition*
JOHNSON and KOTZ · Distributions in Statistics
JOHNSON and KOTZ (editors) · Leading Personalities in Statistical Sciences: From the
    Seventeenth Century to the Present
JOHNSON, KOTZ, and BALAKRISHNAN · Continuous Univariate Distributions,
    Volume 1, *Second Edition*
JOHNSON, KOTZ, and BALAKRISHNAN · Continuous Univariate Distributions,
    Volume 2, *Second Edition*
JOHNSON, KOTZ, and BALAKRISHNAN · Discrete Multivariate Distributions
JOHNSON, KOTZ, and KEMP · Univariate Discrete Distributions,
JUREČKOVÁ and SEN · Robust Statistical Procedures: Asymptotics and Interrelations
JUREK and MASON · Operator-Limit Distributions in Probability Theory
KADANE · Bayesian Methods and Ethics in a Clinical Trial Design

*Now available in a lower priced paperback edition in the Wiley Classics Library.

KADANE and SCHUM · A Probabilistic Analysis of the Sacco and Vanzetti Evidence

KALBFLEISCH and PRENTICE · The Statistical Analysis of Failure Time Data,
Second Edition

KARIYA and KURATA · Generalized Least Squares

KASS and VOS · Geometrical Foundations of Asymptotic Inference

KAUFMAN and ROUSSEEUW · Finding Groups in Data: An Introduction to Cluster Analysis

KEDEM and FOKIANOS · Regression Models for Time Series Analysis

KENDALL, BARDEN, CARNE, and LE · Shape and Shape Theory

KHURI · Advanced Calculus with Applications in Statistics, *Second Edition*

KHURI, MATHEW, and SINHA · Statistical Tests for Mixed Linear Models

*KISH · Statistical Design for Research

KLEIBER and KOTZ · Statistical Size Distributions in Economics and Actuarial Sciences

KLUGMAN, PANJER, and WILLMOT · Loss Models: From Data to Decisions, *Second Edition*

KLUGMAN, PANJER, and WILLMOT · Solutions Manual to Accompany Loss Models:
From Data to Decisions

KOTZ, BALAKRISHNAN, and JOHNSON · Continuous Multivariate Distributions,
Volume 1, *Second Edition*

KOTZ and JOHNSON (editors) · Encyclopedia of Statistical Sciences: Volumes 1 to 9
with Index

KOTZ and JOHNSON (editors) · Encyclopedia of Statistical Sciences: Supplement Volume

KOTZ, READ, and BANKS (editors) · Encyclopedia of Statistical Sciences: Update Volume 1

KOTZ, READ, and BANKS (editors) · Encyclopedia of Statistical Sciences: Update Volume 2

KOVALENKO, KUZNETZOV, and PEGG · Mathematical Theory of Reliability of
Time-Dependent Systems with Practical Applications

KUROWICKA and COOKE · Unceentainty Analysis with High Dimensional
Dependence Modelling

LACHIN · Biostatistical Methods: The Assessment of Relative Risks

LAD · Operational Subjective Statistical Methods: A Mathematical, Philosophical, and
Historical Introduction

LAMPERTI · Probability: A Survey of the Mathematical Theory, *Second Edition*

LANGE, RYAN, BILLARD, BRILLINGER, CONQUEST, and GREENHOUSE ·
Case Studies in Biometry

LARSON · Introduction to Probability Theory and Statistical Inference, *Third Edition*

LAWLESS · Statistical Models and Methods for Lifetime Data, *Second Edition*

LAWSON · Statistical Methods in Spatial Epidemiology, *Second Edition*

LE · Applied Categorical Data Analysis

LE · Applied Survival Analysis

LEE and WANG · Statistical Methods for Survival Data Analysis, *Third Edition*

LePAGE and BILLARD · Exploring the Limits of Bootstrap

LEYLAND and GOLDSTEIN (editors) · Multilevel Modelling of Health Statistics

LIAO · Statistical Group Comparison

LINDVALL · Lectures on the Coupling Method

LINHART and ZUCCHINI · Model Selection

LITTLE and RUBIN · Statistical Analysis with Missing Data, *Second Edition*

LLOYD · The Statistical Analysis of Categorical Data

LOWEN and TEICH · Fractal-Based Point Processes

MAGNUS and NEUDECKER · Matrix Differential Calculus with Applications in
Statistics and Econometrics, *Revised Edition*

MALLER and ZHOU · Survival Analysis with Long Term Survivors

MALLOWS · Design, Data, and Analysis by Some Friends of Cuthbert Daniel

MANN, SCHAFER, and SINGPURWALLA · Methods for Statistical Analysis of
Reliability and Life Data

MANTON, WOODBURY, and TOLLEY · Statistical Applications Using Fuzzy Sets

MARCHETTE · Random Graphs for Statistical Pattern Recognition

MARDIA and JUPP · Directional Statistics

*Now available in a lower priced paperback edition in the Wiley Classics Library.

MARONNA, MARTIN and YOHAI · Robust Statistics: Theory and Methods
MASON, GUNST, and HESS · Statistical Design and Analysis of Experiments with
    Applications to Engineering and Science, *Second Edition*
McCULLOCH and SEARLE · Generalized, Linear, and Mixed Models
McFADDEN · Management of Data in Clinical Trials
McLACHLAN · Discriminant Analysis and Statistical Pattern Recognition
McLACHLAN and KRISHNAN · The EM Algorithm and Extensions
McLACHLAN, DO and AMBROISE · Analyzing Microarray Gene Expression Data
McLACHLAN and PEEL · Finite Mixture Models
McNEIL · Epidemiological Research Methods
MEEKER and ESCOBAR · Statistical Methods for Reliability Data
MEERSCHAERT and SCHEFFLER · Limit Distributions for Sums of Independent
    Random Vectors: Heavy Tails in Theory and Practice
MICKEY, DUNN, and CLARK · Applied Statistics: Anaysis of Variance and
    Regression, *Third Edition*
*MILLER · Survival Analysis, *Second Edition*
MONTGOMERY, PECK, and VINING · Introduction to Linear Regression Analysis,
    Third Edition
MORGENTHALER and TUKEY · Configural Polysampling: A Route to Practical
    Robustness
MUIRHEAD · Aspects of Multivariate Statistical Theory
MURRAY · X-STAT 2.0 Statistical Experimentation, Design Data Analysis, and
    Nonlinear Optimization
MURTHY, XIE, and JIANG · Weibull Models
MYERS and MONTGOMERY · Response Surface Methodology: Process and Product
    Optimization Using Designed Experiments, *Second Edition*
MYERS, MONTGOMERY, and VINING · Generalized Linear Models. With
    Applications in Engineering and the Sciences
†NELSON · Accelerated Testing, Statistical Models, Test Plans, and Data Analyses
†NELSON · Applied Life Data Analysis
NEWMAN · Biostatistical Methods in Epidemiology
OCHI · Applied Probability and Stochastic Processes in Engineering and Physical Sciences
OKABE, BOOTS, SUGIHARA, and CHIU · Spatial Tesselations: Concepts and
    Applications of Voronoi Diagrams, *Second Edition*
OLIVER and SMITH · Influence Diagrams, Belief Nets and Decision Analysis
PALTA · Quantitative Methods in Population Health: Extensions of Ordinary Regressions
PANKRATZ · Forecasting with Dynamic Regression Models
PANKRATZ · Forecasting with Univariate Box-Jenkins Models: Concepts and Cases
*PARZEN · Modern Probability Theory and It's Applications
PEÑA, TIAO, and TSAY · A Course in Time Series Analysis
PIANTADOSI · Clinical Trials: A Methodologic Perspective
PORT · Theoretical Probability for Applications
POURAHMADI · Foundations of Time Series Analysis and Prediction Theory
PRESS · Bayesian Statistics: Principles, Models, and Applications
PRESS · Subjective and Objective Bayesian Statistics, *Second Edition*
PRESS and TANUR · The Subjectivity of Scientists and the Bayesian Approach
PUKELSHEIM · Optimal Experimental Design
PURI, VILAPLANA, and WERTZ · New Perspectives in Theoretical and Applied Statistics
†PUTERMAN · Markov Decision Processes: Discrete Stochastic Dynamic Programming
QIU · Image Processing and Jump Regression Analysis
*RAO · Linear Statistical Inference and Its Applications, *Second Edition*
RAUSAND and HØYLAND · System Reliability Theory: Models, Statistical Methods and
    Applications, *Second Edition*
RENCHER · Linear Models in Statistics
RENCHER · Methods of Multivariate Analysis, *Second Edition*
RENCHER · Multivariate Statistical Inference with Applications

*Now available in a lower priced paperback edition in the Wiley Classics Library.
†Now available in a lower priced paperback edition in the Wiley-Interscience Paperback Series.

RIPLEY · Spatial Statistics

RIPLEY · Stochastic Simulation

ROBINSON · Practical Strategies for Experimenting

ROHATGI and SALEH · An Introduction to Probability and Statistics, *Second Edition*

ROLSKI, SCHMIDLI, SCHMIDT, and TEUGELS · Stochastic Processes for Insurance and Finance

ROSENBERGER and LACHIN · Randomization in Clinical Trials: Theory and Practice

ROSS · Introduction to Probability and Statistics for Engineers and Scientists

ROSSI, ALLENBY and McCULLOCH · Bayesian Statistics and Marketing

ROUSSEEUW and LEROY · Robust Regression and Outlier Detection

RUBIN · Multiple Imputation for Nonresponse in Surveys

RUBINSTEIN · Simulation and the Monte Carlo Method

RUBINSTEIN and MELAMED · Modern Simulation and Modeling

RYAN · Modern Regression Methods

RYAN · Statistical Methods for Quality Improvement, *Second Edition*

SALTELLI, CHAN, and SCOTT (editors) · Sensitivity Analysis

*SCHEFFE · The Analysis of Variance

SCHIMEK · Smoothing and Regression: Approaches, Computation, and Application

SCHOTT · Matrix Analysis for Statistics

SCHOUTENS · Levy Processes in Finance: Pricing Financial Derivatives

SCHUSS · Theory and Applications of Stochastic Differential Equations

SCOTT · Multivariate Density Estimation: Theory, Practice, and Visualization

*SEARLE · Linear Models

SEARLE · Linear Models for Unbalanced Data

SEARLE · Matrix Algebra Useful for Statistics

SEARLE, CASELLA, and McCULLOCH · Variance Components

SEARLE and WILLETT · Matrix Algebra for Applied Economics

SEBER · Multivariate Observations

SEBER and LEE · Linear Regression Analysis, *Second Edition*

SEBER and WILD · Nonlinear Regression

SENNOTT · Stochastic Dynamic Programming and the Control of Queueing Systems

*SERFLING · Approximation Theorems of Mathematical Statistics

SHAFER and VOVK · Probability and Finance: Its Only a Game!

SILVAPULLE and SEN · Constrained Statistical Inference: Inequality, Order, and Shape
    Restrictions

SMALL and McLEISH · Hilbert Space Methods in Probability and Statistical Inference

SRIVASTAVA · Methods of Multivariate Statistics

STAPLETON · Linear Statistical Models

STAUDTE and SHEATHER · Robust Estimation and Testing

STOYAN, KENDALL, and MECKE · Stochastic Geometry and Its Applications,
    Second Edition

STOYAN and STOYAN · Fractals, Random Shapes and Point Fields: Methods of
    Geometrical Statistics

STYAN · The Collected Papers of T. W. Anderson: 1943–1985

SUTTON, ABRAMS, JONES, SHELDON, and SONG · Methods for Meta-Analysis in
    Medical Research

TANAKA · Time Series Analysis: Nonstationary and Noninvertible Distribution Theory

THOMPSON · Empirical Model Building

THOMPSON · Sampling, *Second Edition*

THOMPSON · Simulation: A Modeler's Approach

THOMPSON and SEBER · Adaptive Sampling

THOMPSON, WILLIAMS, and FINDLAY · Models for Investors in Real World Markets

TIAO, BISGAARD, HILL, PEÑA, and STIGLER (editors) · Box on Quality and
    Discovery: with Design, Control, and Robustness

*Now available in a lower priced paperback edition in the Wiley Classics Library.

TIERNEY · LISP-STAT: An Object-Oriented Environment for Statistical Computing and
Dynamic Graphics

TSAY · Analysis of Financial Time Series

UPTON and FINGLETON · Spatial Data Analysis by Example, Volume II:
Categorical and Directional Data

VAN BELLE · Statistical Rules of Thumb

VAN BELLE, FISHER, HEAGERTY, and LUMLEY · Biostatistics: A Methodology for
the Health Sciences, *Second Edition*

VESTRUP · The Theory of Measures and Integration

VIDAKOVIC · Statistical Modeling by Wavelets

VINOD and REAGLE · Preparing for the Worst: Incorporating Downside Risk in Stock
Market Investments

WALLER and GOTWAY · Applied Spatial Statistics for Public Health Data

WEERAHANDI · Generalized Inference in Repeated Measures: Exact Methods in
MANOVA and Mixed Models

WEISBERG · Applied Linear Regression, *Second Edition*

WELSH · Aspects of Statistical Inference

WESTFALL and YOUNG · Resampling-Based Multiple Testing: Examples and
Methods for *p*-Value Adjustment

WHITTAKER · Graphical Models in Applied Multivariate Statistics

WINKER · Optimization Heuristics in Economics: Applications of Threshold Accepting

WONNACOTT and WONNACOTT · Econometrics, *Second Edition*

WOODING · Planning Pharmaceutical Clinical Trials: Basic Statistical Principles

WOOLSON and CLARKE · Statistical Methods for the Analysis of Biomedical Data,
Second Edition

WU and HAMADA · Experiments: Planning, Analysis, and Parameter Design Optimization

YANG · The Construction Theory of Denumerable Markov Processes

*ZELLNER · An Introduction to Bayesian Inference in Econometrics

ZELTERMAN · Discrete Distributions: Applications in the Health Sciences

ZHOU, OBUCHOWSKI, and McCLISH · Statistical Methods in Diagnostic Medicine